# Research for Development

The series Research for Development serves as a vehicle for the presentation and dissemination of complex research and multidisciplinary projects. The published work is dedicated to fostering a high degree of innovation and to the sophisticated demonstration of new techniques or methods. The aim of the Research for Development series is to promote well-balanced sustainable growth. This might take the form of measurable social and economic outcomes, in addition to environmental benefits, or improved efficiency in the use of resources; it might also involve an original mix of intervention schemes. Research for Development focuses on the following topics and disciplines: Urban regeneration and infrastructure, Info-mobility, transport, and logistics, Environment and the land, Cultural heritage and landscape, Energy, Innovation in processes and technologies, Applications of chemistry, materials, and nanotechnologies, Material science and biotechnology solutions, Physics results and related applications and aerospace, Ongoing training and continuing education. Fondazione Politecnico di Milano collaborates as a special co-partner in this series by suggesting themes and evaluating proposals for new volumes. Research for Development addresses researchers, advanced graduate students, and policy and decision-makers around the world in government, industry, and civil society.

More information about this series at http://www.springer.com/series/13084

Maurizio Rossi

# Circadian Lighting Design in the LED Era

Springer

Maurizio Rossi
Laboratorio Luce, Department of Design
Politecnico di Milano
Milan, Italy

ISSN 2198-7300 ISSN 2198-7319 (electronic)
Research for Development
ISBN 978-3-030-11086-4 ISBN 978-3-030-11087-1 (eBook)
https://doi.org/10.1007/978-3-030-11087-1

Library of Congress Control Number: 2018966856



This Springer imprint is published by the registered company Springer Nature Switzerland AG
The registered company address is: Gewerbestrasse 11, 6330 Cham, Switzerland

*To my wife*

# Foreword

The last two decades have seen several revolutions in the field of lighting design: among them, it is worth mentioning on one side the development and more and more growing spread of LED-based lighting systems that allow to obtain dynamic light scenes, on the other the progress in the knowledge of the physiological mechanisms that regulate the circadian rhythms and their connections with light stimuli.

The awareness that luminous environments not only affect visual performances, but have a significant impact on humans owing to the so-called non-visual effects or non-image forming effects as mood, alertness, circadian rhythms and more in general health, is today well confirmed; however, designers cannot yet dispose of specific design rules, based on a quantitative approach. With regard to the circadian effects of light assessment, the difficulties are only in part due to the lack of an established metric accepted by all the scientific community, but mainly because this evaluation requires an approach very different from the visual one: indeed the time, the duration of light exposure and the previous light history, i.e. the light received before the considered period, have an impact on circadian effects, and not only the intensity and spectrum, as it is true for the quite instantaneous visual effects. For this reason, much research was published about both the proposal of "circadian" metrics and how to transfer the scientific findings into the lighting design practice. The most critical environments are the indoor ones, especially those where people spend most of their lives, as work environments, schools, hospitals, residences for aged people, etc. In these environments, daylight access should be maximized, and at the same time, glare and heating effects due to direct sunlight should be avoided. In many cases, daylight illuminance levels are sufficient to perform visual tasks, but not to entrain circadian rhythms and proper electric light integration is necessary.

This book tackles all these topics and shows how to connect visual and non-visual light requirements with innovative lighting systems and controls, without excluding the important role of a proper lighting design specifically adapted to different needs. With this purpose, the important role of a daily light pattern is highlighted, and the potentiality of applications aimed at coming to the aid of people affected by diseases as Alzheimer's syndrome or seasonal depression is illustrated. It must not be forgotten that this scenario is in constant evolution and that in this context research activities help to drive manufacturers and designers in improving their work and to develop new solutions. Besides presenting an updated state of the art of the scientific contributions in this field, the author reports a comprehensive description of research activities carried out at the Lab. Luce of the Politecnico di Milano. The rich excursus of analysed case studies leads the reader towards a new comprehensive approach in the lighting design field demonstrating that, despite some interesting conclusions that can be drawn, there are still things to be investigated and this is really interesting.

Naples, Italy Laura Bellia

# Preface

I decided to write this book inspired by my applied research activities concerning the relationship between artificial lighting and the individual's well-being. These activities, which I have led and coordinated in the last 15 years at the Lab. Luce of the Politecnico di Milano, have been carried out in collaboration with companies and with other research groups.

For over 20 years in the field of physiology, numerous researches have been presented and have shown scientifically that light, in addition to vision, has non-image forming (NIF) effects on the well-being of people. However, these important results have not yet been incorporated into the lighting design. The 2017 Nobel Prize in Medicine and Physiology, awarded on the topic of circadian research, has reawakened the attention of researchers, designers, and stakeholders to the theme of the relationship between lighting design and the human circadian cycle. In Chap. 1, the foundations necessary for understanding the following topics, also regarding the non-image forming (NIF) effects of light, are presented. The science known as chronobiology is introduced and the important impact that the research of the three 2017 Nobel prizes Jeffrey C. Hall, Michael Rosbash (Brandeis University, Boston) and Michael W. Young (Rockefeller University, New York) has had in this context. This chapter presents the historical development of the philosophical and scientific definition of light, up to the description used today in lighting design. Continuing the historical analysis, the theme of the impact of artificial lighting on human society is introduced along with how electric light sources have evolved over the past 150 years. The theme of the relationship between artificial and natural light is introduced in relation to the profound change in the lifestyles of the last five generations following the industrial revolution. The position of legislative bodies, regulators and international trade associations, with regard to the NIF effects and the environmental sustainability of the lighting systems, is also presented.

The analysis of the human circadian system in detail is the subject of Chap. 2. From a lighting design-oriented perspective, the results of the main research are presented, applied to lighting or developed in the physiological medical field, with reference to the NIF effects produced by artificial lighting on the human organism.

These primarily affect the regulation of melatonin production, which is the main marker of the state of the human circadian cycle, but also the central nervous system, with effects on neurophysiological parameters such as cognitive performance, attention level and alertness level. The existing positive and negative relationships between light and some diseases such as seasonal depression, Alzheimer's syndrome, sleep disorders and premature birth are also examined. Furthermore, this chapter provides a general description of the functioning of the human circadian system, the factors that regulate it and its effects on the organism. Finally, it analyses the possible relationships between the circadian cycle and the modern environment, with particular reference to the human interior spaces.

After having illustrated the various relationships existing between light and human physiology, the quantitative aspects of light, which may have NIF effects on the organism, are deepened in Chap. 3. The emphasis is placed on the dose–effect relationships that can come into play in human-centric lighting, as positive or negative elements, and on possible sources of risks, with particular reference to light emitting diode (LED) light. These elements include consolidated requirements of lighting design, such as the control of glare and of the temporal modulation of light, which are now applied to LED lighting. A further topic of reflection, which has seen a lot of interest by the press, concerns the presumed photobiological hazard related to LED lighting. Finally, this chapter introduces the basic elements for the possibility of defining a circadian photometry that can support the activity of the lighting designer, presenting the current state of the research in this area.

Chapter 4 deals with a topic of central importance for the book. Knowing that natural light is the best light for a correct timing of the human circadian cycle, with all the benefits that derive from it, the question is if the real availability of natural light, for people who live and work in interior spaces, is adequate for a correct timing of the circadian cycle as it happens outside. This topic is developed in three contexts of interior design: office spaces, residential areas and training facilities. The question is analysed based on studies conducted at the Lab. Luce Politecnico di Milano. In the latter chapter, these results are analysed in relation to those carried out by other research groups. The survey methods used are presented, and finally, the results are discussed. Only in particular cases, unfortunately limited, the natural light present in interior spaces is able to correctly stimulate the human circadian system. This therefore leads to the consideration that artificial lighting in interior spaces could have NIF effects in the daylight hours, compensating for the lack of natural light.

In order to implement artificial circadian lighting in interior spaces, one must take into account the light requirements of human beings and therefore the technologies that make it possible to meet them. In Chap. 5, the latest lighting technologies that can facilitate the circadian lighting design are presented. A first part introduces the types of LEDs currently available on the market, analysing the main positive and negative aspects of the different technological solutions. The following part focuses on sensors and light management systems that make it possible to receive information on the amount of light and the presence and position of people in interiors. This information must be managed by smart lighting control systems, which are presented both from a theoretical and an applicative standpoint. The study also focuses on the regulatory aspects of lighting products, with reference to energy saving. Lastly, some smart lighting solutions aimed at integrating into the new smart home concept are introduced.

Chapter 6 presents a research conducted at the Laboratorio Luce of the Politecnico di Milano as part of an industrial research project. Research has focused on the creation of a human-centric lighting for residential spaces of the elderly, taking into consideration both the lighting design and the lighting product design based on the new LED technologies. This study has started with a detailed analysis of the needs and problems of the elderly in relation to artificial lighting. With reference to this aspect, the characteristics of a correct lighting for the elderly have been outlined and then translated into requirements to be applied to lighting design and lighting product design, eventually identifying technological solutions and defining guidelines, lighting programs and concepts.

Today there is no internationally accepted standard that establishes rules for designing a circadian lighting system for interior spaces. For several years, in many different contexts, human-centric lighting has been a discussion topic, although both the definition and the practical application of this discipline are controversial and debated. In Chap. 7, through the analysis of the original research results presented in this book and of other sectoral research, a framework of possible guidelines for circadian lighting is proposed. The elements that come into play in this framework are the natural light present in interiors, but also lighting design and lighting product design achievable currently thanks to the availability of LED light sources and the new lighting management systems.

Milan, Italy Maurizio Rossi

# Acknowledgements

Although this text was written by a single author, its writing was possible thanks to many people with whom I have collaborated in the last 20 years. The first thanks is addressed to Prof. Alberto Seassaro who over 20 years ago strongly wanted to realize the Laboratorio Luce and found with Prof. Flaviano Celaschi, Director of the Department, the resources to allow me realizing it in 2002 at the Indaco Department, the actual Design Department of the Politecnico di Milano. A big thanks is also addressed to the other directors who have succeeded and that have favoured the activities of the laboratory: Prof. Marco Gaiani, Prof. Arturo Dell'Acqua Bellavitis and Prof. Silvia Piardi.

A warm thanks is addressed to the collaborators who, in recent years, have participated and contributed to the research and advanced training activities in the laboratory—Andrea Siniscalco for participation in research and organizational activities that supported the laboratory. In addition, a special thanks is addressed to Fulvio Musante and Daria Casciani for active and proactive participation in many of the research activities covered in this book and to Simonetta Fumagalli of ENEA for the 8 years of research carried out together.

A special memory is addressed to Danilo Paleari who, after obtaining the Master's degree in lighting design, continued to collaborate with the Lab. Luce with generosity, dedication and broad expertise in new lighting technologies, until his untimely death in 2016.

# Contents

# About the Author

**Maurizio Rossi M.Sc., Ph.D.** is Associate Professor at the Politecnico di Milano where he teaches lighting design and design methods. At the same university, he is also Director of the Master in Lighting Design and LED Technology and of the Master in Colour Design and Technology and Member of the Ph.D. Faculty in Design.

Starting from 1990, he collaborated with the Eidomatics Laboratory at Università degli Studi di Milano, the CNR, The University of Manchester, the Consorzio Milano Ricerche and the Università degli Studi di Pavia, participating in various national and European research projects in the areas of multimedia and virtual reality applied to lighting and cultural heritage.

Since 1997, he had participated in the design of the Laboratorio Luce at Politecnico di Milano of which he then took care of the implementation. Since 2010, he has been Scientific Head. In this laboratory, since 2002, he has directed 23 financed research projects. In this context, he has been deepening the aspects of design research and new technologies, related to lighting and colour, in the environmental and product design. His research activities are directed to the issues of visual and non-visual perception, aimed, among other objectives, to the psychophysiological well-being of the individual in interior spaces. As part of these research and advanced training activities, at the Politecnico, he has collaborated with over 40 companies and public research organizations, including: Almeco, Artemide, Barbieri Electronic, Barco, Boero Group, Castaldi, Chiron, Clariant, Clay Paky, CNR, Danese, Disano, Eelectron, Elementi Moda, ENEA, ERCO, FLOS, Fontegrafica, General Electric, GEWISS, iGuzzini, IMQ, Konica Minolta, KONE, Lechler, Luceplan, Mantero, MaxMeyer, Missoni, OIKOS Group, Osram, Oxytech, Panzeri, Philips, Prisma-PIL, QLT, Reggiani, Rexel, SHARP, Spotlight, Trilux, Viabizzuno and X-rite.

From 2012 to 2018, he was President of Associazione Italiana Colore. He is Member of the Executive Committee of the AIC (International Colour Association). He is also Member of Società Italiana Design, Associazione Italiana di Illuminazione and International Light Association. Since 2014, he has been Editor-in-Chief of the Color Culture and Science journal. Since 1998, he has written 138 publications: 5 books as author, 16 as editor, 17 essays in books, 36 papers published on scientific journals and 58 papers in conference proceedings.

# Chapter 1
# Introduction: From Chronobiology to Lighting

## 1.1 Chronobiology and Nobel Prize for Circadian Research

The term *circadian* was only recently introduced, but the first scientist asking questions and making observations on the circadian rhythm was Jean Jacques d'Ortous de Mairan already in the eighteenth century. With his experiments on Mimosa pudica, he observed that the opening movements of the flowers of plants during the day and their closing during the night persisted also placing the plants in the dark, erroneously inferring that they felt the presence of the sun even without be placed in the sun. In reality there is instead an internal biological mechanism that acts as a clock in the organism of the plant (de Mairan 1729). Almost a century later, Augustin Pyramus de Candolle repeated the experiments on Mimosa pudica, observing that indeed, in the absence of light, the plant followed a cycle of about 22–23 h instead of the 24 h of the rhythm of the earth's revolution. Based on these observations, he concluded that in the absence of external light stimuli there could be an endogenous clock in the plant, with a duration slightly lower than that of the earth's day (de Candolle 1832). Despite these discoveries, for a long time many scientists rejected the idea of an internal clock in living organisms, while supporting the existence of an exogenous "factor X" linked to the rotation of the Earth; factor that they tried to find empirically until the middle of the twentieth century, when the foundations were laid for a new science called chronobiology (Foster and Kreitzman 2005).

Chronobiology studies the periodic phenomena affecting all living beings, with reference to the revolution cycles of the sun and the moon with respect to the earth's surface. Franz Halberg from the University of Minnesota, considered one of the founders of chronobiology, created the new word *circadian*, from Latin *circa*, which means approximately and *diem* which means day, to describe a periodic biological cycle that lasts roughly one day (Halberg et al. 2003). These studies focus on the description of the molecular mechanisms that are involved in the natural physiological cycles of living organisms. This also includes research on genetics, physiology, comparative anatomy, molecular biology and the ethology of animals and humans. Other aspects taken into consideration also concern the new epigenetics, reproduc-



M. Rossi, *Circadian Lighting Design in the LED Era*, Research for Development,
https://doi.org/10.1007/978-3-030-11087-1_1

tion, evolution and the relationship with the environmental sciences. In the first phase, chronobiology has described the circadian phenomena, but the biological and genetic aspects underlying the internal physiological clock remained to be demonstrated.

The 2017 Nobel Prize in Physiology or Medicine was awarded to scientists Jeffrey C. Hall, Michael Rosbash and Michael W. Young for their findings on the molecular mechanisms that control the circadian rhythm.

Jeffrey Connor Hall is professor emeritus of biology at the Brandeis University in Massachusetts. His contribution has been essential for the development of this field of investigation, as he developed a strategy for the isolation and neurological characterisation of mutants of the *Drosophila melanogaster*, also known as fruit fly. His studies have led to the discovery of the mechanisms that underlie the biological clock in all living beings. In particular, he succeeded in obtaining a vast collection of mutants that presented very significant changes in the sleep/wake cycles, and showed that these alterations were genetically transmitted to the following generations. The collaboration with Michael Rosbash has made it possible to carry out molecular studies on these mutants, which have led to the identification of some of the genes involved in these processes.

Michael Rosbash, professor of genetics at Brandeis, used the then nascent technologies of recombinant DNA to characterise the mutants isolated by Hall, identifying many genes involved in these circadian clocks: this has made possible, also in collaboration with Jeff Hall, to demonstrate that circadian clocks are present in all living beings, from bacteria to plants and animals. Among the genes identified by Rosbash, one gene emerges above all: the *period*, with the protein it encodes, PER. This periodic PER protein plays an essential role, as it is able to suppress its transcription from DNA to RNA. Biological transcription is the fundamental mechanism with which the DNA reaches the production of functional proteins. This self-suppression mechanism causes PER production to occur with a cyclic mode defined as a transcription-translation feedback loop model (TTFL), which therefore plays a central role in the circadian rhythm. In addition to having worked with Hall on the discovery and definition of the TTFL, he has identified the gene *clock* and the gene *cycle* in arrhythmic mutants of the fruit fly. Furthermore, he has discovered that the photoreceptor cytochrome plays a fundamental role in the circadian synchronisation with the light and dark cycles (Siwicki et al. 1988; Liu et al. 1992). He later collaborated with other research groups identifying, in cells of various parts of the organism subjected to light and dark cycles, the genetic processes that are at the base of the TTFL mechanism, coming to the discovery of the pigment dispersing factor (PDF) protein and of the neural cells that control its production, as an element of synchronisation that contributes to the control of the circadian cycle of the whole organism (Zehring et al. 1984).

Also Michael Warren Young, a geneticist and biologist, initially conducted studies on the circadian synchronisation of the fruit fly during his research at the Rockefeller University. Young first investigated the *period* gene responsible for producing the PER protein and later discovered the gene *timeless* responsible for the TIM protein, which facilitates the entry of PER into the cell nucleus and therefore has functional connections with the *period* gene and with the PER protein. Subsequent studies conducted by his research group have also discovered that light acts on TIM cycles through the *double-time* gene and therefore on the correct timing of the circadian rhythm (Bargiello et al. 1984; Hardin et al. 1990; Vosshall et al. 1994; Price et al. 1998). In his Nobel Lecture, he mainly presented very interesting data concerning sleep alterations in humans, proving that there are hundreds of genes that have cycles of activation/repression with circadian rhythms, and that alterations to the expression of these genes are at the basis of many sleep disorders, with serious clinical consequences.

Research conducted by Hall, Rosbash and Young, although they were carried out at first on the fruit fly, is at the base of the explanation of the functioning of the circadian rhythm also of plants, mammals and humans. In our body there is therefore a timed system, which lasts about a day, through which all physiological processes of the human body are managed, such as sleep, nutrition, hormone production, blood pressure and body temperature. It is important to stress that all these factors act on our immune system. However, this clock needs to be synchronised and the exogenous element that generates this synchronism is light. In fact, in the absence of the natural rhythms of light and darkness, following the earth's rotation, the clock can go out of phase, with multiple implications for our health and well-being (Evans and Davidson 2013; Hatori et al. 2017; Yadav et al. 2017).

## 1.2 The Historical Description of Light

In this text we do not cover the basics of lighting technology. However, in order to introduce some of the parameters useful for a quantitative evaluation of the effects of light on human health, it is necessary to recall some fundamental concepts that make it possible to describe the perception of light by the human visual system.

Since ancient Greece, for centuries, humanity has been wondering about the nature of light. Until the Renaissance, the concept of light was more philosophical than physical. Pythagoras and Euclid supported the emission theory according to which light rays, or visual rays, spread from the eyes to explore the surrounding environment. On the other hand, the atomists, among whom Democritus, supported the input theory, according to which light moved from the objects towards the eyes, generating vision. According to Plato, the vision was born from the soul, the visual focus, and it spread out of the body where it interacted with the daylight, which was also indispensable for the vision, and this was the reason why, in his theory, one could not see in the dark. Aristotle stated that light was a property of the transparent medium that carried it, whether it be indeterminate, like air, or determined, like ice.

Leaving the field of philosophy and taking into consideration the context of physics and the scientific method, we recall the famous dispute that, from the seventeenth to the mid-nineteenth, opposed the corpuscular theory of the English scientist Isaac Newton and the undulatory theory of the Dutch Christiaan Huygens. The British scientist Michael Farady, who lived between 1791 and 1867, was the first to highlight the relationship between the oscillations of the electric field and the magnetic field and halfway through the nineteenth century he hypothesised that light was an electromagnetic wave. In the second half of the nineteenth century, James Clerk Maxwell defined the equations of the electromagnetic field with which he succeeded in giving a unitary description of the electric, magnetic and optical phenomena. According to this theory, light is composed of electromagnetic waves having a wavelength between about 380 and 780 nm and different wavelengths correspond to different perceived colours. The most accurate definition available to modern science comes from the description of the photoelectric effect by the scientist Albert Einstein, based on which we came to define the light through the light quanta, or photons: wave packets that have partly undulatory and partly corpuscular behaviour. However, such an accurate description of light is superfluous for design purposes. In fact, in our daily macroscopic reality, the phenomena of light propagation and its interaction with the environment can be adequately described by the theory of electromagnetism, which considers light an electromagnetic radiation. The electromagnetic definition of light comes into crisis if we move at speeds close to those of light or in the presence of very strong gravitational fields, situations that never concern lighting design.

The artificial and natural light we use every day is never composed of a single monochromatic electromagnetic wave, but is the sum of many monochromatic waves with different frequencies and amplitudes. Exactly as it happens for the surface of the sea that is disturbed by waves of different length and width superimposed on each other. The set of electromagnetic radiation of all wavelengths is called the electromagnetic spectrum. Usually electromagnetic waves are classified, according to their decreasing wavelength, as: radiofrequency waves, microwaves, infrared rays (IR), light, ultraviolet rays (UV), X-rays, gamma rays and cosmic rays. The lower the wavelength, the greater the energy associated with the wave and the damage it can cause to living organisms. Within the electromagnetic spectrum, non-ionizing radiations are defined as those which, while having the ability to act on electrons, do not carry enough energy to extract electrons from atoms. This category includes the range between radio waves and ultraviolet rays nearing light in the spectrum (UVA from 315 to 400 nm).

## 1.3 Social Impact of Artificial Lighting

We know that life on our planet depends on sunlight and this is true both on land and in the oceans. The sun is indeed fundamental in the mechanisms of chlorophyll photosynthesis in plants and phytoplankton. These are at the base of the food chain and the production of oxygen in the earth's atmosphere. Even for humans, the best lighting is natural light, which has been the main source of light for thousands of years (Mead 2008). However, the rhythms and lifestyles have changed profoundly over the course of a few generations, as a consequence of the second phase of the industrial revolution in the second half of the nineteenth century, thanks to important discoveries and the development of inventions and multiple technological innovations, such as introduction of chemicals, oil and electricity.

Electricity has made possible the development of artificial lighting, with all its connected benefits, but it also has a negative biological impact on our planet. Indeed, there is evidence that artificial lighting at night in the exteriors, due to street lighting, advertising signs, shops and other sources of uncontrolled light that contribute to light pollution (Aubé et al. 2013; Luginbuhl et al. 2014), may have a negative effect on nocturnal animals and on plants (Rich and Longcore 2006; Gaston et al. 2014, 2015).

In the last 150 years, human activities have undergone a radical transformation and have been centralised in the cities, with the consequent mass migration from the countryside to the new urban centres and the formation of the industrialised societies. We therefore moved from a working life mainly conducted outdoors, in the countryside, to a working life mainly conducted in closed environments, with a limited supply of natural light and subjected instead to artificial light that has characteristics that are different from those of natural light. To understand the extent of this change in Europe, for example, in 1800 only 2% of the population lived in a city. In 1900 the share of the city population rose to 15%. In this century, most of the European population lives in cities (Fig. 1.1, UNICEF 2008) and works in closed spaces; even those who do not live in urban centres often do not work outdoors anyway. In Italy the percentage of agricultural workers in 2017 was 5.5% and that of construction workers was 9%, but this does not mean that all these people spend a significant number of hours outdoors during their day. In fact, most people live and work indoors with a limited supply of natural light. It is estimated that today in industrialised societies people spend between 80 and 90% of their time indoors (Boubekri 2008; Evans 2003; Hui et al. 2008).

Based on this radical change in current lifestyles, we can however observe that 150 years are nothing compared to the evolution of human beings and, from this point of view, our exposure to artificial lighting must be considered a factor of very recent

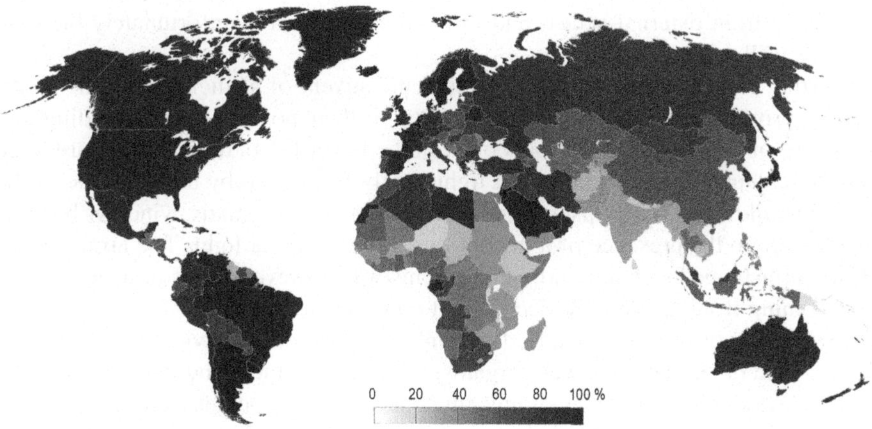


**Fig. 1.1** Percentage of urbanised population worldwide in 2006

introduction (Stevens 1987). We have evolved over millions of years as hunters and gatherers in an upright position, active during the day, in natural light, and resting at night, in the absence of natural light. Our organism is made to work and synchronise according to the rhythm of the continuous variations of natural light. Our physiology would require to be exposed to natural light by day and to complete darkness at night, in order to facilitate sleep, with its regenerative cycles that have a fundamental role for our happiness, health and fitness. There is no doubt therefore that exposure to natural light is beneficial for our well-being and our health (CIE 139-2001). The fundamental questions that we ask ourselves from the design standpoint are: how much exposure to natural light is sufficient or excessive and how should artificial light compensate for the lack of natural light?

In the history of human constructions, the presence of light has influenced and was an integral part in the development of the project, helping to shape the functional and artistic aspects of the structures. In the Roman villa, urban or rustic, where the windows towards the urban outside were limited, the centre of the activity took place around the *impluvium*, where the natural light, direct or diffused by the clouds, penetrated and then spread to the adjacent living rooms. The purpose of the impluvium, in addition to collecting rain, was precisely to capture and spread light inside the house. In the Middle Ages the different historical-cultural situation saw a new example of the use of light in the large polychrome windows of the Gothic cathedrals that, crossed by coloured light rays, contributed to creating the mystical and imposing climate of the structure. On the contrary, in residential spaces little attention was paid to light, as if to underline the obscurantism and the closure of an era in which the need to protect the occupants from the outside prevailed. In the Renaissance, the need for fortification and defence of living spaces finally decreases, leaving space for the development of windows facing outwards, the streets of the city centre, and inwards, the courtyard used as a vegetable garden or garden according to the social status. This openness to light on both exterior facades of the homes reflected the cultural and social development of the era; heavy metal gratings, however, remained in defence from external access and in many places this is unfortunately the case today as well.

Starting from the Renaissance and until the advent of artificial light, the architectural project sees windows closely related to their position in the building and to the use of the different interior rooms. In the lower floors the openings are large to try to capture a greater amount of light, partly obstructed by the presence of the adjacent buildings. Going upstairs and going down in social status, windows become smaller due to the presence of a greater quantity of natural light. The arrangement of the furnishings inside the individual rooms was however aimed at using natural light coming from the window according to the various visual tasks: activities such as sewing and reading were carried out only in the immediate vicinity of the windows, while other activities were organised in the innermost and dark areas of the room. Towards the end of the eighteenth century, in Europe gas lighting spread in the most important areas of the main inhabited centres with the fundamental purpose of facilitating people's recognition and safety. Gas lighting replaced the faint public lighting that until then had been mainly obtained with torches and lanterns, while in the interiors candles and oil lanterns prevailed. Prior to the advent of electricity, artificial lighting was therefore mainly based on devices that produce a limited luminous flux. Attempts to increase lighting capacity were inevitably destined to clash with the problems of fuel supply, heat development, oxygen consumption and with the practical difficulty of conveying and directing the luminous flux. Candles were very expensive and were used on special occasions by the most affluent social classes. At the end of the nineteenth century, the incandescent electric light bulb was invented. Created in 1878 by Thomas Alva Edison, these bulbs began to spread and be sold in large quantities by General Electric (GE) only after 1911.

The lighting of the interior at the beginning of the twentieth century was divided between the functional and basic in the workplace and the decorative and artistic in the homes of the more affluent classes. The development of rationalist thinking and the relationship between function and form found an unquestionable technological application in the creation of luminaires for workplaces: simple, streamlined forms of suspended luminaires for the projection of the luminous flux in the area used for visual tasks. The same shapes that can still be found today in mass production.

In the decorative and artistic production of indoor lamps, the incandescent source found a ready application in the production of unique and precious pieces, enriched by the use of materials such as alabaster and mother-of-pearl combined with coloured glass and gold and silver metals, such as the Louis Comfort Tiffany table lamps (Eidelberg et al. 2005). The form was not dictated by functional needs, but by artistic and scenic choices that took inspiration from the pictorial and sculptural productions of the artistic culture of the time. On the contrary, the more economic productions still saw for many years, even up to the present day, the repetition of forms prior to the advent of electric light such as, for example, the candlestick, the lampshade and the lantern.

In the context of the lighting project applied to work activities, the development of industrial production and the assembly line, activities that are forcedly carried out indoors, thanks to electrical lighting could finally achieve constant production processes independent of the geographical and environmental conditions in which these activities are carried out. The development of the service sector and office activities are also favoured by artificial lighting, as well as the commercial activities that, from the shop window to the vast areas of the department stores, have needs related to product display and the accessibility and safety of people protected from external atmospheric agents and with controllable and comfortable lighting.

## 1.4 The Progress of Electric Light Sources Before LEDs

The first source of electrical light experimentally created with the purpose of lighting was invented by Humphry Davy in the first decade of 1800. He used an enhanced version of Alessandro Volta's pile to produce an electric arc between two carbon electrodes. However, his invention did not find a practical application until 1870, since the arc bulb needed direct current, which became more available only after

**Fig. 1.2** Incandescent lamp with tungsten filament, halogen lamp, compact fluorescent lamp

the invention of the dynamo. Furthermore, during its operation the carbon electrodes were quickly consumed and therefore required a clockwork mechanism bringing them closer together so as to keep the distance constant. This type of lamp could produce a lot of light but it was also inefficient, it needed continuous maintenance and could cause fires. It was used until the first decade of 1900 in street lighting and large factories.

The first real revolution in electrical lighting came with the invention in 1878 of the incandescent lamp with carbonized cotton thread by Thomas Edison. Thanks to the improvements introduced by William David Coolidge who invented a method for working tungsten, to be able to use it as a filament of the lamp, in 1911 the mass marketing of incandescent lamps began (Fig. 1.2 left). Thus began the sale to the large masses of the classic light bulb for the home. This product was also one of the first subject to a multinational cartel to manage its planned obsolescence, control its sale and maximize profits by agreeing on sales prices. The Phoebus cartel was an agreement made in December 1924 between GE, Osram, Philips and other companies to establish that light bulbs should not last more than 1,000 h and control their prices and sales worldwide (MacKinnon 2016). The agreement should have lasted until 1955 but was interrupted by the Second World War and a legal action initiated in 1942 by the US government against General Electric and other companies of the cartel for the violation of antitrust regulations. Long-life incandescent lamps were made before the Phoebus agreement was launched. As reported by the Guinness Book of World Records, at the firehouse in Livermore, California, there is an incandescent lamp, handmade at the end of the nineteenth century, which has been lit almost uninterruptedly since 1901 and has exceeded one million hours of operation (GWR 2015).

In 1959, GE patented the halogen incandescent lamp that was perfected in subsequent phases until the 80s, when it began to spread in the interiors thanks to the dichroic spotlights. Most incandescent lamps are no longer sold today because they are not very energy efficient, but there are still many in the residences around the world (Fig. 1.2 centre). Their definitive elimination from the European market, except for those for special uses, happened in September 2018.

Another type of lamp is that in which light is produced by an electric discharge, in a gas contained in a transparent ampoule that can be made of glass, quartz or ceramic. In 1901, Peter Cooper Hewitt invented the mercury vapour discharge lamp, which was perfected in 1930 by GE, which decreased the amount of mercury and started a wide dissemination of this product for general lighting. Based on previous research by other scientists in 1911, Georges Claude created the first neon discharge lamp for practical use. This lamp did not have an illuminating function but was used for signalling and is still used today in advertising signs. In 1920, the Nobel Prize laureate for physics Arthur Holly Compton, while working at the Westinghouse Lamp Company, invented the sodium vapour lamp. This was later perfected by Marcello Pirani in Osram and in 1932, Philips began marketing sodium vapour lamps for lighting large areas in the Netherlands. In 1962, Robert Reiling perfected his previous research creating the first metal halide discharge lamp, whose light offers superior quality performances compared to those of mercury and sodium vapours.

A particular category of discharge lamps, which is still the most widespread in the world for interior lighting, is the one in which the coloured light, or the ultraviolet radiation, produced by the discharge in the gases, is modified by fluorescent materials that change the spectral composition to obtain white light. Also based on the research previously carried out by other scientists, in 1926 Edmund Germer patented the fluorescent lamp containing mercury vapour, which was perfected in the following years and whose commercialisation began about ten years later by GE. It is estimated, however, that only in 1951 the lighting produced by fluorescent lamps succeeded in overcoming the one produced by incandescent lamps in terms of dissemination. In 1976, also as a result of the 1973 oil and energy crisis, Edward E. Hammer created the compact fluorescent lamp (CFL) in the General Electric research laboratories. However, Philips launched the first compact fluorescent lamp on the market, the SL*18, in 1984. It was the first non-incandescent lamp proposed as an alternative to incandescent lamps: with a consumption of 18 W, it produced a flux equivalent to that of a 75 W incandescent lamp. Since the 90s the CFL, thanks to a further increase in efficiency and duration and the reduction in size, weight and prices, has begun to seriously affect the market for residential lighting until then dominated by incandescent lamps (Fig. 1.2 right).

## 1.5 Relationship Between Artificial Light and Natural Light

In the last century, the availability and the wide dissemination of artificial light made it possible to change the styles, activities and quality of life of an ever-wider segment of the population and in a very short period of time compared to the evolution of human beings. This today certainly applies to industrialised countries, where artificial lighting is now comparable to primary consumer goods such as water and bread. Most modern work, manual and intellectual activities, as well as recreational leisure time, often avoid direct sunlight because it is too intense, dazzling and difficult to control. On the other hand, the light diffused by the atmosphere, that of the sky, and that filtered by the clouds, is more pleasant. For many centuries, curtains and blinds have been commonly used to control the entry of natural light in interiors.

We also know that window glass reduces the part of solar radiation necessary for vitamin D synthesis by about 95% (Boubekri 2008). If on the one hand direct sunlight is therefore essential for the synthesis of vitamin D, on the other hand many information campaigns have been promoted, especially in the last 20 years, to warn white populations against the risks of excessive exposure to sunlight, because it can cause long-term skin cancers (ISO/CIE 28077:2016).

But we must distinguish between direct exposure of the skin to sunlight and natural light understood in a more general sense as the sum of the light of the sky and the light of the sun. The sky can have extremely variable climatic conditions that modify the spectral content of the light diffused by the earth's atmosphere and that can obscure direct sunlight. The light of the sky is a form of indirect lighting, because the gases present in the atmosphere profoundly modify the spectrum of the radiations present in the sunlight. For this reason, we see a yellow sun and a blue sky, or a white sky when it is covered by clouds. Ultimately, with the term "natural light" we mean all light that is certainly produced by the sun, which changes going through the atmosphere, but also by the integration with the light diffused by the sky and the light reflected from the ground, water and objects present on the territory, be it the prairies, the trees or the buildings.

The CIE (Commission Internationale de l'Éclairage) has described the spectral content of natural light with the standard D-series illuminants (CIE 015:2004) that are defined according to the correlated colour temperature (CCT) of natural light, which may vary from the neutral white light of the overcast sky (D50) to that of the sun plus the sky at midday at European latitudes (D65), to that of the northward sky characterised by a colder white light, tending towards blue (D75). The relative spectral power distribution (SPD) of D65 natural light is shown in Fig. 1.3. In the graph the chromatic content of the light at midday is visible in the situation in which the direct sunlight and the diffused light from the sky are added.

It is important to distinguish the effects of direct sunlight on the skin from the physiological effects of light, whether natural or artificial, when it enters our eyes and is understood in the context of the illumination that can be achieved in the exteriors and interiors of buildings (CIE 158:2009; CIE 2015).

As we have seen, human activities are mainly carried out indoors, in environments where the use of natural light can be made difficult by chaotic and unreasonable urban development, geographical location, climatic conditions or even by design needs for an efficient use of urban spaces. Since the end of the nineteenth century, there has been a profound change in society (Bell 2006) and human activities no longer follow the rhythms of natural light, both work and leisure time take place also during the evening and night hours (Brox 2011). In a well-known epidemiological research, Stevens and Rea (2001) have analysed the relationship between the lack of natural

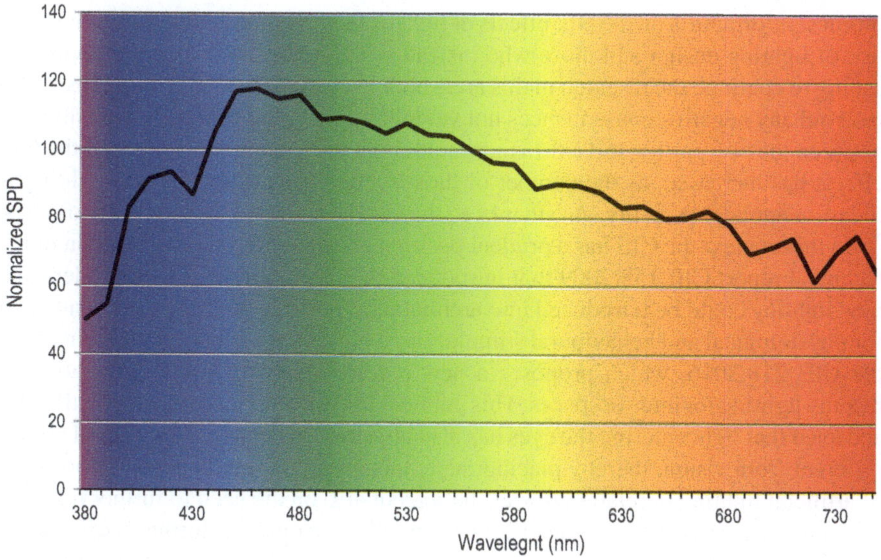


**Fig. 1.3** Relative spectral power distribution (SPD) of D65 natural light

light in the interiors and the endocrine disruptors, which can be the cause of health problems that include the reduction of melatonin, such as stress, sleep disorders, changes in the menstrual and estrogen cycle and the impact of these elements on breast cancer risk.

The duration of what, in the common sense, we call night, has in fact been reduced. Artificial light integrates and indeed, more often than not, replaces natural light in our visual perception, both indoors and outdoors, when this is not available due to the reduction of the duration of solar exposure for climatic, urban or seasonal reasons. If, from the standpoint of visual perception, artificial lighting can positively replace or supplement natural light indoors, this does not apply to the circadian aspects. Indeed, the physiological effects of artificial lighting, now present in almost all living spaces of human beings, are profoundly different from those certainly more beneficial provided by natural lighting (Boyce 2014).

## 1.6 The Position of International Bodies on NIF Effects

Regarding the non-image forming (NIF) effects of light, there is no common consensus on the applicability of scientific discoveries concerning the relationship between artificial light and the circadian rhythm in the practice of lighting design. Even before the discovery of melanopsin and intrinsically photosensitive retinal ganglion cells (ipRGCs), there was a fundamental difference in opinion among those who wanted this new information on the NIF effects of light to be immediately inserted in the practice of lighting design and those who instead advocated a more cautious approach, asking to continue the research in this area, both to evaluate the beneficial effects and to avoid any negative consequences not yet highlighted by the studies. This division between the two parties still exists today, although the scientific evidence of the NIF effects has increased, as the number of those who believe that circadian lighting is today an applicable reality, also thanks to the new LED-based lighting technologies.

In this context the CIE has a prudent position, because after the publication of the technical report CIE 158:2009 that introduced the first studies useful to understand how lighting could be introduced into architectural projects optimised for vision and for physiological and behavioural stimulation, it presented a subsequent publication, the CIE 218:2016, which proposes a new research path for the identification of healthy lighting for interior spaces. This publication, however, considers scientifically accepted that light entering the eyes has also physiological and psychological effects different from vision, thereby placing the emphasis on the fundamental importance of artificial lighting in relation to well-being, but also on the need to further research this area. The aspects that, according to the CIE, should be further explored by the research concern the fundamental processes, the patterns of daily exposure to light, the analysis of longer exposure times, the relationship with the lighting design applications and the natural differences between individuals. In addition, the JTC9 Joint Technical Committee is working on the preparation of the "CIE system for Metrology of ipRGC influenced light response" standard (CIE JTC9 2018), which aims at defining the spectral sensitivity functions, the quantities and the metrics, to describe the ability of light to stimulate the five types of photoreceptors, S cones, M cones, L cones, rods and ipRGCs, which can contribute to the non-visual effects of light in humans.

In parallel, however, some national and supranational bodies have published in recent years more detailed rules, which provide the first guidelines at the scientific and applicative level. This for example has been the case in the United States, with the rule on the relationship between light and health of human beings (IES TM-18-18 2011) and with the one that defines the guidelines for lighting in hospitals (IES/ANSI RP-29-16 2017). Also in Germany a standard has been published that defines and quantifies the ocular effects of light on melatonin (DIN SPEC 5031-100 2015).

In Europe there is the CEN/TC 169 technical committee, which includes the working group "WG13 Non-visual effects of light on human beings", which has already published the technical report CEN/TR 16791:2017, that actually anticipates at European level the possible contents of the CIE-JT9 "CIE system for metrology of ipRGC influenced light response". This technical report proposes metrics that can be used to evaluate lighting on the human eye with respect to the possibility of generating NIF effects. To this end, the spectral sensitivity of the five photoreceptors on the retina is considered in relation to the optical radiations visible in the wavelength range from 380 to 780 nm.

Also at the European level, the Scientific Committee on Emerging and Newly Identified Health Risks (SCENIHR) has expressed an opinion on the relationship between artificial lighting and human health (SCENIHR 2012). This body has been established by the European Commission to provide scientific advice on emerging or newly identified health and environmental risks, on broad, complex or multidisciplinary issues requiring a global risk assessment regarding consumer safety and public health. In this document, ample space has been dedicated to a review of epidemiological research that has shown that inadequate artificial lighting can alter the circadian rhythm with consequent reduction of melatonin. These two joint effects have been also evaluated as a factor leading to a moderately increased risk of breast cancer, gastrointestinal problems, cardiovascular disorders and affective disorders. Further evaluated effects concern sleep disturbances caused by the short wavelength (blue) light component, possibly also present in the cool white light, and the effects this may have on the mood, the level of vigilance and the cognitive performance. However, they emphasized that these effects depend very much on the characteristics of the light sources used and primarily on the lighting project affecting individuals in their everyday life. It is interesting to note that the opinion expressed by this committee of experts focuses on what harmful effects artificial lighting may cause, with reference to design parameters such as the amount of light, its spectral composition and its daily exposure pattern, but makes no assessment of the lighting project in the environmental context and in relation to the individual.

Subsequently, the European Commission has set up a new committee, the Scientific Committee on Health, Environmental and Emerging Risks (SCHEER), which is working on the issue of possible risks to human health in the use of LEDs and has published a document on this topic (SCHEER 2018). In this document particular attention is paid to the question of short wavelengths of light, which may be present in artificial lighting. With reference to this aspect, SCHEER has concluded that there is no scientific evidence that the blue light component present in LEDs can have harmful effects on human health. The committee however pointed out that the short wavelengths of light produced by all light sources, including television, computer and tablet screens, if used in the late evening can have an impact on the human circadian system and cause possible sleep disorders, effects that have been already proven for some time in many epidemiological researches also for artificial light sources other than LEDs. Furthermore, for children under three years of age, who have a higher sensitivity to blue light, the incorrect use of very bright cool white light sources, which contain a high amount of blue light, could induce retinopathies in the long run.

SCHEER has also highlighted that there are studies done on animal cells that have shown adverse effects due to the blue light component. However, these researches have been carried out using conditions and levels of exposure far in excess of the normal conditions of exposure of humans to artificial lighting produced by LEDs or other light sources. These surveys therefore state that, given the fact that the technology of light sources is rapidly evolving, further long-term epidemiological research is certainly desirable.

An important design consideration, derived from the analysis of the documents produced by SCENIHR and SCHEER, is the fact that a correctly designed, installed and used circadian lighting is not harmful; on the contrary, some harm could be caused by the artificial lighting, aimed exclusively at the objectives of vision, which is still used in working and residential areas, in the absolute unawareness of the non-visual effects of artificial light on the human well-being. In particular, the availability on the market of fluorescent light sources or LEDs, with cool CCT, used unconsciously in the evening hours, can affect the correct human circadian rhythm. Another harmful factor could be, for the normal sleep cycle, the use of modern backlit screens with white LEDs, which have a considerable content of blue light radiations, present in modern television, computer and tablet screens (Cajochen et al. 2011; Chang et al. 2015; Green et al. 2017). It is for this reason that recently the manufacturers of the operating systems of these devices have introduced a feature that can be activated to automatically decrease the blue light component emitted from the screen from dusk to dawn; in this operating mode the screens acquire a warm white colour due to the decrease of the blue light component.

The International Association of Lighting Designers (IALD), by its very nature close to the evidence and needs of professional practice, has published a white paper (Ladopoulos and Shaw 2017) in which it introduces the theme of a holistic lighting design aimed at the objectives of human centric lighting (HCL). Among these objectives the theme of circadian lighting plays a central role, among other topics, such as LEDs innovation, flickering, photobiological safety of light, the colour rendering index and the regulatory situation in the sector. This document introduces the topic starting from the scientific researches in the sector but maintains a strong applicative and informative connotation. It is noted that, with the advent of LEDs, industries have gradually proposed new products for circadian lighting or more generally for the well-being of people indoors. However, the use of these products should not be considered equivalent to a portentous remedy, but rather a new lighting tool, more flexible and able, potentially, to adapt to the needs of designers and users, becoming an interesting prospect for lighting designers in the very near future. Emphasis is placed on the fact that any benefit can only derive from the correct use of the products, which no longer serve only a function useful for vision but also for well-being, provided they are correctly installed and used. The risk to consider is that inadequate use could instead produce potentially harmful effects on health. But this consideration is even more true for other types of technological products or products in the food sector that are misused or used excessively in industrialised countries. The IALD therefore provocatively poses the question of whether the lighting designer of the future should also be equipped with a degree in medicine, concluding that the designer must be responsible and always informed about the scientific and technological innovations of the sector and the new rules and recommendations that are proposed by international bodies.

Also the International Light Association (ILA 2003), active since 2003 in the therapeutic context, has emphasised the importance of a joint evaluation of light and colour in the treatment of some pathologies. This association organises a yearly conference gathering light and chromotherapy professionals, healthcare and alternative medicine professionals, scientists, educators, companies, artists, architects and interior designers who are interested in applying the latest methods and technologies for using light and colour to promote well-being, improve performance, facilitate learning and increase people's moral, social and political awareness.

## 1.7 The Position of the Manufacturers

In the middle of the first decade of 2000, on the basis of the first research on the relationship between artificial lighting and the well-being of individuals, some companies introduced interior lighting products, defined as biodynamic, and designed to synchronize the human circadian cycle (van Bommel 2006, Centro studi e ricerca iGuzzini 2007; Rossi et al. 2009). However, those early experiments were based on linear fluorescent lamps that are not suitable to be used to create the circadian lighting, precisely because of their technological characteristics, their size and the difficulty in managing their dimming with the power systems. In the last ten years, these biodynamic products have mostly gone out of production, due to high costs and management difficulties, but above all because they have not been understood and accepted by designers and users. In addition, in the last 10 years, companies in the sector have committed themselves to accept and implement the new solid-state lighting technology, i.e. LEDs, which have largely replaced the other artificial light sources. LEDs are photoelectronic components and the manufacturing companies, which had operated in the electromechanical sector for decades, had to convert into companies with a greater focus in the electronics sector (Romm 2016). This process, indispensable for the survival of the companies themselves, has absorbed most of the resources available for research and development, relegating the circadian theme to the background.

Today the situation has radically changed. The LEDification process has reached a stage of maturity in which the fundamental topic is no longer that of incorporating the new solid state lighting technology, but is to explore all possible lighting applications that make it possible to exploit the potential of this new lighting technology in the best possible way. The LightingEurope association, which gathers over a thousand manufacturers in the lighting sector in the European Union, declares four areas of research and development as fundamental:

- LEDification
- Intelligent Lighting Systems
- Human Centric Lighting
- Circular Economy

The term LEDification refers to the process by which LEDs have been progressively replacing traditional light sources such as incandescent lamps, halogen lamps, gas discharge lamps and fluorescent lamps over the last few years. LEDs make it possible to implement energy-efficient solutions and at the same time also improve the quality of light. New operators, coming from the electronics sector, are entering this market. This process has also changed the total value chain in the lighting products sector. Indeed, luminaire manufacturers can now integrate LED light sources directly into luminaires and in many cases there is no longer a physical separation between the light source and the luminaire; the life cycle of products has increased compared to traditional light sources. Finally, lighting systems can and must become easier to configure and install, in order to be called smart lighting.

The introduction of LED technology is changing the lighting products from analogue to digital, implementing the new concept of Intelligent Lighting System (ILS). An ILS is part of an interconnected system. As buildings are becoming more intelligent and must operate as connected ecosystems, lighting can become the backbone of an intelligent building. From the operational standpoint, lighting is an autonomous system but it is connected to other building components such as the local network, video surveillance systems, sensors, natural light control, air conditioning systems and other systems in the building. Therefore, an ILS will be able to provide greater control over the quality, flexibility and adaptability of lighting in the building.

The theme of Human Centric Lighting is considered of fundamental importance by LightingEurope, jointly with the LEDification process, which should therefore not be understood as aimed at energy saving only. It is important to note that in this context, indeed, the definition of HCL is very similar to the concepts of circadian lighting or biodynamic illumination introduced in previous years. While according to a more recent vision, expressed by other bodies and authors, circadian lighting would be one of the key elements and actually the fundamental part of the HCL, but along with other design factors (CIE TN 003:2015; IALD 2017).

It is important to highlight that a product cannot simply be defined as an HCL enabler, because the concept of this new lighting must always be associated with the context, which also includes people, time and interior spaces, i.e. the application of the lighting project as a whole. LightingEurope therefore emphasises that artificial lighting can do much more than simply enabling a correct visual function. Light can energise, promote relaxation, improve mood levels, increase the level of subjective vigilance and cognitive performance, but also improve the sleep-wake cycle of human beings. That is, it can support the health, well-being and performance of human beings, combining the benefits of a correct vision with the biological and emotional benefits of light. The true value of the artificial lighting of the future will therefore be in the combination of excellent visual, biological and emotional benefits. LEDs can do much more than facilitate simple, albeit important, energy savings. In this sector, the association of European producers has also supported the development of the Lighting for People portal (SSL-erate 2016) developed as part of the SSL-erate research funded by the European Commission under the 7th Framework Program.

Finally, LightingEurope shares the principles of the circular economy and believes that with the advent of LEDs and the goal of increasing the life of LED lighting systems for end users, the lighting industry can be among the leaders in the area of energy efficiency for industrial products. In this sector there is a great deal of experience in prolonging product life as well as in collecting, recycling and reducing hazardous substances. The goal is to further reduce the environmental footprint of this industrial sector.

The concept of circular economy is used to define an economic system that regenerates itself and in which the incoming material resources, waste, emissions and energy dispersions are kept to the minimum, trying to maintain the energy consumption circuits and the materials circuits as much as possible self-sufficient. These objectives are achieved through a long-term design, the enhancement of maintenance and repair processes, reuse, regeneration of products and systems and recycling. It is therefore a development model that is different from the classical one, defined linear, in which there is an input, a production and sale of products, and an uncontrolled disposal of the same at the end of their life cycle (Geissdoerfer et al. 2017).

## 1.8 Building Certifications and NIF Effects

Since lighting is one of the fundamental systems in a building, another aspect to consider is the relationship between circadian lighting and building certification.

With regard to the energy certification of buildings, there are various national and international standards, such as the Leadership in Energy and Environmental Design (LEED) in the United States (USGBC 2014), the Building Research Establishment Environmental Assessment Method (BREEAM) in the United Kingdom (BRE 2018), but also adopted in other countries, the Green Star in Australia (GBCA 2003), the Excellence in Design for Greater Efficiencies (EDGE) promoted by the World Bank Group in developing countries (IFC 2012) and various national certifications in European Union countries that were based on Directive 2002/91/EC, now replaced by Directive 2010/31/EU on the energy performance of buildings. In the various national contexts of the EU, the European directives are then implemented through national laws and applied thanks to the sectoral legislation. Although some of these are being revised to also consider aspects of people's well-being in relation to lighting, to date, the certification of buildings only considers lighting aspects in relation to energy saving and possibly the management of natural light.

There is also the recent Well Building Standard certification (IWBI 2014), which strongly focuses on the health and psychological-physical well-being of people in buildings. This certification, based on scientific research in various sectors, takes into consideration the characteristics of the building that have a direct impact on human health and well-being, such as the quality of air and water, the possibility of promoting proper nutrition and fitness activities, comfort in general and lighting.

As far as lighting is concerned, the guidelines laid down by the WELL standard aim at minimising the negative interference that the lack of natural light or inadequate artificial light can produce on the normal human circadian rhythm. They also aim at improving productivity, promoting a good quality of sleep and ensuring an appropriate and comfortable visual function. Starting from these objectives, among the main characteristics of light, natural and artificial, which are controllable through an adequate design, the guidelines consider some fundamental elements. Firstly, a lighting design aimed at the needs of visual perception, by defining the lighting levels needed for the various activities and environments, ensuring glare containment in artificial sources, with the aim of ensuring a comfortable vision. Secondly, the circadian lighting design assessed according to the type of environments and with a metric that is alternative to classical photometry. This type of lighting must be able to adapt dynamically in order to avoid circadian stimulation in the evening and night hours, while it must favour it during the day in the absence of an adequate amount of natural light. It is also important to manually and automatically control the contribution of natural light in order to optimise its use and minimize the energy consumption of artificial light, and to avoid glare, caused by uncontrolled natural light, during its normal daily variability and thus facilitating the availability of an adequate dose of natural light to the occupants. Equally important is a high colour rendering, aimed at a natural visual perception of colours as a factor contributing to visual comfort. And, last but not least, a surface and colour design properly related to lighting design by controlling the reflection factors of the surfaces used in the interior design, in order to contain energy consumption while maintaining good visual comfort. These surfaces in fact contribute decisively to the amount of indirect lighting.

The WELL standard is quite recent, but it is raising a lot of interest worldwide. The International WELL Building Institute (IWBI), which has defined this standard, has immediately started collaborating with the LEED, BREEAM and Green Star standards to identify common working methods and provide guidelines to mutually implement the Well and the other standards mentioned. In particular, also the latest version of the BREEAM standard (BRE 2018), for new buildings, highlights the importance of natural light for the correct stimulation of the circadian system and the well-being of individuals. The many topics covered by the WELL recall, in turn, the concepts of holistic design mentioned in the IALD white paper (Ladopoulos and Shaw 2017).

## References


2010/31/EU Directive of the European Parliament and of the Council of 19 May 2010 on the Energy Performance of Buildings. Available at: https://eur-lex.europa.eu/legal-content/EN/ALL/?uri=CELEX%3A32010L0031. Accessed May 20, 2017.

Aubé, M., Roby, J., & Kocifaj, M. (2013). Evaluating potential spectral impacts of various artificial lights on melatonin suppression, photosynthesis, and star visibility. *PLoS ONE, 8*(7), e67798. https://doi.org/10.1371/journal.pone.0067798.

Bargiello, T. A., Jackson, F. R., & Young, M. W. (1984). Restoration of circadian behavioural rhythms by gene transfer in Drosophila. *Nature, 312*(5996), 752–754. https://doi.org/10.1038/312752a0.

Bell, D. (2006). In J. Hollows (Ed.), *Historicizing lifestyle: Mediating taste, consumption and identity from the 1900s to 1970s* (1st ed). Aldershot, England; Burlington, VT: Routledge.

Boubekri, M. (2008). *Daylighting, architecture and health: Building design strategies*. Oxford: Elsevier.

Boyce, P. (2014). *Human factors in lighting* (3rd ed.). Boca Raton: CRC Press.

BRE. (2018). *Building Research Establishment Environmental Assessment Method (BREEAM)*. Available at: https://www.breeam.com/. Accessed May 14, 2018.

Brox, J. (2011). Artificial light: How man-made brightness has changed the way we live. *The Independent*. Available at: http://www.independent.co.uk/life-style/gadgets-and-tech/features/artificial-light-how-man-made-brightness-has-changed-the-way-we-live-and-see-forever-2282563.html. Accessed April 18, 2018.

Cajochen, C., et al. (2011). Evening exposure to a light-emitting diodes (LED)-backlit computer screen affects circadian physiology and cognitive performance. *Journal of Applied Physiology* (Bethesda, MD.: 1985), *110*(5), 1432–1438. https://doi.org/10.1152/japplphysiol.00165.2011.

CEN/TR 16791:2017 Quantifying irradiance for eye-mediated non-image-forming effects of light in humans.

Centro studi e ricerca iGuzzini. (2007). *La ricerca Sivra. In More than vision* (1st ed, pp. 50–59). Recanati: iGuzzini - Editoriale Domus (ilibriGuzzini).
Chang, A.-M., et al. (2015). Evening use of light-emitting eReaders negatively affects sleep, circadian timing, and next-morning alertness. *Proceedings of the National Academy of Sciences of the United States of America, 112*(4), 1232–1237. https://doi.org/10.1073/pnas.1418490112.
CIE 015:2004 Colorimetry, 3rd edition.
CIE 139-2001 The Influence of Daylight and artificial light variations in humans. A bibliography.
CIE 158:2009 Ocular lighting effects on human physiology and behaviour.
CIE 2015 Statement on non-visual effects of light—Recommending proper light at the proper time (2015, June 28). Available at: http://www.cie.co.at/publications/statement-non-visual-effects-light-recommending-proper-light-proper-time-june-28-2015. Accessed April 21, 2017.
CIE 218:2016 Research roadmap for healthful interior lighting applications.
CIE JTC9. (2018). *CIE system for metrology of ipRGC influenced light response*. Available at: http://www.cie.co.at/technicalcommittees/cie-system-metrology-iprgc-influenced-light-response. Accessed March 20, 2018.
CIE TN 003:2015 Report on the First International Workshop on Circadian and Neurophysiological Photometry, 2013. Available at: http://www.cie.co.at/publications/report-first-international-workshop-circadian-and-neurophysiological-photometry-2013. Accessed February 20, 2017.
de Candolle, A. P. (1832). *Physiologie végétale. Ou exposition des forces et des fonctions vitales des végétaux, pour servir de suite a l'organographie végétale, et d'introduction a la botanique géographique et agricole*. Paris: Béchet jeune.
de Mairan, J.-J. (1729). *Observation botanique*. Bibnum Education. Available at: https://www.bibnum.education.fr/sciencesdelavie/biologie/observation-botanique. Accessed January 27, 2018.
DIN SPEC 5031-100. (2015). Optical radiation physics and illuminating engineering—Part 100: Melanopic effects of ocular light on human beings—Quantities, symbols and action spectra.
Eidelberg, M., et al. (2005). *The lamps of Louis Comfort Tiffany*. New York: Vendome Press.
Evans, G. W. (2003). The built environment and mental health. *Journal of Urban Health: Bulletin of the New York Academy of Medicine, 80*(4), 536–555. https://doi.org/10.1093/jurban/jtg063.
Evans, J. A., & Davidson, A. J. (2013). Health consequences of circadian disruption in humans and animal models. *Progress in Molecular Biology and Translational Science, 119*, 283–323. https://doi.org/10.1016/B978-0-12-396971-2.00010-5.
Foster, R., & Kreitzman, L. (2005). *Rhythms of life. The biological clocks that control the daily lives of every living thing*. Yale University Press.
Gaston, K. J., et al. (2014). Human alteration of natural light cycles: Causes and ecological consequences. *Oecologia, 176*(4), 917–931. https://doi.org/10.1007/s00442-014-3088-2.
Gaston, K. J., Visser, M. E., & Hölker, F. (2015). The biological impacts of artificial light at night: The research challenge. *Philosophical Transactions of the Royal Society B, 370*(1667). https://doi.org/10.1098/rstb.2014.0133.
GBCA. (2003). *Green Star*. Available at: https://new.gbca.org.au/green-star/. Accessed May 14, 2018.
Geissdoerfer, M., et al. (2017). The circular economy—A new sustainability paradigm? *Journal of Cleaner Production, 143*, 757–768. https://doi.org/10.1016/j.jclepro.2016.12.048.
Green, A., et al. (2017). Evening light exposure to computer screens disrupts human sleep, biological rhythms, and attention abilities. *Chronobiology International, 34*(7), 855–865. https://doi.org/10.1080/07420528.2017.1324878.
GWR. (2015). *Longest burning light bulb, Guinness World Records*. Available at: http://www.guinnessworldrecords.com/world-records/longest-burning-light-bulb. Accessed July 29, 2018.
Halberg, F., et al. (2003). Transdisciplinary unifying implications of circadian findings in the 1950s. *Journal of Circadian Rhythms, 1*(0). https://doi.org/10.1186/1740-3391-1-2.
Hardin, P. E., Hall, J. C., & Rosbash, M. (1990). Feedback of the Drosophila period gene product on circadian cycling of its messenger RNA levels. *Nature, 343*(6258), 536–540. https://doi.org/10.1038/343536a0.

Hatori, M., et al. (2017). Global rise of potential health hazards caused by blue light-induced circadian disruption in modern aging societies. *npj Aging and Mechanisms of Disease, 3*(1), 9. https://doi.org/10.1038/s41514-017-0010-2.

Hui, P.-S., Mui, K.-W., & Wong, L.-T. (2008). Influence of indoor air quality (IAQ) objectives on air-conditioned offices in Hong Kong. *Environmental Monitoring and Assessment, 144*(1–3), 315–322. https://doi.org/10.1007/s10661-007-9994-6.

IALD. (2017, February). *Joint position paper by LightingEurope and the International Association of Lighting Designers (IALD) on Human Centric Lighting*. Available at: https://www.iald.org/Advocacy/Publications. Accessed September 30, 2018.

IES TM-18-18. (2011). Light and human health: An overview of the impact of light on visual, circadian, neuroendocrine and neurobehavioral.

IES/ANSI RP-29-16. (2017). Lighting for hospitals and healthcare facilities.

IFC. (2012). *Excellence in design for greater efficiencies (EDGE)*. Available at: https://www.edgebuildings.com. Accessed May 14, 2018.

ILA. (2003). *International Light Association*. Available at: https://www.international-light-association.org/. Accessed May 20, 2018.

ISO/CIE 28077:2016 Photocarcinogenesis action spectrum (non-melanoma skin cancers).

IWBI. (2014). *The WELL Building Standard*. International WELL Building Institute. Available at: https://www.wellcertified.com/en/explore-standard. Accessed May 14, 2018.

Ladopoulos, I., & Shaw, K. (2017). *IALD white paper: Lighting design for health, wellbeing and quality of light, a holistic approach on human centric lighting*. IALD. Available at: http://iald.org/News/Reflections-Newsletter/IALD-REFLECTIONS-24-February-2017. Accessed May 8, 2018.

Liu, X., et al. (1992). The period gene encodes a predominantly nuclear protein in adult Drosophila. *The Journal of Neuroscience: The Official Journal of the Society for Neuroscience, 12*(7), 2735–2744.

Luginbuhl, C. B., Boley, P. A., & Davis, D. R. (2014). The impact of light source spectral power distribution on sky glow. *Journal of Quantitative Spectroscopy & Radiative Transfer, 139*, 21–26. https://doi.org/10.1016/j.jqsrt.2013.12.004.

MacKinnon, J. B. (2016). *The L.E.D. quandary: Why there's no such thing as "Built to Last", The New Yorker*, 14 July. Available at: https://www.newyorker.com/business/currency/the-l-e-d-quandary-why-theres-no-such-thing-as-built-to-last. Accessed July 29, 2018.

Mead, M. N. (2008). Benefits of sunlight: A bright spot for human health. *Environmental Health Perspectives, 116*(4), A160–A167.

Price, J. L., et al. (1998). double-time is a novel Drosophila clock gene that regulates PERIOD protein accumulation. *Cell, 94*(1), 83–95.

Rich, C., & Longcore, T. (2006). *Ecological consequences of artificial night lighting*. Island Press.

Romm, J. (2016). *5 Charts that illustrate the remarkable LED lighting revolution*. Available at: https://thinkprogress.org/5-charts-that-illustrate-the-remarkable-led-lighting-revolution-83ecb6c1f472/. Accessed October 13, 2017.

Rossi, M., Siniscalco, A., & Zanola, F. (2009). From physiology to a new sustainable lighting design: The "My White Light" case study. In: Multiple Ways to Design Research Symposium 2009, pp. 272–278. Lugano: Swiss Design Network.

SCENIHR. (2012). *Health effects of artificial light*. European Commission.

SCHEER. (2018). *Opinion on potential risks to human health of Light Emitting Diodes LEDs*. European Commission.

Siwicki, K. K., et al. (1988). Antibodies to the period gene product of Drosophila reveal diverse tissue distribution and rhythmic changes in the visual system. *Neuron, 1*(2), 141–150.

SSL-erate. (2016). *Lighting for people—Web-based platform for high quality information on solid state lighting in Europe*. Available at: http://lightingforpeople.eu/. Accessed October 13, 2017.

Stevens, R. G. (1987). Electric power use and breast cancer: A hypothesis. *American Journal of Epidemiology, 125*(4), 556–561.

Stevens, R. G., & Rea, M. S. (2001). Light in the built environment: Potential role of circadian disruption in endocrine disruption and breast cancer. *Cancer Causes & Control: CCC, 12*(3), 279–287.

UNICEF. (2008). *State of the world's children 2008—Child survival*. Available at: https://www.unicef.org/sowc08/. Accessed April 18, 2018.

USGBC. (2014). *Leadership in energy and environmental design*. Available at: https://new.usgbc.org/leed. Accessed May 14, 2018.

van Bommel, W. (2006). Dynamic Lighting at work-both in level and colour. In *2nd CIE expert symposium "Lighting and Health"*. Ottawa:CIE.

Vosshall, L. B., et al. (1994), Block in nuclear localization of period protein by a second clock mutation, timeless. *Science (New York, N.Y.), 263*(5153), 1606–1609.

Yadav, A., Verma, P., & Singh, S. (2017). Going beyond the limits: Effect of clock disruption on human health. *Biological Rhythm Research, 48*(5), 693–700. https://doi.org/10.1080/09291016.2017.1345428.

Zehring, W. A., et al. (1984). P-element transformation with period locus DNA restores rhythmicity to mutant, arrhythmic Drosophila melanogaster. *Cell, 39*(2 Pt 1), 369–376.

# Chapter 2
# The Human Circadian System

## 2.1 Introduction on the Effects of Artificial Light on Human Beings

For many years, the research in the lighting sector has asked the question whether artificial light should, and for what reasons, try to reproduce some of the qualitative aspects of natural light. It is by now a shared opinion, in the contexts of research on the design of artificial lighting, that the electric light used in the last century in the interiors, while fulfilling in an increasingly effective manner the needs of human visual perception, is however lacking in the aspects that concern the physiological well-being, the influence on mood and the effects it can have on human behaviour (CIE TN 003:2015; IALD 2017). In this context, we talk about non-visual effects or non-image forming (NIF) effects of light.

Some research conducted in the past, before the discovery of melanopsin (Freedman et al. 1999), have produced results that are in part contradictory and difficult to interpret. Both because these results have to be evaluated based on human psychology and physiology, and for the absence of shared protocols useful for quantifying light and evaluating the effects that it has with respect to the human NIF system (Lucas et al. 2014). However, in the last few years, more defined results have been achieved that allow us to outline some useful guidelines for lighting design. In these guidelines the idea of wanting to reproduce some of the effects of natural light is an undoubted guiding thread, highlighted by scientific research in the fields of medicine and physiology.

Research in this area has had a remarkable development at the beginning of this millennium, also as a consequence of the discovery of a new category of photoreceptors on the human retina (Provencio et al. 2000), containing the melanopsin protein and called intrinsic photosensitive Retinal Ganglion Cells (ipRGCs). These photoreceptors are not involved in vision like cones and rods. In a first phase, it was observed that the ipRGCs, thanks to their photosensitive capability, plus the information they collect from the other photoreceptors, have an effect on the regulation of the melatonin hormone and therefore on the human circadian system; new research



M. Rossi, *Circadian Lighting Design in the LED Era*, Research for Development,
https://doi.org/10.1007/978-3-030-11087-1_2

is highlighting further physiological effects and other effects on human behavioural responses.

There are non-visual effects of light that affect the human body through mechanisms other than those that act on the regulation of melatonin, through direct action on the human nervous system and with consequences that affect the alertness level (Cajochen 2007), the mood (LeGates et al. 2014), the behaviour (Chellappa et al. 2011; LeGates et al. 2014) and other human physiological parameters such as heart rate and body temperature. However, empirical studies in this area have sometimes been vitiated by basic observation conditions, or surrounding conditions, not always consistent and sometimes by generic descriptions of light, rather than being based on the use of real objective photometric and radiometric parameters, which would have led to the construction of dose-response curves, descriptive of the effects considered (Souman et al. 2018).

In this context, it is worth introducing an element used in neurophysiology to evaluate the activity of the central nervous system: brain waves. These are electrical brain activities, mostly periodic, which can be recorded via an electroencephalogram (EEG); they identify different levels of activity of the human brain and therefore have a significant importance in cognitive processes and in the identification of the awareness and consciousness stages. A detailed study of the cerebral electrical activity goes beyond the scope of this book, but, broadly speaking, we can divide brain waves into five main categories, according to their frequency intervals:

- Gamma waves have a frequency ranging between 30 and 100 Hz. These waves arise when the brain is subject to demanding processing activities and high cognitive functions. They are believed to be important for learning, memory and information processing in general. It is believed that the 40 Hz frequency is important for the connection between our senses, the perception and the learning of new notions. It has been shown that individuals with learning difficulties tend to have a lower gamma activity than the average.
- Beta waves can be aperiodic, but generally have a frequency between 13 and 30 Hz. They are observed in a wake person who is engaged in intense brain activity, such as mathematical calculation.
- Alpha waves have a frequency between 8 and 13 Hz and are typical of the transition between our conscious thought and the subconscious mind. They are observed when a person is awake with eyes closed, in the phases of relaxation and in the moments before falling asleep. They are therefore the waves that identify the transition between the waking and sleeping states.
- Theta waves range from 4 to 8 Hz and are observed in a subject in the rapid eye movements (REM) light sleep.
- Delta waves have a very low frequency, between 1 and 4 Hz and arise in the brain in deep sleep stages.

Therefore, along with the aspects related to research on the production and operation of lighting, there is research conducted in the physiological field on the positive or negative effects that light can have on human beings. The methods of investigation of these areas of research are different from one another and the results obtained are different as well. Let us look at some aspects concerning the effects of light on human physiology, highlighted by scientific research.

## 2.2 Effects of Light on the Brain

In the field of neuropsychology, cognitive performance is within the human mind's ability to acquire, store and process information, in order to solve problems of any kind; from the simplest everyday life needs, like transporting three objects if I have only two hands, to the most complex research activities, e.g. how to identify which genes are involved in a specific disease. It is in this type of activity that the human cerebral cortex is particularly adept. Then there are other factors such as the level of subjective vigilance and the level of alertness that involve the brain. Light can affect all of these factors.

### *2.2.1 Main Effects*

In a research conducted by Badia et al. (1991) the immediate psycho-physiological and behavioural effects of light stimulation have been tested on a sample of 43 human subjects, evaluating the differences between an intense illuminance of 5,000 lx and one of only 50 lx, both used after sunset. The alertness level, assessed with EEG through beta brainwaves, has shown a significant increase in the strong light condition, which also had the effect of decreasing drowsiness and increasing body temperature.

The research carried out by Okamoto and Nakagawa (2016) has studied the effects of the different wavelengths of light on the brain waves associated with memorisation processes, on a sample of twelve subjects aged between 20 and 31. Cortical activity has been assessed using magnetoencephalography (MEG), while subjects have been submitted to an auditory Sternberg memory task. The experiment has been carried out by exposing the subjects for 30 min, during the day, to two different lighting conditions: with green light with a peak wavelength of 520 nm and with blue light with a peak wavelength of 470 nm. The results have shown that blue light increases the cortical activity related to active maintenance of working memory and attention to auditory stimuli. Similar results have been also observed in the research of Noguchi and Sakaguchi (1999) in which, instead of monochromatic lights, they has used two white lights with a correlated colour temperature (CCT) of 3,000 and 5,000 K, observing that the 5,000 K light stimulates the central nervous system more.

Working memory in the human cognitive system is a part of memory, with a time limited capacity, which supports the temporary storage of information available for brain processing (Miyake and Shah 1999). This cognitive structure is important in reasoning and in guiding decision-making and behavioural processes. It should not be confused with short-term memory, which merely stores information in a temporary way but is not directly related to the processing of information by the brain (Cowan 2008). Working memory is a central theoretical concept in cognitive psychology, neuropsychology and neuroscience in general. Work memory performance is essential for learning activities and problem solving. The relationship between artificial lighting and working memory performance is extremely important for the illumination of all internal work, study, learning, research and development spaces, where learning and problem-solving skills are an essential factor for competitiveness and also for the balanced development of human activities with respect to workloads. These effects of light also seem to be confirmed by another research that evaluated the amount of light (Huiberts et al. 2016) and has been conducted, during the day, on a sample of thirty-nine people subjected to white light with a CCT of 4,700 K, with three different levels of illuminance, 165, 600 and 1,700 lx, directly on the eye for 90 min, in the morning and in the afternoon, to evaluate the heartbeat and working memory. After an hour of exposure to the stronger illuminance condition, 1,700 lx, an increase in working memory performance and in heart rate has been recorded, while the illuminance condition of 600 lx has not shown significant effects. In addition, participants in the experiment evaluated the two stronger levels of illuminance as significantly more stimulating.

In the study carried out by Cajochen et al. (2005), attention has been placed on the chromatic component of light. Ten subjects were exposed in the evening to two hours of blue light with a peak at 460 nm and yellow light at 550 nm. In the blue light condition, in addition to observing a reduction of melatonin, there has been an increase in body temperature, heart rate and the level of vigilance of the subjects involved. This did not happen with the yellow light exposure. Similar results were also observed in the research of Smolders et al. (2012) conducted during normal office work on 32 people. A 4,000 K light source and two different levels of illuminance have been on the eye, 200 and 1,000 lx, for one hour of exposure, respectively in the morning and the afternoon. The results have shown that with the more intense illumination the subjects had shorter reaction times, a higher level of vigilance and an increase in heart rate, especially towards the end of the hour of exposure to the stronger light. These results would show that even under normal daytime conditions, not in the dark and temporally far from rest periods, i.e. in physiological conditions different from those in which melatonin regulation comes into play, a more intense light can improve the sensations of vigilance and vitality, as well as the subjective performance and the physiological activation level.

The research conducted by Revell et al. (2006) has aimed at verifying the NIF effects of light on twelve human subjects, to evaluate the subjective mood and the level of vigilance, according to the exposure to coloured lights of four different wavelengths, respectively of 420, 440, 470 and 600 nm. The light exposure has lasted four hours, starting at 7:15. During these four hours, vigilance and mood have been evaluated every 30 min. A higher level of vigilance has been observed for light at 420 nm compared to that at 470 nm and even higher than that at 600 nm. In general, the level of vigilance has increased with the increase in light exposure. It follows that the level of subjective vigilance has also increased as the wavelength has decreased, while no significant differences in mood have been noted. A similar research was carried out by Lockley et al. (2006), using two almost monochromatic light sources with wavelengths of 460 and 550 nm, on sixteen adults, for a period of six hours and thirty minutes at night. The subjects have been monitored continuously by EEG and have undergone performance tests at intervals of 30–60 min. The results have shown unequivocally that the subjects exposed to blue light (460 nm) have had a lower subjective drowsiness, a higher level of attention, a decrease in delta and theta waves in favour of an increase in alpha waves between 9.5 and 10.5 Hz.

The researches mentioned above therefore clearly show that the wavelengths in the central area of the spectrum, typical of the sensitivity of the visual system, do not influence the alert level, the cognitive performance capabilities and the ability to maintain attention. These brain states depend instead on the NIF effects of light that are more sensitive to the short wavelengths of the light spectrum, those that we normally perceive as blue and azure light.

The human being is an endothermic organism, i.e. the adjustment of body temperature is mainly due to endogenous phenomena such as sweating and vasodilatation. The ability to make this adjustment therefore depends on the body surface involved and the difference in temperature with the external environment. A study by Kulve et al. (2015) has analysed the researches that relate the quantity of light and its CCT to thermoregulation and the subjective sensation of thermal comfort. In some cases, the light with colour shades turning towards red induces a feeling of greater environmental heat compared to a light that turns towards blue. However, the results of this last type of research are sometimes contradictory because the context, the experimentation modes and the results detection are different. In this perspective, psychological factors can also come into play, leading to different results in different social and cultural contexts.

### 2.2.2 *Effects of Light on SAD*

Winter depression or seasonal affective disorder (SAD) is a mood disorder that affects during winter people who are normally healthy in the other seasons of the year (Rosenthal 2006). The symptoms are typical of depression, like a lack of energy, the tendency to sleep a lot, but also obesity, asthenia, insomnia and difficulty in concentrating. It has been also observed that the same subjects in the summer can show symptoms of intensified anxiety. For this reason, today this disease is no longer considered as an exclusively winter mood disorder but as a disorder with seasonal patterns that can recur in several periods of the year with different symptoms. Some studies have shown that the probability of contracting this disease is greater with increasing latitude and particularly in Nordic countries, especially when exposure to solar radiation is lower (Rosenthal et al. 1987; Rosen et al. 1990; Magnusson et al. 2000).

The causes of this disease could also be of an evolutionary nature, since many animals tend to hibernate or to behave in a way that saves energy during the winter months when food shortages occur. This phenomenon must also have affected human beings in their prehistory. Among the biological causes there would be a diminished production of serotonin (Johansson et al. 2001) but another possible cause could be an excess of melatonin production by the pineal gland at night and also in daytime with little or no natural light. To reinforce this thesis, it has been discovered that by subjecting a human being affected by SAD in winter to treatments with strong artificial lighting there is a remission of the disease after a few days (Lewy et al. 1982).

For years, white artificial light with a CCT of 5,000 K, with eye illuminance levels of approximately 10,000 lx for 30 min or 2,500 lx for two hours, has been used as a therapy for SAD in Nordic countries. Some people, however, consider the treatment of SAD by a therapy using these high levels of illuminance to be bothersome. In a research carried out on 22 patients with SAD, it has been observed that the use of a cold white light with a CCT of 17,000 K, with a high blue light content, also leads to positive therapeutic effects with an illuminance of only 750 lx compared to 10,000 lx needed in the case of a white light with a CCT of 5,000 K (Meesters et al. 2011). In another research conducted on 48 patients suffering from subsyndromal SAD, a milder form of the disease, it has been observed that, using an illuminance of only 100 lx of narrow-band blue light with a peak wavelength of 470 nm for 20 min, the same therapeutic effect as with a white light exposure of 10,000 lx is achieved (Meesters et al. 2016). The lighting system used in this case is the goLITE BLU HF3320/01 by Philips, which has been proposed for light therapy as a possible substitute for the blue light of the clear summer sky. This product has now been replaced by the Philips EnergyUp HF3420/03, which is proposed as an energising product, which produces the equivalent of natural light and helps improve the mood with a 20–30 min exposure a day. The product is guaranteed UV free and anti-glare, with a level of illuminance to the eye of only 200 lx. The term therapy is not present in its descriptive notes (Philips 2018). In many countries the term "therapy" in reference to a product can only be used after having passed a series of protocols provided by the respective ministries of health, as is the case with medical products. For this reason, manufacturers today are very wary of proposing products of this type by associating them with the therapy concept.

### 2.2.3 *Effects of Light on Alzheimer's Disease*

Another example of artificial lighting that is hypothesised to be used for therapeutic purposes is in support of the treatment of the severe Alzheimer's disease. Alzheimer's disease is a form of progressive degenerative dementia that can usually occur after 65 years of age. Among the various forms of senile dementia, Alzheimer's is the most common, around 60–70% of cases. In 2015, 47.5 million cases of dementia were estimated worldwide and 7.7 million new cases are estimated each year. Dementia is one of the main causes of disability and dependence of older people worldwide (WHO 2015).

Alzheimer's disease primarily affects memory in the short term and increasingly serious problems arise in relation to language and mood, with anger and orientation problems and finally the inability to take care of oneself and the tendency to isolate oneself from others. The Alzheimer's patient, if managed at home, is also a strong cause of stress for caregivers. In fact, the disease has a strong physical, psychological, social and economic impact on caregivers, families and society. These patients often show a poor quality of night-time sleep with frequent awakenings, night wandering and many daytime naps, which worsen the quality of life of patients and caregivers.

In a well-known research, the patients' resting cycle has been evaluated through the use of actigraphs on 22 patients suffering from Alzheimer's disease with a mean age of 79 years (Van Someren et al. 1997). Ceiling-mounted white lighting was used, with an illuminance on the eye varying between 790 and 2,190 lx, depending on the position in the common areas of hospitalisation, for an observation period of one week. It has been observed that patients subjected to intense lighting have increased the stability of the sleep-wake cycle in the following days, thus recovering a more correct circadian rhythm in their organism, but only if normal-sighted. Similar results have been deduced by another research (Sloane et al. 2007), conducted in the public areas of a nursing home for long-term residents, on 66 patients with senile dementia, whose activity was recorded through actigraphs over a period of three weeks. In this experimentation, luminaires were installed with a low glare index, capable of generating ambient lighting of approximately 2,500 lx through a gradual increase in lighting over a period of 30 min. In three phases of the experiment, the over-illumination has been applied all day for over 8 h, or for 3 h in the morning or in the evening. The results have shown that patients exposed to intense light in the morning or all day had a significant increase in sleep duration at night.

### *2.2.4 Effects of Light on Learning*

The effects that light can have on learning environments is a topic of research investigated for some time. Three studies conducted in Sweden (Küller and Lindsten 1992) and in the USA (Salt and Karmel 1967; Heschong et al. 2002) have demonstrated the importance of the presence of windows and natural light on psychophysiological well-being and student performance. Moreover, two of them have also proved that these positive psychophysiological and performance effects are absent in windowless classrooms.

Some studies have verified the effects of different artificial lighting on elementary school pupils. In the study conducted by Berman et al. (2006) they have shown that with constant levels of illuminance on school desks, a CCT of 5,500 K is more effective, than a 3,600 K one, on the near visual acuity in 24 children out of 27. In the study of Barkmann et al. (2012) they have experimented with variable lighting to evaluate the levels of concentration, reading speed and comprehension, in two classrooms over a period of 9 months. This study has shown that with an illuminance of 1,060 lx with a CCT of 5,800 K, pupils have made a smaller number of errors of omission, have improved reading speed and there has been also a slight improvement in reading comprehension compared with an illuminance of 300 lx and 4,000 K. These results have also been confirmed by the research of Sleegers et al. (2013) in which a dynamic lighting system has been experimented, with an illuminance from 350 to 1,000 lx and a CCT in the interval 3,000–12,000 K, finding an improvement in pupils' concentration. Furthermore, the fact that a higher level of lighting, with a cool CCT, stimulates cognitive processes for learning performances of pupils has also been confirmed by two recent studies conducted in Korea (Choi and Suk 2016) and in the USA (Pulay 2016).

The effects of light on students' performance have also been investigated in the context of academic education. In a study conducted in the month of November in Finland by Rautkylä et al. (2010), a higher level of students' alertness has been found while attending lectures, in particular after lunch, with a CCT of 17,000 K compared to a 4,000 K one. The same study repeated in spring has not detected the same effects as the level of alertness is already higher than in autumn, also due to a greater exposure of the students to outdoor daylight. Teixeira et al. (2013) have conducted a study on university working students attending evening courses in Brasil. The results have shown that the use of a high level of illuminance, up to 8,000 lx of white light with a CCT of 4,000 K, has increased alertness level in all the students. This effect has been attributed to a decrease in the production of the evening melatonin, due to the bright light. Finally the characteristics of lighting can have effects on the students' performance but this depends marginally on the technology of the light sources used, as it has been highlighted in a study conducted in Sweden comparing lighting systems based on fluorescent lamps and LEDs lighting (Gentile et al. 2018). Nevertheless, the classrooms fitted with LED lighting systems have been slightly preferred and achieved also a small energy savings.

## 2.3 Light and Sleep Disorders

Many people suffer from sleep disorders, which may have different causes and different effects, even of serious pathological nature, which hinder a healthy daily life both from a physiological and psychological standpoint, interfering with emotional states and social relations.

### *2.3.1 Insomnia*

The most known disorder is that of insomnia, which occurs when the individual fails to fall asleep despite being tired and having an effective physiological need for sleep. In other cases, insomnia manifests itself with repeated awakenings during the night or with very early awakenings in the morning and the impossibility of going back to sleep. The typical symptoms of insomnia include the inability to find a comfortable position in bed, anxiety, the feeling of strong tiredness when waking up, but also night walks with the inability to go back to bed. The effects of insomnia also have repercussions on the quality of daily life, as the affected individuals often have difficulties in concentration and learning, linked to the malfunction of working memory, but also chronic fatigue and irritable mood. In a person, insomnia can emerge without any apparent triggers or as a consequence of external factors such as, for example, the abuse of caffeine, nicotine and alcohol, but also heart problems, hypertension, heartburn, menopause, psychological stress and sleep apnoea. It is estimated that about 30% of the world's population reports symptoms of insomnia in their lifetime, with a greater incidence in women and people over 65. In general, the main discomforts, which occur in 40% of cases of people suffering from insomnia, are psychiatric disorders, especially depression and anxiety (Roth 2007).

### *2.3.2 Bruxism*

Another form of sleep disturbance is nocturnal bruxism, which has very severe negative effects: tooth erosion down to dentin, with a greater probability of fracture and to develop caries and tongue inflammation. It can also induce headaches in the temporal zones, with possible disturbances to the ears caused by the repeated nocturnal tension of the muscles that act on the jaw. It has been estimated that nocturnal bruxism occurs in a percentage ranging from 8 to 16% of the adult population, but the percentages are even higher in young people. Unfortunately, only 3% of the people who suffer from bruxism are aware of being affected by this disorder (Shetty et al. 2010; Manfredini et al. 2013). The issue of awareness is very important because most people suffer from various types of illness without being aware that they are a direct consequence of this sleep disorder. It has been observed that among the main causes of nocturnal bruxism there are sleep disorders that fall into the categories of parasomnias. This type of disorder involves a restless sleep with nightmares, restless leg syndrome, increased heartbeat and various other types of disorders, including bruxism. It has been shown that about 86% of episodes of nocturnal bruxism occur during restless sleep (Lavigne et al. 2007).

### *2.3.3 Nigh Terror*

Another discomfort affecting the quality of night rest is the sleep terror disorder that predominantly affects children, 3–12 years old, with a percentage of less than 6%, and to a lesser extent adults, with a percentage of 1%, mainly in the age group between 20 and 30. This disturbance should not be confused with nocturnal nightmares and sleepwalking that can affect all age groups (Szelenberger et al. 2005). The subject suffering from this type of disorder is affected by real attacks, which in the most serious cases can also repeat several times in the same night, during which he/she suddenly wakes up screaming and with wide open eyes, but does not consciously respond to those looking to reassure him/her and does not even respond to environmental stimuli. The duration of these attacks is usually a few minutes, but in rare cases they can last up to half an hour. Then the subject suffering from night terror goes back to sleep and almost never remembers anything from the episode. The causes of nocturnal terror disorder are numerous and not yet fully explained: hereditary factors, situations of prolonged psychological and physical stress and many others, including the lack of sleep and any light or sound disturbances during sleep.

### 2.3.4 *Visually Impaired People*

Older people often have degenerative problems in the visual system, such as cataracts or degradation of transparency and consistency of the vitreous body. These degenerations of the visual system mean that the elderly tend to avoid strong light sources because they increase the luminance of the internal veil of the eye, which has the effect of obscuring the sight. These degenerations also decrease the blue light component that reaches the retina in a normal white daylight condition, with the result that this limited contribution of short wavelengths on the retina also affects the circadian system. However, as highlighted in another research (Figueiro et al. 2014a), with a careful control of the daily exposure patterns and a smaller amount of comfortable white light, characterised by a spectral distribution rich in blue light, positive effects on sleep-wake cycles can be achieved by a large percentage of patients, without unwanted effects of over-illumination of patients and caregivers. This trial has been carried out in a nursing home (NH) for long-term patients on a sample of 14 normal-sighted patients, who however had problems with agitation or sleep disorders and for whom a medium-moderate level of senile dementia was diagnosed. In addition, the participants were chosen because they tended to spend most of the day in their rooms. In this case, indirect lighting has been installed in the hospital rooms with a CCT of 9,325 K. The average illuminance measured at the patients' cornea was about 324 lx. The system has been timed to turn on at the time scheduled for waking up and turn off at 18:00. In order to limit the energy consumption, by means of an infrared passive presence sensor (PIR), the system automatically shut off the light in the case of absence of the patient from his or her room. The results have shown an increase in the duration and quality of sleep, better timing of the circadian rhythm and a reduction in the level of depression and agitation of the patients.

### 2.3.5 *Effects and Causes of Sleep Disorders*

Regardless of the cause, be it lighting or other factors, it has now been scientifically proven that poor quality and reduced duration of sleep can be associated with a worsening of cognitive performance, physical and mental fatigue, low level of vigilance, changes in mood and health risks. For example, Christoffersson et al. (2014) have shown that sleep deprivation in young people leads to a decrease in blood neutrophils, the immune cells already present in the circulatory system and the first reacting in case of infections. While Engle-Friedman (2014) has highlighted that the loss of sleep is accompanied by a decrease in physical and mental performance, due to a decreased ability to access the body's energy resources. Furthermore, Cedernaes et al. (2015) have analysed the relationship between sleep disorders and the increased possibility of occurrence of diseases such as obesity and type 2 diabetes mellitus.

Many sleep disorders are caused by the alteration of the circadian rhythm. These are divided into two main categories: intrinsic disorders, caused by physiological malfunctioning factors within the organism, and extrinsic factors, which may be caused, for example, by intercontinental flights that cross multiple time zones or the need to work in night shifts. Alteration of the circadian rhythm of sleep may occur late (delayed sleep phase disorder—DSPD) or early (advanced sleep phase disorder—ASPD). Actually, many individuals may experience delayed sleep, also known as owl behaviour, or early sleep, also known as lark behaviour (Phillips 2009). A delayed shift of the circadian rhythm is the one affecting individuals who struggle to wake up in the morning and have problems working or studying early in the day. Conversely, an advanced shift is typical of people who wake up early and are very active, both intellectually and physically already in the morning, but tend to be very tired and sleepy in the late afternoon or immediately after dinner, resulting in poor propensity for post-working social activities or evening work.

Some well-known researches have shown that the use of bright light in the morning can help with typical sleep disturbances with a delayed shift (Rosenthal et al. 1990; Figueiro 2016), while a strong illuminance of 2,500 lx after 20:00 can help resetting the rhythm in disturbances typical of the advanced shift (Lack and Wright 1993). All these studies analysed sleep hours, as well as other parameters such as body temperature and muscle activity through the use of actigraphs. In another field survey conducted on 23 people working in the office in Switzerland, special glasses were used with sensors able to record, during the working day, the level of illuminance and the irradiance of the blue component of the light on the eyes (Hubalek et al. 2010). The results of the research have shown that during the working days the exposure to light had a regular trend, while on holidays the exposure was much more irregular. Additionally, a positive effect of the light received during the day, in terms of duration, quantity and spectrum, has been associated with sleep quality the following night.

An epidemiological study carried out by Sloane et al. (2008) has reviewed twenty-one researches that have investigated the use of light as a therapy for sleep disorders, which are much more common in the elderly population. The elements that have been taken into consideration are the amount of light, the spectral power distribution (SPD) of light, the time of exposure to light, the duration of the exposure and the direction of light. The light therapy proposed by this study, for a correct timing of the sleep for the elderly, can be summarised by these three main points:

- In daylight hours, it is best to use 7,500 K cool white light sources that provide 1,000 lx illuminance. If these levels are considered too high, it is appropriate to

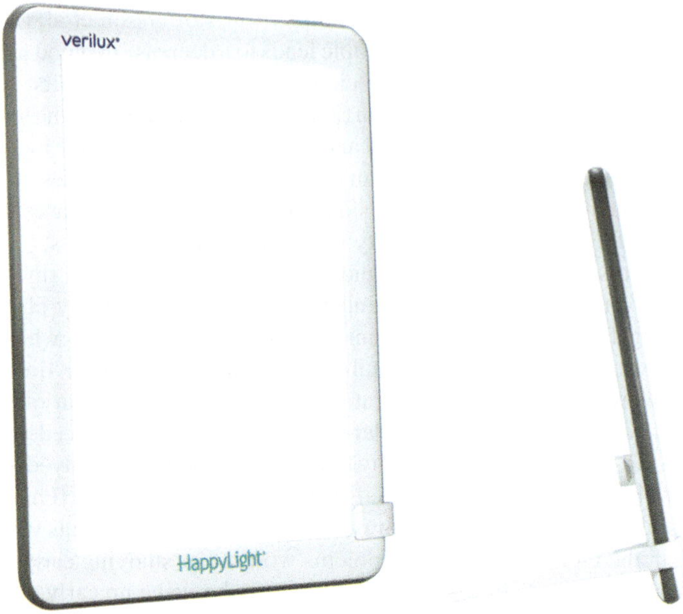


**Fig. 2.1** The HappyLight Lucent LED. *Courtesy* Verilux Inc.

integrate the normal lighting with low-intensity blue light sources that are visible to the subjects involved in the periods provided for the therapy.

- The timing of light therapy should be provided according to the patient's chronotype; generally, the elderly suffer from an advanced shift of the sleep phase, but each case must be assessed individually. It is therefore important to check with the patient if he/she tends to be more tired in the morning or late afternoon, consequently providing light therapy during this time interval.
- For the evening and night hours it is always better to use a light of low intensity and low CCT, which is not able to affect the circadian system of the elderly in the phase that precedes sleep and in the case of nocturnal awakenings to perform physiological functions.

The positive effects of light on the correct timing and quality of sleep have also been verified by other subsequent researches, in which the relationships with the different effects that different light wavelengths have in this context have also been highlighted (Figueiro et al. 2011; Figueiro et al. 2014b). Thus excluding the need for excessively bright and annoying lighting not applicable in the design reality, and using instead normal comfortable lighting systems with white light and a suitable content of short wavelengths (blue). There is now a wide scientific consensus on the fact that light can be used as a therapy to improve the quality of sleep of all individuals (Montgomery and Dennis 2002; Figueiro 2015). To this end, for years light therapy products have been available on the market, also valid in the treatment of SAD. Figure 2.1 shows one of the light therapy products on the market that use high levels of white light, up to 10,000 lx, the Verilux HappyLight Lucent LED.

However, as mentioned above, in some countries the term "therapy" cannot be associated with these products. Among the leading manufacturers in the industry we can mention Philips, Verilux, Nature Bright and Carex.

## 2.4 Effects of Light on Premature Infants

According to the World Health Organization (WHO), preterm infants are those born before the completion of the 37th week of gestation (WHO 2018). It is estimated that around 15 million children worldwide are born in these conditions each year and that this number is increasing. According to an analysis conducted in 184 countries, the percentage of premature births varies between 5 and 18%. Complications due to premature birth are one of the two leading causes of death among children under 5 years of age (Liu et al. 2016), but three quarters of deaths related to premature birth could be avoided with the aid of adequate medical-surgical devices and with the use of neonatal intensive care units (NICU). According to an analysis conducted in 2010, the 10 countries with the highest number of premature births are (Blencowe et al. 2012):

1. India: 3,519,100
2. China: 1,172,300
3. Nigeria: 773,600
4. Pakistan: 748,100
5. Indonesia: 675,700
6. United States of America: 517,400
7. Bangladesh: 424,100
8. Philippines: 348,900
9. Democratic Republic of the Congo: 341,400
10. Brazil: 279,300

### *2.4.1 Lighting Trials in the NICU*

A well-known research conducted by Miller et al. (1995) has assessed the lighting condition of the NICU. The subjects considered were 41 premature infants weighing less than 2.5 kg at birth and with a gestation of less than 37 weeks. This research has aimed at experimenting two different artificial lighting conditions: a fixed lighting over the 24 h and a cyclic lighting on the model of natural light variation. For research purposes, 20 premature infants underwent a cyclically variable artificial lighting condition between day and night, while 21 premature infants underwent a constant 24-h illumination condition. The cyclic lighting condition provided for an average amount of light on the horizontal plane of about 200 lx from 7:00 to 18:00 and 32 lx the rest of the day. The constant lighting condition provided for a quantity of light in the horizontal plane between 232 and 274 lx over 24 h. These values have been measured by luxmeters placed next to the baby cots, and both NICU rooms were shielded from direct sunlight. After the intensive care phase, the new-borns were moved to the normal nursing rooms where they received a cyclic day-night light. The results presented by this research have shown that in the intensive care the 20 premature infants undergoing cyclic lighting, compared to the other 21 undergoing constant lighting, had a weight increase on average and could be fed orally before the other group. They also spent less time in forced ventilation and showed faster improvement also in motor coordination.

Another trial was conducted by Brandon et al. (2002) on 60 premature infants born before 31 weeks of gestation, excluding those with neurological diseases or visual problems. The main parameter taken into consideration was weight gain, because it is the factor that can have a greater impact on the length of the recovery period. The NICU rooms with incubator cradles used for experimentation, as often happens, did not have natural light contributions. The lighting conditions experienced in this case have seen a comparison between a cyclic lighting and an almost absent lighting. The lamps used for lighting were cool white linear fluorescent lamps produced by Philips, equipped with a filter to eliminate the ultraviolet component of the emitted radiation. The level of cyclic lighting provided for 11 h of operation, with an illuminance between 200 and 225 lx, and 11 h of shutdown with 1 h of time for gradual transitions between the two levels. On the other hand, the almost absent lighting condition, experimented over 24 h, was implemented with a lighting between only 5 and 10 lx, with the exception of 6:30–7:30 and 18:30–19:30 during which the medical operators were active on new-borns. The results proposed by this research, evaluated through a statistical analysis, argue that cyclic illumination has short term advantages on the health of premature babies compared to low continuous illumination. In fact, the cyclic lighting condition has provided, over the same time, a greater increase in weight of new-borns compared to those subjected to almost absent lighting. This research has also assessed the risk of onset of retinopathy of prematurity (ROP), as some established practices suggested keeping premature infants almost in the dark to reduce the risk of this disease. The research therefore also aimed at assessing whether cyclic lighting could cause an increased risk of this disease. On the contrary, the results have shown that the incidence of ROP was greater in infants constantly kept almost in the dark, deducing that cyclic illumination would facilitate the normal development of the retina provided that the light level is kept within the limits used in this experimentation. In fact, also a previous research (Glass et al. 1985) had shown an increase in the incidence of ROP in premature infants undergoing a high constant illuminance of approximately 645 lx compared to those illuminated with approximately 270 lx.

### 2.4.2 *Possible Guidelines for the NICU*

Recently, the Cochrane non-profit association, which deals with organising medical research with direct implications on people's health for promoting professional and political interventions, has conducted an epidemiological review on the studies published in the last twenty years in this field, on a total of 544 premature babies (Morag and Ohlsson 2016). The conclusions expressed in this report are that some of the published research are not considered high-level, due to the small number of new-borns considered or due to the fact that the medical staff who collected the data were aware of the research and therefore were directly involved. The report expresses the hope that research in this area will be further explored, but it nevertheless acknowledges that several studies have highlighted positive aspects of cyclical lighting compared to almost absent lighting or strong lighting always present. This is evident from the general decrease in hospitalisation days for premature babies undergoing cyclic lighting.

The topic of correct lighting for premature infants is an open field of research in which over the last thirty years many studies have been carried out to assess which lighting condition, between cyclic, constant bright or near darkness is better (Engwall et al. 2014). The elements that are considered are the speed in weight growth, the required ventilation time, the possible damage to the retina and other neurological diseases. In addition, we must take into consideration the health of the paramedics who work in those environments for many hours and also in night shifts. The case of the NICU is therefore an example of a situation in which the lighting design has to deal with different needs in the same environmental context (Rea and Figueiro 2016).

## 2.5 Light and Production of Human Melatonin

Even before the research on the circadian cycle by the 2017 Nobel laureates Jeffrey C. Hall, Michael Rosbash and Michael W. Young, it was known that light has an effect on the production of the melatonin hormone in humans.

Until 1980, an effect of light on the production of melatonin in other mammals was observed, but it was not believed that this could happen also in human beings and it was also erroneously hypothesised that this could have been an evolutionary advantage with respect to other animals (Perlow et al. 1980).

On the contrary, a research published in Science by Lewy et al. (1980) has demonstrated that the use of bright artificial light at night has suppressed the production of melatonin in the six human beings subjected to the experiment. Although the tests of this research have been carried out on six subjects only, the importance of this discovery for research conducted in the following years is also evidenced by the high number of citations received, about 1200, catalogued by both Web of Science and Scopus. In the experiment, 3 types of artificial lighting have been used. A white fluorescent lamp producing about 500 lx of lighting, one incandescent lamp producing about 1,500 lx and one incandescent lamp producing about 2,500 lx. The amount of lighting has been measured at eye level. The melatonin level has been measured in the blood at regular time intervals throughout the night. The subjects were sent to sleep at 23:00 and were woken up at 2:00 and then exposed to lighting for two hours, after which they were sent back to sleep. In subjects exposed to 2,500 lx, melatonin concentration has shown a decrease in blood within 10–20 min, going down to levels similar to those measurable by day. Removing the lighting, the melatonin levels have returned to the typical nocturnal ones, before waking up, within 40 min. Using the 500 lx light source no decrease has occurred in melatonin. With the 1,500 lx source there has been a melatonin reduction of about half compared to that obtainable with 2,500 lx.

In this research, great attention has been paid to the method of measuring melatonin in the blood. In commenting on the results, the authors have also hypothesised that the failures of previous research, aimed at demonstrating that artificial light can have an effect on the adjustment of human melatonin, have been due to two possible reasons. Firstly, the melatonin measuring method was not sufficiently correct and, secondly, levels of artificial lighting were used, in line with the requirements of office lighting standards, which would be insufficient for the purpose. With this last observation they have laid the base for a question of fundamental importance in the context of interior design, which only in recent times has received a partial answer: whether the artificial light we have in interior spaces is able to facilitate a normal adjustment of melatonin as it happens in outdoor areas under natural light. The results of this research have been very important, because they have paved the way for a series of more in-depth studies in the following years. In particular, from the standpoint of medical-physiological research, it has been tried to investigate what were the human physiological structures able to control this melatonin response to light. From the application and design standpoint, the focus has been on understanding what characteristics of light are able to influence the human melatonin cycle, with the aim of being able to define the guidelines for lighting design of interior spaces.

### *2.5.1 Dose-Response Relationship*

It has been demonstrated that this action of light on melatonin occurs with a dose-effect relationship, as evidenced in a research by Brainard et al. (1988) that has used a monochromatic light with a 509 nm wavelength for an hour at night on six human subjects. Five different levels of illumination have been used, measured on the eyes in terms of irradiance: 0.01, 0.3, 1.6, 5 and 13 $\mu W/cm^2$. It has been observed that melatonin suppression is greater with the increase of the irradiance level, with a dose-response curve showing that the average threshold necessary for the melatonin suppression is between 1.6 and 13 $\mu W/cm^2$. This dose-response principle has also been verified in an experimentation conducted by McIntyre et al. (1989b) with a full spectrum white light source. They used, for one hour at night, five different illuminance levels, 3,000, 1,000, 500, 350 and 200 lx. On average, they measured levels of melatonin suppression of 71, 67, 44, 38 and 16% respectively. These researches have shown the existence of a fundamental principle: the amount of light influences a dose-effect response curve, of a non-linear type, which was also confirmed by other subsequent researches (McIntyre et al. 1989a; Rea et al. 2001).

### 2.5.2 *Phase Shift in Melatonin Production*

The research proposed by Zeitzer et al. (2000) has also confirmed the existence of a dose-effect curve in the fast suppression of nocturnal melatonin by light, but another element has been also verified, which is important for the application aspects of interior lighting. This is the effect that different levels of lighting in the late evening may have in the timing of the production of melatonin over the 24 h. It was already known indeed, in a research published in Nature (Boivin et al. 1996), that in human beings an exposure to moderate light levels of white light, 180 lx, repeated for three consecutive mornings, can phase advance the human circadian system, but the possible shifting effect of a moderate evening light had not been evaluated yet. Twenty-three healthy human subjects undergoing a 9-day lighting protocol have been involved in this experiment, which included a 6.5 h light exposure from 23:00 onwards. The duration of the research was therefore motivated by the purpose of evaluating how the light could also shift the entire melatonin cycle in 24 h over several days. The endogenous timing of the melatonin in the subjects has been evaluated under controlled conditions before the experiment. It has been observed that illuminance levels lower than 15 lx cause irrelevant shifts, while at 500 lx there is a delay shift of the melatonin cycle of about 2.5 h in the next day. Just as the dose-response curve of melatonin suppression does not have a linear trend, also the dose-response curve of the melatonin shift effect is not linear, so much so that approximately 9,000 lx result in a 3-h shift, while only 100 lx may suffice to cause half of this shift (1.5 h). These results clearly show that even moderate levels of artificial lighting in late evening hours can significantly shift the amount and the melatonin cycle over the 24 h.

### 2.5.3 *Human Sensors for Melatonin Control*

From a physiological standpoint, researchers have also started to ask questions about the physiological structures responsible for the adjustment of melatonin according to lighting, first with experiments on animals and then with direct observations on human beings.

In a research conducted by Czeisler et al. (1995), the attention has been focused on the production of melatonin in total blind patients. In most cases, these patients experience a loss of synchronisation of the circadian rhythm in the 24 h (Sack et al.

1992) and suffer from insomnia. However, this rule does not apply to all blind people. Eleven blind subjects and six normal-sighted subjects have been involved in the experimentation. All have undergone a high illumination protocol with eye illuminance values between 6,000 and 13,700 lx, comparable to exterior lighting levels immediately after sunrise. Brilliant lighting has been used at night for a time between 90 and 100 min, during which the melatonin level was checked every 5–10 min. It has been observed that in three of the eleven blind subjects there has been a suppression of nocturnal melatonin similar to that of the 6 normal-sighted subjects. It is also important to note that these three blind patients, in their daily lives, had not reported sleep disturbances or circadian rhythm shifts. It has been also observed that, if the eyes of these blind patients were covered, there was no suppression of melatonin. This research has led to the conclusion that the mechanism of melatonin adjustment follows physiological routes different from those of the visual perception that are described by photometry, but still passes through the eyes. These results have been equally confirmed in a subsequent research conducted on two blind subjects whose organism, however, has correctly responded to light stimuli for the suppression of melatonin and the pupillary contraction (Zaidi et al. 2007).

A fundamental aspect to be highlighted is also the following: it is now scientifically established that it is exclusively the retina's task to translate light into nerve stimulations that affect the adjustment of melatonin. In fact, all the researches conducted to highlight effects of this type deriving from extraocular stimulations have given a negative result. Using the same light levels that have an effect on melatonin suppression, or even higher levels, lighting tests have been carried out behind the knees (Lockley et al. 1998; Eastman et al. 2000) and on the chest and abdomen (Lindblom et al. 2000). These experiments have led to the conclusion that the light reaching our skin has no effect on the melatonin cycle, and that only the light entering our eyes can have significant effects in this sense.

### *2.5.4 The Fifth Photoreceptor: ipRGC*

After a series of experiments on mice, at the beginning of the new millennium it was discovered that another photoreceptor is also present in the retina of the human being, different from the three types of cones and the rods that are present on the retina. The cones and rods are the photosensitive cells on the retina, thanks to which we receive the light signals that, in the processes of visual perception, are transformed into images and colours in our brain. Light-sensitive proteins, called opsins, are present in photosensitive cells. This additional photoreceptor has been located in some retinal ganglion cells and is characterised by containing a particular protein called melanopsin (Provencio et al. 2000).

When the cones and rods receive the light, depending on its spectral content they generate neurotransmitters that, through the synapses, are transmitted to the bipolar neural cells, which in turn transmit these signals to other neural cells present on the retina, ganglion cells and amacrine cells. The ganglion cells can collect information from a single photoreceptor or from sets of photoreceptors, and this information is then transmitted via their axon to the optic nerve. Only some of the ganglion cells contain melanopsin, which is a different protein from the other 4 human opsins. The signal collected by the other photoreceptors and generated by the ipRGCs towards the suprachiasmatic nucleus (SCN) would not influence the process of visual perception but would instead act on the pupillary reflex to light and on the melatonin level.

Alongside these discoveries, it was then demonstrated physiologically, by two well-known studies published in Science (Berson et al. 2002; Hattar et al. 2002) that the photosensitive cells that act on the adjustment of melatonin are the ipRGCs, which have a slow response compared to the other photoreceptors, and whose axons send signals to the retina-hypothalamus tract of the optical nerve. In the hypothalamus there is the SCN that plays a pacemaker role on the circadian rhythm in human beings as in the other mammals (Klein et al. 1991). The SCN is therefore the intrinsic timer that is however synchronised thanks to the bright extrinsic signals of day and night (Duffy and Czeisler 2009).

### *2.5.5 Sensitivity to Wavelengths*

A subsequent research, which involved healthy subjects with normal sleep-wake rhythms, subjected at night to monochromatic lights, has verified that actually the cones, which are the cells involved in the photopic vision, would not be the basis of the melatonin adjustment (Brainard et al. 2001b). Another research, conducted on 72 normal-sighted human subjects, that did not present problems in the sleep-wake cycle, has also assessed the level of suppression of nocturnal melatonin according to the wavelength of eight monochromatic lights, within the interval from 440 to 600 nm, obtaining a response curve with a maximum of 464 nm, which is different from the sensitivity curves of the rods and the three types of cones (Brainard et al. 2001a). These results confirm that, in humans, there is another photopigment, different from cones and rods, which acts as the main transducer of light information in the suppression of melatonin and with a greater sensitivity for the blue area of the spectrum. These results have been uphold by a similar research (Thapan et al. 2001), conducted in the same period, on 22 normal-sighted human subjects who did not present sleep disorders, subjected to sessions of monochromatic lights with wavelengths of 424, 456, 472, 496, 520 and 548 nm. The results obtained by this research also confirm a spectral sensitivity curve for the melatonin suppression different from those of cones and rods, with a maximum sensitivity for the short wavelengths estimated at 459 nm.

## 2.6 The Human Biological Clock

The circadian cycle of human beings lasts approximately 24 h and determines the moment and duration of biological functions, such as sleep and nutrition, whose balancing is fundamental for the well-being of the human body. In this cycle there are periodic variations in body temperature, hormonal production, brain activity, cell regeneration and other biological activities, with daily frequency (Dijk and Archer 2009). The detailed description of all physiological, cellular and genetic mechanisms that govern and contribute to the functioning of the human circadian system goes beyond the scope of this book.

For the purposes of lighting design, we can summarise that our internal biological clock is composed of the SCN of the hypothalamus located in its central part, the most primitive part of the brain, which controls the circadian cycle (Klein et al. 1991). The neurons present in the dorsomedial SCN are believed to have an endogenous rhythm of about 24 h, which can persist also in the perduring absence of light, as it occurs in other cells of the human body. In fact, at the cellular level there is an autonomous molecular oscillator based on the CLOCK-BMAL1 protein heterodimer that acts on the transcription of the Period and Cryptochrome genes. The proteins produced by this transcription accumulate in the cell nucleus to the point of repressing this transcription process, until they decrease again and the transcription process resumes. However, this oscillatory process can be modified within the cell by post-translational modifications of proteins, which may depend on external factors (Takahashi 2017). The latter property can generate different chronotypes in people, also depending on factors external to the organism, such as light.

The neurological stimuli produced by light on the photoreceptors, collected by the ipRGCs, from the retina reach the SCN through a branch of the optic nerve, which does not reach the parts of the cerebral cortex responsible for visual perception (Morin 1994). This branch is called retinohypothalamic tract. The SCN intrinsically generates an oscillation cycle of about 24 h in the neural activity, known as circadian rhythm (Czeisler et al. 1999) but, in order to synchronise itself exactly over the 24 h, it needs an external signal that is constituted by natural light, in the alternating rhythm of day and night. Its natural 24-h phase can however be altered if light stimuli are received with the characteristics sufficient to act on the SCN. This can happen, for example, in the case of intercontinental flights between multiple time zones that modify the daily lighting pattern, or in the case of exposure to artificial light, with circadian characteristics, after sunset. As we have seen in the previous paragraphs, the SCN can be disrupted also in the absence of adequate circadian light during the daylight hours, because in this case the natural pattern of exposure to the normal cycle of natural light is lost.

The SCN also controls the functioning of the pineal gland that produces the melatonin hormone and releases it into the bloodstream (Reiter 1991). Melatonin can be detected in the blood, saliva and urine as a factor that can highlight the status of the internal biological clock (Rosenthal 1991). The presence of melatonin in the blood has effects on many internal organs and promotes relaxation and sleep, while its suppression, or its absence, makes the individual more alert and ready to work both physically and intellectually.

Figure 2.2 illustrates the extrinsic factor that synchronises the circadian rhythm: light enters the eyes and reaches the photoreceptors and the ipRGCs cells; from the latter, a neural signal is sent to the retina-hypothalamus tract of the optic nerve

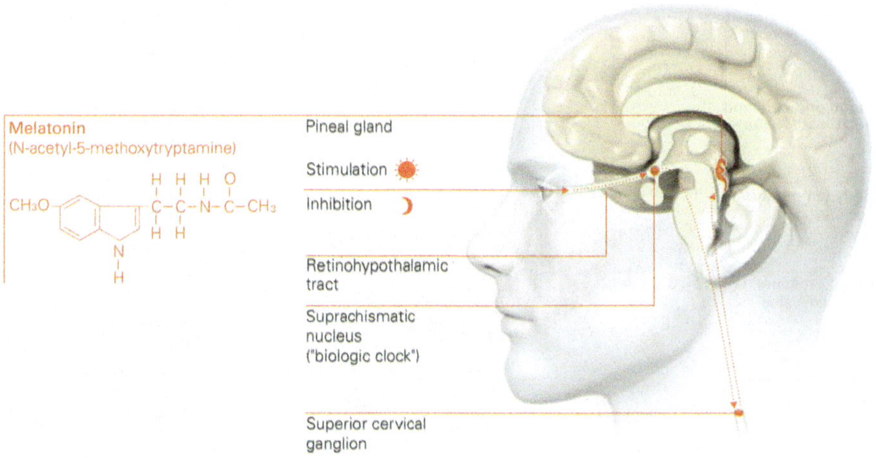


**Fig. 2.2** The effect of light on melatonin. *Courtesy* iGuzzini Illuminazione

reaching the SCN. The SCN acts on the brain and on the pineal gland that produces melatonin.

Among the main human physiological elements that are influenced by the circadian cycle with a 24-h periodicity, we can list:

- Sleep cycle.
- Change in body temperature.
- Change in blood pressure.
- Defecation.
- Testosterone production.
- Activities of the immune system.
- Alertness level.
- Muscle coordination.
- Reaction times.
- Level of muscle strength.

For years melatonin has been available as a drug or food supplement, taken by people who have sleep problems and want to restore their circadian rhythm (Pévet et al. 2002). Many argue that the intake of melatonin, in appropriate doses, times and modalities, helps improving the psychophysiological well-being of the individual. Figure 2.3 illustrates some of the physiological cyclical activities of the human circadian rhythm in the 24 h.

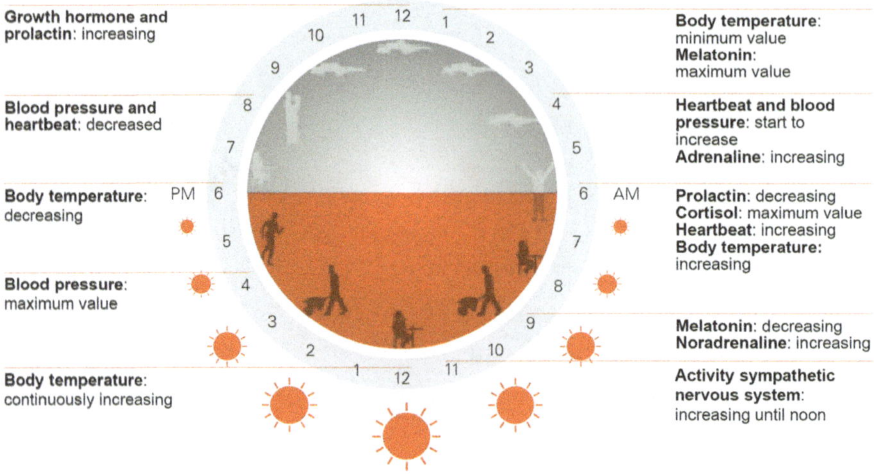


**Fig. 2.3** The physiological activities of the human circadian rhythm. *Courtesy* iGuzzini Illuminazione

## *2.6.1 Markers of the Circadian Rhythm*

Normally, the melatonin level increases during the night and decreases at the end of the dark. However, this level is not controlled only by the light and dark cycle, since it starts to decrease towards the end of the night even in the absence of light. Therefore, the melatonin level is the main marker that can be used to evaluate a person's circadian timing but it is not the only element that comes into play in the circadian rhythm.

An aspect highlighted by van Bommel and van den Beld (2004) is the relationship between the daily cycles of melatonin, cortisol and body temperature that in a healthy individual can vary by about 0.4 °C during the 24 h. Melatonin decreases in the morning, causing drowsiness and sleepiness and increases in the evening to promote relaxation and restful sleep, while cortisol increases in the morning. Figure 2.4 shows the changes in levels of cortisol, melatonin, alertness and body temperature in a human subject normally exposed to natural light.

Cortisol, also known as stress or activity hormone, promotes the metabolism of fats, proteins, carbohydrates, helps to increase the sugar available in the blood and therefore the energy available in the muscles. Cortisol is produced by the adrenal glands on stimulation of the pituitary, which is itself controlled by the hypothalamus. Cortisol has the property of inducing an increase in blood sugar, has an anti-insulin effect, protects from inflammation and supports the efficiency of the immune system. For this reason, some anti-inflammatory drugs are based on synthetic cortisol precursors, such as cortisone discovered by the Nobel laureate for Medicine 1950, the American Edward Calvin Kendall. Cortisone is a precursor of cortisol, but the effects of cortisone are not the same as cortisol. On the one hand, cortisone controls

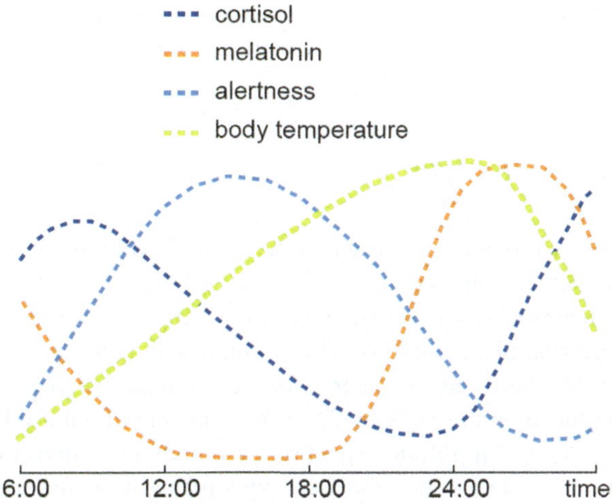


**Fig. 2.4** The markers of the human circadian rhythm

inflammatory reactions, but on the other hand it mitigates the effects of autoimmune diseases by decreasing the effectiveness of the immune system.

It is interesting to note that the secretion of cortisol is, in practice, opposite to that of melatonin. However, when cortisol remains high for too long, the body tends to lose efficiency and energy, becoming exhausted. Cortisol increases in the morning, thus preparing the human being to the physical and mental activity and acts on the immune system to protect the body from pathogens that could attack it during the day. Although cortisol has this counter-phase relationship with melatonin, two clinical studies have shown that the morning production of cortisol is more related to the time of awakening and less to the biological clock governed by the suprachiasmatic nucleus (Wilhelm et al. 2007; Rea et al. 2012).

### *2.6.2 Effects of the Disruptions of the Circadian Rhythm*

Many individuals suffer from more or less severe phase shifts of the circadian phase, which can also be caused by exposure to inadequate artificial lighting. These disorders affect sleep, digestion and intestinal activities; they can cause migraines (van Oosterhout et al. 2018), headaches (Pringsheim 2002), irritability (Evans et al. 2013), seasonal depression (Rosenthal 2006), deficiencies of the immune system (Christoffersson et al. 2014), chronic fatigue (Bonsall and Harrington 2013), obesity (Reiter et al. 2012) and diabetes mellitus (Cedernaes et al. 2015). An increase in the probability of developing some tumors has been hypothesised as a consequence of circadian circulatory disruption, which affects hormone production and the efficiency of the immune system (Stevens and Rea 2001; Yadav et al. 2017). It has also been observed that in the absence of the daily variations of natural light, the circadian rhythm in most of the individuals is prolonged beyond 24 h, with a consequent misalignment of the whole organism's biological clock.

### 2.6.3 *Jet Lag*

Other problems occur when the normal cycle of natural light is altered, for example in subjects that make intercontinental flights crossing multiple time zones, causing a series of disorders such as fatigue, digestive problems and a generalised psychophysiological condition that is commonly called jet lag. This occurs due to the mismatch between the sleep cycle, the internal organs and the new circadian rhythm induced on the SCN by the different cycle of natural lighting (Boulos et al. 1995).

A very important aspect of the human circadian rhythm is the fact that the melatonin cycle can be influenced by shifting it or timing it, thanks to appropriate light stimuli, but it is not influenced by the time one goes to sleep or by the time one is forced to wake up (Burgess and Eastman 2005; Crowley et al. 2006). This is the fundamental reason why all people who suffer from a delayed circadian rhythm (owl syndrome) and therefore still have high levels of melatonin in the morning, have cognitive and motor difficulties if they are forced to get up before the natural dropping of melatonin. As often happens for social, study or work reasons, these people must wake up at a different time from their biological clock and remain in a state of torpor until their melatonin level drops to daytime levels. This can also be favoured by an adequate exposure, in the awakening phases and later, to natural light, if present, or even to an artificial light that is able to quickly decrease the melatonin level in the blood. Artificial light, if repeated for several days, may be used to correct in the right way the circadian cycle with respect to the schedule of activity and rest expected during the period and in the specific social context.

### 2.6.4 *Foods and Drugs*

The widespread use of caffeine, as a method to counteract the drowsiness induces by melatonin, is not however a valid substitute for a correct exposure to adequate light in the morning and during the day for the timing of our circadian rhythm. Indeed, as shown by several studies (Shilo et al. 2002; Snel and Lorist 2011; Watson et al. 2016) the excessive use of caffeine has the counter-effect of reducing the quantity and quality of sleep. In turn, reducing the quality and quantity of sleep has the effect of causing drowsiness during the day, which is contrasted with more caffeine. In fact, caffeine is a psychoactive substance like other drugs and, if taken in excessive quantities, leads to a whole series of negative effects, long recognised in physiological research. From a physical standpoint, it causes an increase in blood pressure and vasoconstriction of the vascular system (Mahmud and Feely 2001). In women after menopause, a high consumption of caffeine can accelerate the weakening of the bones due to osteoporosis (Rapuri et al. 2001). It can also alter intestinal motility and gastric acid secretion (Boekema et al. 1999). From a psychological standpoint, the abuse of caffeine has less harmful and acute effects, compared to the physical ones, but not less important. It can in fact induce anxiety, nervousness and reduction of motor coordination (Winston et al. 2005; Smith 2002).

A scientific review carried out by Versteeg et al. (2016) has also suggested that exposure to an inappropriate evening light could have a direct negative effect on the relationship between food consumption and body weight. Artificial lighting in the evening could in fact facilitate a consumption of food in moments less favourable from the metabolic standpoint when the body is not predisposed to digestion and the proper metabolization of nutrients.

## 2.7 The Influence of the Environment on the Human Circadian Cycle

For a human being exposed to a natural life in the open air, the circadian cycle is synchronised by the cycle of natural light with sunrise and sunset. When the normal circadian cycle is shifted, exposure to a high amount of light in the early morning can be useful, because it helps to correctly trigger the normal biological clock. At the same time, the use of light having effect on the circadian system at night should be avoided because, in addition to decreasing or slowing down the production of melatonin, it also has the effect of disrupting the natural human circadian rhythm.

If it is true that light affects the circadian system, it is also true that the way in which this action can take place is different in the exteriors than in the interiors.

### *2.7.1 Outdoor*

In the exteriors there is natural light. It should be remembered that the term natural light means not only direct sunlight, also because the sun can be covered by clouds, or not directly visible for orographic, vegetative or urbanistic issues. Natural light is understood in a more general sense as the sum of the light of the sky and that of the sun. The sky is extremely changeable depending on the time, the latitude and the meteorological situation, factors that continuously change the direction and the spectral content of natural light. With the term natural light or daylight, we mean all the light that comes from the sun, but it is modified by the atmosphere, and the light reflected from the ground, the seas, the woods, and the orographic and urban structures. In outdoor areas, the amount of light is always much higher than the natural or artificial one measurable in interior spaces. But above all, natural light has two fundamental characteristics: it has both the timing and the quantity suitable for a correct phasing of the human circadian system. As stated by Mark Rea: "…the circadian system can be considered a blue-sky detector" (Rea 2007). It is highly probable that the circadian rhythm has evolved to allow the organism to adapt and react optimally to the variability of daily activities. This is to be able to anticipate expected and periodic events, such as meals and the regenerating rest at night, with the aim of ensuring that the bodily processes are temporally synchronised to better adapt to the surrounding environment.

### 2.7.2 *Indoor*

In interior spaces the quantity and quality of the light is different from that found outside. The amount of natural light in interior spaces is much lower than that found outside. Furthermore, the amount of natural light depends on many architectural factors and the urban context. Often, the optimisation of the use of interior spaces dedicated to work, but also of residential spaces, reduces the amount of natural light by various orders of magnitude, particularly when considered in relation to the use of spaces, to the distance from the windows and the orientation of the view. In addition, direct sunlight is also perceived as too intense in the interior and is therefore masked and reduced with the aid of blinds or other anti-glare systems. Often in the workplace, even during the day, regardless of the presence of natural light, artificial light is switched on anyway.

So, in buildings there is a lower amount of light and this may not be able to correctly stimulate the circadian system for the duration of the stay of the subject in the interior spaces. This fundamental topic will be explored in chapter 4 of this book. In the case of typical office work, this can lead to incorrect light stimulation eight hours a day for five days a week. Another question concerning the light present in interior spaces is the correct evaluation of the light spectrum that reaches the eyes of human beings and how this is spatially distributed in the images we see. We know that the human visual system spatially distinguishes the contents of the images thanks to the contrasts, brightness and colour of the light that arrives with different intensities and spectra in the various positions of the retina. Given that the ipRGCs cells are distributed throughout the retina, it is legitimate to ask whether the spatial distribution of intensity and light spectra that we receive can also affect the circadian system. Various studies have been carried out on this aspect.

Another aspect to consider is that interior spaces are characterised by furnishings and surfaces that can be coloured and that therefore influence the spectrum of lighting they receive, as was also shown in the research by Bellia et al. (2015). Moreover, human gaze is almost never directed towards the sources of light but rather towards the other surfaces of the environment and therefore a correct evaluation of the characteristics of lighting, from the application standpoint, presupposes experimentations done in the field with the possibility of monitoring the light to which human subjects are actually subjected (Figueiro 2013) or other methods of investigation based on the subjective evaluation of additional parameters such as comfort, mood, performance and satisfaction level (Veitch and Newsham 1998).

### 2.7.3 Directions of Light

However, some studies have shown that with the same energy, spectrum, timing and duration, there is no dependence of the circadian system on the spatial distribution of the received radiation. That is, the retina would act as a sensor that spatially integrates the information useful for stimulating the biological clock regardless of the information content of the received image. This has been experimented with light boxes arranged in various configurations (McIntyre et al. 1989a, b; Adler et al. 1992). However, regardless of the visual message contained in the image, recent research has nevertheless highlighted the fact that the number of photoreceptors useful to the circadian system is greater in the lower part of the retina (Lasko et al. 1999). In one of these studies (Glickman et al. 2003) special glasses have been used to mask the upper or lower part of the retina. The results have shown that with the same light stimulation, the suppression of melatonin, due to the lower part of the retina, is practically identical to that due to the stimulation of the entire retina, while the stimulation of the upper part of the retina does not change significantly with respect to the absence of light stimulation. These researches would therefore suggest that the light radiation useful for the stimulation of the circadian system is mainly that coming from the visual field in front of us and from above. There are probably evolutionary reasons that justify this phenomenon, since human beings have always been accustomed to receiving light from above: from the sun and from the sky. This aspect is particularly interesting for the research regarding lighting design. In fact, the light stimulation that can affect physiologically the circadian system would be mainly due to the light coming from the horizon up. These results are also confirmed by a research conducted on a group of thirteen elderly people (Smith et al. 2002).

### 2.7.4 The Social Jet Lag

The social context can also influence the evening light exposure of people, consciously or not, contributing to a delayed shift of the circadian rhythm defined as social jet lag, very common among young people. The person is in fact systematically in a state of shifted circadian rhythm compared to social standards that generally provide for a morning commitment of work and/or study. The research conducted by Joo et al. (2017) has demonstrated a high probability that in individuals with delayed chronotype, the exposure to unnatural light patterns, with an intense evening light, contributes to maintaining a delayed shift of the circadian rhythm, leading to poor quality of sleep and mood disorders, typical of these individuals. The social shifting of circadian rhythm has been associated with possible problems such as excess consumption of stimulants such as caffeine, alcohol and nicotine (Wittmann et al. 2006). Furthermore, this social shifting has also been linked to an increased risk of heart disease and metabolic diseases such as obesity and diabetes (Parsons et al. 2015).

## References

Adler, J. S., et al. (1992). Peripheral vision suppression of melatonin. *Journal of Pineal Research, 12*(2), 49–52.

Badia, P., et al. (1991). Bright light effects on body temperature, alertness, EEG and behaviour. *Physiology & Behavior, 50*(3), 583–588.

Barkmann, C., Wessolowski, N., & Schulte-Markwort, M. (2012). Applicability and efficacy of variable light in schools. *Physiology & Behavior, 105*(3), 621–627. https://doi.org/10.1016/j.physbeh.2011.09.020.

Bellia, L., Pedace, A., & Fragliasso, F. (2015). Indoor lighting quality: Effects of different wall colours. *Lighting Research and Technology, 49*. https://doi.org/10.1177/1477153515594654.

Berman, S., et al. (2006). A comparison of traditional and high colour temperature lighting on the near acuity of elementary school children. *Lighting Research & Technology, 38*(1), 41–49. https://doi.org/10.1191/1365782806li155oa.

Berson, D. M., Dunn, F. A., & Takao, M. (2002). Phototransduction by retinal ganglion cells that set the circadian clock. *Science, 295*(5557), 1070–1073. https://doi.org/10.1126/science.1067262.

Blencowe, H., et al. (2012). National, regional, and worldwide estimates of pre-term birth rates in the year 2010 with time trends since 1990 for selected countries: A systematic analysis and implications. *Lancet (London, England), 379*(9832), 2162–2172. https://doi.org/10.1016/S0140-6736(12)60820-4.

Boekema, P. J., et al. (1999). Coffee and gastrointestinal function: Facts and fiction. A review. *Scandinavian Journal of Gastroenterology* (Supplement 230), 35–39

Boivin, D. B., et al. (1996). Dose-response relationships for resetting of human circadian clock by light. *Nature, 379*(6565), 540–542. https://doi.org/10.1038/379540a0.

Bonsall, D. R., & Harrington, M. E. (2013). Circadian rhythm disruption in chronic fatigue syndrome. *Advances in Neuroimmune Biology, 4*(4), 265–274. https://doi.org/10.3233/NIB-130074.

Boulos, Z., et al. (1995). Light treatment for sleep disorders: Consensus report: VII. Jet lag. *Journal of Biological Rhythms, 10*(2), 167–176. https://doi.org/10.1177/074873049501000209.

Brainard, G. C., et al. (1988). Dose-response relationship between light irradiance and the suppression of plasma melatonin in human volunteers. *Brain Research, 454*(1), 212–218. https://doi.org/10.1016/0006-8993(88)90820-7.

Brainard, G. C., et al. (2001a). Action spectrum for melatonin regulation in humans: Evidence for a novel circadian photoreceptor. *Journal of Neuroscience, 21*(16), 6405–6412. https://doi.org/10.1523/JNEUROSCI.21-16-06405.2001.

Brainard, G. C., et al. (2001b). Human melatonin regulation is not mediated by the three cone photopic visual system. *Journal of Clinical Endocrinology and Metabolism, 86*(1), 433–436. https://doi.org/10.1210/jcem.86.1.7277.

Brandon, D. H., Holditch-Davis, D., & Belyea, M. (2002). Preterm infants born at less than 31 weeks' gestation have improved growth in cycled light compared with continuous near darkness. *The Journal of Pediatrics, 140*(2), 192–199. https://doi.org/10.1067/mpd.2002.121932.

Burgess, H. J., & Eastman, C. I. (2005). The dim light melatonin onset following fixed and free sleep schedules. *Journal of Sleep Research, 14*(3), 229–237. https://doi.org/10.1111/j.1365-2869.2005.00470.x.

Cajochen, C. (2007). Alerting effects of light. *Sleep Medicine Reviews, 11*(6), 453–464. https://doi.org/10.1016/j.smrv.2007.07.009.

Cajochen, C., et al. (2005). High sensitivity of human melatonin, alertness, thermoregulation, and heart rate to short wavelength light. *The Journal of Clinical Endocrinology and Metabolism, 90*(3), 1311–1316. https://doi.org/10.1210/jc.2004-0957.

Cedernaes, J., Schiöth, H. B., & Benedict, C. (2015). Determinants of shortened, disrupted, and mistimed sleep and associated metabolic health consequences in healthy humans. *Diabetes, 64*(4), 1073–1080. https://doi.org/10.2337/db14-1475.

Chellappa, S. L., Gordijn, M. C. M., & Cajochen, C. (2011). Can light make us bright? Effects of light on cognition and sleep'. *Progress in Brain Research, 190,* 119–133. https://doi.org/10.1016/B978-0-444-53817-8.00007-4.

Choi, K., & Suk, H.-J. (2016). Dynamic lighting system for the learning environment: Performance of elementary students. *Optics Express, 24*(10), A907–A916. https://doi.org/10.1364/OE.24.00A907.

Christoffersson, G., et al. (2014). Acute sleep deprivation in healthy young men: Impact on population diversity and function of circulating neutrophils. *Brain, Behavior, and Immunity, 41,* 162–172. https://doi.org/10.1016/j.bbi.2014.05.010.

CIE TN 003:2015 Report on the First International Workshop on Circadian and Neurophysiological Photometry, 2013. Available at: http://www.cie.co.at/publications/report-first-international-workshop-circadian-and-neurophysiological-photometry-2013. Accessed February 20, 2017.

Cowan, N. (2008). What are the differences between long-term, short-term, and working memory? In W. S. Sossin, et al. (Eds.), *Progress in brain research.* Elsevier (Essence of Memory), pp. 323–338. https://doi.org/10.1016/s0079-6123(07)00020-9.

Crowley, S. J., et al. (2006). Estimating dim light melatonin onset (DLMO) phase in adolescents using summer or school-year sleep/wake schedules. *Sleep, 29*(12), 1632–1641.

Czeisler, C. A., et al. (1995). Suppression of melatonin secretion in some blind patients by exposure to bright light. *New England Journal of Medicine, 332*(1), 6–11. https://doi.org/10.1056/NEJM199501053320102.

Czeisler, C. A., et al. (1999). Stability, precision, and near-24-hour period of the human circadian pacemaker. *Science (New York, N.Y.), 284*(5423), 2177–2181.

Dijk, D.-J., & Archer, S. N. (2009). Light, sleep, and circadian rhythms: Together again. *PLoS Biology, 7*(6), e1000145. https://doi.org/10.1371/journal.pbio.1000145.

Duffy, J. F., & Czeisler, C. A. (2009). Effect of light on human circadian physiology. *Sleep Medicine Clinics, 4*(2), 165–177. https://doi.org/10.1016/j.jsmc.2009.01.004.

Eastman, C., Martin, S. K., & Hébert, M. (2000). Failure of extraocular light to facilitate circadian rhythm reentrainment in humans. *Chronobiology International, 17,* 807–26. https://doi.org/10.1081/cbi-100102116.

Engle-Friedman, M. (2014). The effects of sleep loss on capacity and effort. *Sleep Science, 7*(4), 213–224. https://doi.org/10.1016/j.slsci.2014.11.001.

Engwall, M., et al. (2014). Let there be light and darkness: Findings from a prestudy concerning cycled light in the intensive care unit environment. *Critical Care Nursing Quarterly, 37*(3), 273–298. https://doi.org/10.1097/CNQ.0000000000000031.

Evans, J. A., et al. (2013). Health consequences of circadian disruption in humans and animal models. *Progress in Molecular Biology and Translational Science, 119,* 283–323.

Figueiro, M. G., Lesniak, N. Z., & Rea, M. S. (2011). Implications of controlled short-wavelength light exposure for sleep in older adults. *BMC Research Notes, 4,* 334. https://doi.org/10.1186/1756-0500-4-334.

Figueiro, M. G. (2013). Why field measurements of circadian light exposure are important. *Lighting Research & Technology, 45*(1), 6. https://doi.org/10.1177/1477153512473709.

Figueiro, M. G. (2015). Light therapy for better sleep. *Sleep Review, 5.* Available at: http://www.sleepreviewmag.com/2015/05/light-therapy-better-sleep/. Accessed March 14, 2018.

Figueiro, M. G., et al. (2014a). Tailored lighting intervention improves measures of sleep, depression, and agitation in persons with Alzheimer's disease and related dementia living in long-term care facilities. *Clinical Interventions in Aging, 9,* 1527–1537. https://doi.org/10.2147/CIA.S68557.

Figueiro, M. G., Plitnick, B., & Rea, M. S. (2014b). The effects of chronotype, sleep schedule and light/dark pattern exposures on circadian phase. *Sleep Medicine, 15*(12), 1554–1564. https://doi.org/10.1016/j.sleep.2014.07.009.

Figueiro, M. G. (2016). Delayed sleep phase disorder: Clinical perspective with a focus on light therapy. *Nature and Science of Sleep, 8,* 91–106. https://doi.org/10.2147/NSS.S85849.

Freedman, M. S., et al. (1999). Regulation of mammalian circadian behaviour by non-rod, non-cone, ocular photoreceptors. *Science, 284*(5413), 502–504. https://doi.org/10.1126/science.284.5413.502.

Gentile, N., et al. (2018). A field study of fluorescent and LED classroom lighting. *Lighting Research & Technology, 50*(4), 631–650. https://doi.org/10.1177/1477153516675911.

Glass, P., et al. (1985). Effect of Bright light in the hospital nursery on the incidence of retinopathy of prematurity. *New England Journal of Medicine, 313*(7), 401–404. https://doi.org/10.1056/NEJM198508153130701.

Glickman, G., et al. (2003). Inferior retinal light exposure is more effective than superior retinal exposure in suppressing melatonin in humans. *Journal of Biological Rhythms, 18*(1), 71–79. https://doi.org/10.1177/0748730402239678.

Hattar, S., et al. (2002). Melanopsin-containing retinal ganglion cells: Architecture, projections, and intrinsic photosensitivity. *Science, 295*(5557), 1065–1070. https://doi.org/10.1126/science.1069609.

Heschong, L., Wright, R. L., & Okura, S. (2002). Daylighting impacts on human performance in school. *Journal of the Illuminating Engineering Society, 31*(2), 101–114. https://doi.org/10.1080/00994480.2002.10748396.

Hubalek, S., Brink, M., & Schierz, C. (2010). Office workers' daily exposure to light and its influence on sleep quality and mood. *Lighting Research & Technology, 42*(1), 33–50. https://doi.org/10.1177/1477153509355632.

Huiberts, L. M., Smolders, K. C. H. J., & de Kort, Y. A. W. (2016). Non-image forming effects of illuminance level: Exploring parallel effects on physiological arousal and task performance. *Physiology and Behavior, 164*, 129–139. https://doi.org/10.1016/j.physbeh.2016.05.035.

IALD. (2017). Joint position paper by *LightingEurope and the International Association of Lighting Designers (IALD) on Human Centric Lighting* (February 2017). Available at: https://www.iald.org/Advocacy/Publications. Accessed September 30, 2018.

Johansson, C., et al. (2001). Seasonal affective disorder and serotonin-related polymorphisms. *Neurobiology of Disease, 8*(2), 351–357. https://doi.org/10.1006/nbdi.2000.037.

Joo, E. Y., et al. (2017). Timing of light exposure and activity in adults with delayed sleep-wake phase disorder. *Sleep Medicine, 32,* 259–265. https://doi.org/10.1016/j.sleep.2016.09.009.

Klein, D. C., Moore, R. Y., & Reppert, S. M. (Eds.). (1991). *Suprachiasmatic nucleus: The mind's clock* (1st ed.). New York: Oxford University Press.

Küller, R., & Lindsten, C. (1992). Health and behavior of children in classrooms with and without windows. *Journal of Environmental Psychology, 12*(4), 305–317. https://doi.org/10.1016/S0272-4944(05)80079-9.

Lack, L., & Wright, H. (1993). The effect of evening bright light in delaying the circadian rhythms and lengthening the sleep of early morning awakening insomniacs. *Sleep, 16*(5), 436–443.

Lasko, T. A., Kripke, D. F., & Elliot, J. A. (1999). Melatonin suppression by illumination of upper and lower visual fields. *Journal of Biological Rhythms, 14*(2), 122–125. https://doi.org/10.1177/074873099129000506.

Lavigne, G. J., et al. (2007). Genesis of sleep bruxism: Motor and autonomic-cardiac interactions. *Archives of Oral Biology (Special Issue: International Mastication Symposium, Brisbane, June 2006), 52*(4), 381–384. https://doi.org/10.1016/j.archoralbio.2006.11.017.

LeGates, T. A., Fernandez, D. C., & Hattar, S. (2014). Light as a central modulator of circadian rhythms, sleep and affect. *Nature Reviews Neuroscience, 15*(7), 443–454. https://doi.org/10.1038/nrn3743.

Lewy, A. J., et al. (1980). Light suppresses melatonin secretion in humans. *Science, 210*(4475), 1267–1269. https://doi.org/10.1126/science.7434030.

Lewy, A. J., et al. (1982). Bright artificial light treatment of a manic-depressive patient with a seasonal mood cycle. *The American Journal of Psychiatry, 139*(11), 1496–1498. https://doi.org/10.1176/ajp.139.11.1496.

Lindblom, N., et al. (2000). Bright light exposure of a large skin area does not affect melatonin or bilirubin levels in humans. *Biological Psychiatry, 48*(11), 1098–1104. https://doi.org/10.1016/S0006-3223(00)00905-7.

Liu, L., et al. (2016). Global, regional, and national causes of under-5 mortality in 2000–15: An updated systematic analysis with implications for the Sustainable Development Goals. *Lancet (London, England), 388*(10063), 3027–3035. https://doi.org/10.1016/S0140-6736(16)31593-8.

Lockley, S. W., et al. (1998). Extraocular light exposure does not suppress plasma melatonin in humans. *The Journal of Clinical Endocrinology and Metabolism, 83*(9), 3369–3372. https://doi.org/10.1210/jcem.83.9.5244.

Lockley, S. W., et al. (2006). Short-wavelength sensitivity for the direct effects of light on alertness. *Vigilance, and the Waking Electroencephalogram in Humans Sleep, 29*(2), 161–168. https://doi.org/10.1093/sleep/29.2.161.

Lucas, R. J., et al. (2014). Measuring and using light in the melanopsin age. *Trends in Neurosciences, 37*(1), 1–9. https://doi.org/10.1016/j.tins.2013.10.004.

Magnusson, A., et al. (2000). Lack of seasonal mood change in the Icelandic population: Results of a cross-sectional study. *The American journal of psychiatry, 157*(2), 234–238. https://doi.org/10.1176/appi.ajp.157.2.234.

Mahmud, A., & Feely, J. (2001). Acute effect of caffeine on arterial stiffness and aortic pressure waveform. *Hypertension, 38*(2), 227–231. https://doi.org/10.1161/01.HYP.38.2.227.

Manfredini, D., et al. (2013). Epidemiology of bruxism in adults: A systematic review of the literature. *Journal of Oral & Facial Pain and Headache, 27*(2), 99–110. https://doi.org/10.11607/jop.921.

McIntyre, I. M., et al. (1989). Quantal melatonin suppression by exposure to low intensity light in man. *Life Sciences, 45*(4), 327–332. https://doi.org/10.1016/0024-3205(89)90142-2.

McIntyre, I. M., et al. (1989). Human melatonin suppression by light is intensity dependent. *Journal of Pineal Research, 6*(2), 149–156. https://doi.org/10.1111/j.1600-079X.1989.tb00412.x.

Meesters, Y., et al. (2011). Low-intensity blue-enriched white light (750 lux) and standard bright light (10 000 lux) are equally effective in treating SAD. *A random-ized controlled study. BMC Psychiatry, 11,* 17. https://doi.org/10.1186/1471-244X-11-17.

Meesters, Y., et al. (2016). The effects of low-intensity narrow-band blue-light treatment compared to bright white-light treatment in sub-syndromal seasonal affective disorder. *BMC Psychiatry, 16*. https://doi.org/10.1186/s12888-016-0729-5.

Miller, C. L., et al. (1995). The effects of cycled versus noncycled lighting on growth and development in preterm infants. *Infant Behavior and Development, 18*(1), 87–95. https://doi.org/10.1016/0163-6383(95)90010-1.

Miyake, A., & Shah, P. (Eds). (1999). Models of working memory: Mechanisms of active maintenance and executive control. Cambridge University Press.

Montgomery, P., & Dennis, J. A. (2002). Bright light therapy for sleep problems in adults aged 60+. *Cochrane Database of Systematic Reviews, 2*. https://doi.org/10.1002/14651858.cd003403.

Morag, I., & Ohlsson, A. (2016). Cycled light in the intensive care unit for pre-term and low birth weight infants. *The Cochrane Database of Systematic Reviews, 8*, CD006982. https://doi.org/10.1002/14651858.cd006982.pub4.

Morin, L. P. (1994). The circadian visual system. *Brain Research. Brain Research Reviews, 19*(1), 102–127.

Noguchi, H., & Sakaguchi, T. (1999). Effect of illuminance and colour temperature on lowering of physiological activity. *Applied Human Science: Journal of Physiological Anthropology, 18*(4), 117–123.

Okamoto, Y., & Nakagawa, S. (2016). Effects of light wavelength on MEG ERD/ERS during a working memory task. *International Journal of Psychophysi-ology, 104,* 10–16. https://doi.org/10.1016/j.ijpsycho.2016.03.008.

Parsons, M. J., et al. (2015). Social jetlag, obesity and metabolic disorder: Investigation in a cohort study. *International Journal of Obesity (2005), 39*(5), 842–848. https://doi.org/10.1038/ijo.2014.201.

Perlow, M. J., et al. (1980). Photic regulation of the melatonin rhythm: Monkey and man are not the same. *Brain Research, 182*(1), 211–216. https://doi.org/10.1016/0006-8993(80)90848-3.

Pévet, P., et al. (2002). The chronobiotic properties of melatonin. *Cell and Tissue Research, 309*(1), 183–191. https://doi.org/10.1007/s00441-002-0584-1.

Philips. (2018). *EnergyUp Energy light HF3420/03 Energy light*. Available at: https://www.philips.co.uk/c-p/HF3420_03/energyup-energy-light Accessed June 10, 2018.

Phillips, M. L. (2009). Circadian rhythms: Of owls, larks and alarm clocks. *Nature, 458*(7235), 142–144. https://doi.org/10.1038/458142a.

Pringsheim, T. (2002). Cluster headache: Evidence for a disorder of circadian rhythm and hypothalamic function. *Canadian Journal of Neurological Sciences, 29*(1), 33–40. https://doi.org/10.1017/S0317167100001694.

Provencio, I., et al. (2000). A novel human opsin in the inner retina. *Journal of Neuroscience, 20*(2), 600–605. https://doi.org/10.1523/JNEUROSCI.20-02-00600.2000.

Pulay, A. (2016). Examining student behavior under two correlated color temperature levels of lighting in an elementary classroom. *Educational Planning, 23*(3), 57–69.

Rapuri, P. B., et al. (2001). Caffeine intake increases the rate of bone loss in elderly women and interacts with vitamin D receptor genotypes. *The American Journal of Clinical Nutrition, 74*(5), 694–700. https://doi.org/10.1093/ajcn/74.5.694.

Rautkylä, E., et al. (2010). Effects of correlated colour temperature and timing of light exposure on daytime alertness in lecture environments. *Journal of Light & Visual Environment, 34*(2), 59–68. https://doi.org/10.2150/jlve.34.59.

Rea, M. S., Bullough, J. D., & Figueiro, M. G. (2001). Human melatonin suppression by light: A case for scotopic efficiency. *Neuroscience Letters, 299*(1), 45–48. https://doi.org/10.1016/S0304-3940(01)01512-9.

Rea, M. S. (2007). Oltre la visione. In *More than vision* (1st edn, pp. 4–18). Recanati: iGuzzini - Editoriale Domus (ilibriGuzzini).

Rea, M. S., et al. (2012). Relationship of morning cortisol to circadian phase and rising time in young adults with delayed sleep times. *International Journal of Endocrinology*. https://doi.org/10.1155/2012/749460.

Rea, M. S., & Figueiro, M. G. (2016). The NICU lighted environment. *New-born and infant nursing reviews: NAINR, 16*(4), 195–202. https://doi.org/10.1053/j.nainr.2016.09.009.

Reiter, R. J. (1991). Pineal melatonin: Cell biology of its synthesis and of its physiological interactions. *Endocrine Reviews, 12*(2), 151–180. https://doi.org/10.1210/edrv-12-2-151.

Reiter, R. J., et al. (2012). Obesity and metabolic syndrome: Association with chronodisruption, sleep deprivation, and melatonin suppression. *Annals of Medicine, 44*(6), 564–577. https://doi.org/10.3109/07853890.2011.586365.

Revell, V. L., et al. (2006). Alerting effects of light are sensitive to very short wavelengths. *Neuroscience Letters, 399*(1–2), 96–100. https://doi.org/10.1016/j.neulet.2006.01.032.

Rosen, L. N., et al. (1990). Prevalence of seasonal affective disorder at four latitudes. *Psychiatry Research, 31*(2), 131–144. https://doi.org/10.1016/0165-1781(90)90116-M.

Rosenthal, N. E., et al. (1987). Disturbances of appetite and weight regulation in seasonal affective disorder. *Annals of the New York Academy of Sciences, 499*(1), 216–230. https://doi.org/10.1111/j.1749-6632.1987.tb36213.x.

Rosenthal, N. E., et al. (1990). Phase-shifting effects of bright morning light as treatment for delayed sleep phase syndrome. *Sleep, 13*(4), 354–361. https://doi.org/10.1093/sleep/13.4.354.

Rosenthal, N. E. (1991). Plasma melatonin as a measure of the human clock. *The Journal of Clinical Endocrinology & Metabolism, 73*(2), 225–226. https://doi.org/10.1210/jcem-73-2-225.

Rosenthal, N. E. (2006). *Winter blues: Everything you need to know to beat seasonal affective disorder* (4th ed.). New York: Guilford Press.

Roth, T. (2007). Insomnia: Definition, prevalence, etiology, and consequences. *Journal of Clinical Sleep Medicine: JCSM: Official Publication of the American Academy of Sleep Medicine, 3*(5 Suppl), S7–S10.

Sack, R. L., et al. (1992). Circadian rhythm abnormalities in totally blind people: Incidence and clinical significance. *The Journal of Clinical Endocrinology and Metabolism, 75*(1), 127–134. https://doi.org/10.1210/jcem.75.1.1619000.

Salt, S., & Karmel, L. J. (1967). The windowless school. *The Clearing House: A Journal of Educational Strategies, Issues and Ideas, 42*(3), 176–178. https://doi.org/10.1080/00098655.1967.11477476.

Shetty, S., et al. (2010). Bruxism: A literature review. *Journal of Indian Prosthodontic Society, 10*(3), 141–148. https://doi.org/10.1007/s13191-011-0041-5.

Shilo, L., et al. (2002). The effects of coffee consumption on sleep and melatonin secretion. *Sleep Medicine, 3*(3), 271–273. https://doi.org/10.1016/S1389-9457(02)00015-1.

Sleegers, P. J. C., et al. (2013). Lighting affects students' concentration positively: Findings from three Dutch studies. *Lighting Research & Technology, 45*(2), 159–175. https://doi.org/10.1177/1477153512446099.

Sloane, P. D., et al. (2007). High-intensity environmental light in dementia: Effect on sleep and activity. *Journal of the American Geriatrics Society, 55*(10), 1524–1533. https://doi.org/10.1111/j.1532-5415.2007.01358.x.

Sloane, P. D., Figueiro, M. G., & Cohen, L. (2008). Light as therapy for sleep disorders and depression in older adults. *Clinical geriatrics, 16*(3), 25–31.

Smith, A. (2002). Effects of caffeine on human behaviour. *Food and Chemical Toxicology, 40*(9), 1243–1255. https://doi.org/10.1016/S0278-6915(02)00096-0.

Smith, J. S., et al. (2002). Illumination of upper and middle visual fields produces equivalent suppression of melatonin in older volunteers. *Chronobiology Inter-national, 19*(5), 883–891.

Smolders, K. C. H. J., de Kort, Y. A. W., & Cluitmans, P. J. M. (2012). A higher illuminance induces alertness even during office hours: Findings on subjective measures, task performance and heart rate measures. *Physiology & Behavior, 107*(1), 7–16. https://doi.org/10.1016/j.physbeh.2012.04.028.

Snel, J., & Lorist, M. M. (2011). Chapter 6—Effects of caffeine on sleep and cognition. In Van Dongen, H. P. A. & Kerkhof, G. A. (Eds.), *Progress in brain research* (pp. 105–117). Elsevier (Human Sleep and Cognition Part II). https://doi.org/10.1016/b978-0-444-53817-8.00006-2.

Souman, J. L., et al. (2018). Acute alerting effects of light: A systematic literature review. *Behavioural Brain Research, 337,* 228–239. https://doi.org/10.1016/j.bbr.2017.09.016.

Stevens, R. G., & Rea, M. S. (2001). Light in the built environment: Potential role of circadian disruption in endocrine disruption and breast cancer. *Cancer Causes and Control, 12*(3), 279–287. https://doi.org/10.1023/A:1011237000609.

Szelenberger, W., Niemcewicz, S., & DĄbrowska, A. J. (2005). Sleepwalking and night terrors: Psychopathological and psychophysiological correlates. *International Review of Psychiatry, 17*(4), 263–270. https://doi.org/10.1080/09540260500104573.

Takahashi, J. S. (2017). Transcriptional architecture of the mammalian circadian clock. *Nature Reviews Genetics, 18*(3), 164–179. https://doi.org/10.1038/nrg.2016.150.

te Kulve, M., et al. (2015). The influence of light on thermal responses. *Acta Physiologica, 216*(2), 163–185. https://doi.org/10.1111/apha.12552.

Teixeira, L., et al. (2013). Exposure to bright light during evening class hours increases alertness among working college students. *Sleep Medicine, 14*(1), 91–97. https://doi.org/10.1016/j.sleep.2012.08.017.

Thapan, K., Arendt, J., & Skene, D. J. (2001). An action spectrum for melatonin suppression: Evidence for a novel non-rod, non-cone photoreceptor system in humans. *Journal of Physiology, 535*(1), 261–267. https://doi.org/10.1111/j.1469-7793.2001.t01-1-00261.x.

van Bommel, W., & van den Beld, G. (2004). Lighting for work: A review of visual and biological effects. *Lighting Research & Technology, 36,* 255–269. https://doi.org/10.1191/1365782804li122oa.

van Oosterhout, W., et al. (2018). Chronotypes and circadian timing in migraine. *Cephalalgia: An International Journal of Headache, 38*(4), 617–625. https://doi.org/10.1177/0333102417698953.

Van Someren, E. J., et al. (1997). Indirect bright light improves circadian rest-activity rhythm disturbances in demented patients. *Biological Psychiatry, 41*(9), 955–963. https://doi.org/10.1016/S0006-3223(97)89928-3.

Veitch, J. A., & Newsham, G. R. (1998). Lighting quality and energy-efficiency effects on task performance, mood, health, satisfaction, and comfort. *Journal of the Illuminating Engineering Society, 27*(1), 107–129. https://doi.org/10.1080/00994480.1998.10748216.

Versteeg, R. I., et al. (2016). Nutrition in the spotlight: Metabolic effects of environmental light. *The Proceedings of the Nutrition Society, 75*(4), 451–463. https://doi.org/10.1017/S0029665116000707.

Watson, E. J., et al. (2016). Caffeine consumption and sleep quality in Australian adults. *Nutrients, 8*(8). https://doi.org/10.3390/nu8080479.

WHO. (2015). *Dementia Fact sheet N° 362*. World Health Organization. Available at: https://web.archive.org/web/20150318030901/http://www.who.int/mediacentre/factsheets/fs362/en. Accessed March 15, 2018.

WHO. (2018). *Preterm birth*. World Health Organization. Available at: http://www.who.int/news-room/fact-sheets/detail/preterm-birth. Accessed June 11, 2018.

Wilhelm, I., et al. (2007). Is the cortisol awakening rise a response to awakening? *Psychoneuroendocrinology, 32*(4), 358–366. https://doi.org/10.1016/j.psyneuen.2007.01.008.

Winston, A. P., Hardwick, E., & Jaberi, N. (2005). Neuropsychiatric effects of caffeine. *Advances in Psychiatric Treatment, 11*(6), 432–439. https://doi.org/10.1192/apt.11.6.432.

Wittmann, M., et al. (2006). Social jetlag: Misalignment of biological and social time. *Chronobiology International, 23*(1–2), 497–509. https://doi.org/10.1080/07420520500545979.

Yadav, A., Verma, P., & Singh, S. (2017). Going beyond the limits: Effect of clock disruption on human health. *Biological Rhythm Research, 48*(5), 693–700. https://doi.org/10.1080/09291016.2017.1345428.

Zaidi, F. H., et al. (2007). Short-wavelength light sensitivity of circadian, pupillary, and visual awareness in humans lacking an outer retina. *Current Biology, 17*(24), 2122–2128. https://doi.org/10.1016/j.cub.2007.11.034l.

Zeitzer, J. M., et al. (2000). Sensitivity of the human circadian pacemaker to nocturnal light: Melatonin phase resetting and suppression. *The Journal of Physiology, 526*(Pt 3), 695–702. https://doi.org/10.1111/j.1469-7793.2000.00695.x.

# Chapter 3
# Evaluation of Artificial Light with Respect to Human Health

## 3.1 Visual Evaluation of Light: Radiometry and Photometry

In lighting design, photometric quantities are used to measure the values of the requirements that come into play in the project. In the case of the evaluation of the non-image forming (NIF) effects, in addition to photometric quantities, it is necessary to consider also other parameters deriving from the theory of electromagnetism defined by the Scottish physicist James Clerk Maxwell in 1864. The fundamental equations of electromagnetism are vector differential equations. Treating electromagnetic waves in vectorial terms leads to their accurate description, but presents notable complications from the notational and the practical calculation standpoint. For these reasons, lighting applications have never treated light in vector form, but with spectral functions and real numbers in the field of radiometry (Table 3.1) and photometry (Table 3.2).

**Table 3.1** List of the main radiometric quantities, symbols and units of measurement

| Context | Quantity | SPD | | Total value | |
|---|---|---|---|---|---|
| | | Symbol | Unit | Symbol | Unit |
| Energy | Radiant energy | $Q_{e,\lambda}$ | J/m | $Q_e$ | J |
| Radiation sources | Radiant flux | $\Phi_{e,\lambda}$ | W/m | $\Phi_e$ | W = J/s |
| | Radiant intensity | $I_{e,\lambda}$ | W/(sr m) | $I_e$ | W/sr |
| Surfaces | Irradiance | $E_{e,\lambda}$ | W/(m$^2$ m) | $E_e$ | W/m$^2$ |
| | Radiant exitance | $M_{e,\lambda}$ | W/(m$^2$ m) | $M_e$ | W/m$^2$ |
| Observer | Radiance | $L_{e,\lambda}$ | W/(sr m$^2$ m) | $L_e$ | W/(sr m$^2$) |

 
M. Rossi, *Circadian Lighting Design in the LED Era*, Research for Development,
https://doi.org/10.1007/978-3-030-11087-1_3

Table 3.2 List of the main photometric quantities, symbols and units of measurement

| Context | Quantity | Symbol | Unit | |
|---|---|---|---|---|
| Light energy | Luminous energy | $Q_V$ | T | talbot |
| Light sources | Luminous flux | $\Phi_V$ | lm | lumen |
| | Luminous intensity | $I_V$ | cd | candle |
| Surfaces | Illuminance | $E_V$ | lx | lux |
| | Luminous exitance | $M_V$ | lx | lux |
| Observer | Luminance | $L_V$ | cd/m$^2$ | candle/m$^2$ |
| Performance | Luminous efficacy | $\eta$ | lm/W | lumen/watt |
| Performance | Luminous efficacy of optical radiation | LER | lm/W | lumen/watt |

### 3.1.1 Radiometry

Radiometry is that part of the optical science that defines the measurement and description of electromagnetic waves through scalar rather than vector fields, contextualizing the direction and orientation of radiation through the use of different quantities with an appropriate nomenclature. Radiometry uses some fundamental quantities defined on the basis of the main quantity, which is the radiant energy associated with an electromagnetic radiation. These quantities are used to describe the power of electromagnetic radiation that are considered with respect to the three fundamental elements that govern lighting design, i.e. the sources, the surfaces of objects and the observers or sensors. The name of each of these quantities therefore is used to identify the direction and the context with respect to a source, a surface or an observer.

When considering a radiation source, the radiant power or flux describes the amount of energy exiting the source in the time unit, while the radiant intensity describes the flux coming out of a point source relative to a precise direction in the space around the source.

In the case of the radiation that interacts with the objects, the irradiance is the flux falling on the surface unit and is therefore always understood as a radiation that arrives on the surface from a source or is reflected from other surfaces. While the radiant exitance describes the outgoing flux from the surface unit. The flux can come out of a surface because this reflects the radiation, or because it transmits it, or because the surface is a geometrically extended source.

Considering an observer, a sensor, a camera or another type of receptor, the radiance describes the flux received from the receptor with respect to the solid angle unit (measured in steradians) and to the area unit projected on a plane orthogonal to the direction between source and observer.

When we consider a radiometric quantity, this can be expressed in two different ways. The first one as a function of the wavelength and in this case it is called spectral power distribution (SPD) that describes all the monochromatic waves that compose

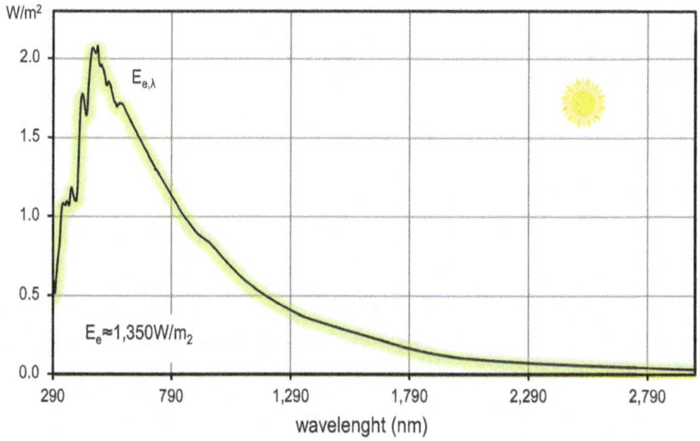


**Fig. 3.1** The SPD of irradiance received from the sun outside the earth's atmosphere

it. The second one as a single value, said integral or total value corresponding to the integral extended to all the wavelengths of the SPD, or to the area subtended by its curve. The total value, being the area subtended by the spectral function, is a measure of the total energy or power associated with the radiation regardless of the distribution of the energy in the various wavelengths. The SPD of the irradiance $E_{e,\lambda}$ received from the sun outside the earth's atmosphere (Mecherikunnel and Richmond 1908) is shown in Fig. 3.1, the area subtended by the curve is the total value $E_e$ of the irradiance. Being a real number, the total value is certainly a more usable representation of the considered magnitude; however, it loses any chromatic information associated with the radiation, given that infinite shapes of the spectra can correspond to the same underlying area, and it is the shape of the SPD that helps to determine the chromatic perception of a luminous radiation.

### *3.1.2 Photometry*

The radiometric quantities so far defined in reality are applicable to all electromagnetic radiations, even those that are not visible. For the purposes of lighting design, from radiometry we move on to photometry, which deals with the measurement and description of electromagnetic radiation according to their perception, in the form of light radiations, by a human observer. Indeed, our visual sensitivity to electromagnetic radiation changes according to the wavelength, but also to the amount of light reaching the eyes. There are three types of vision: the photopic or diurnal vision, which occurs when the luminance levels in the eye are greater than about 10 cd/m$^2$; the scotopic or nocturnal vision, in which it is not possible to distinguish colours

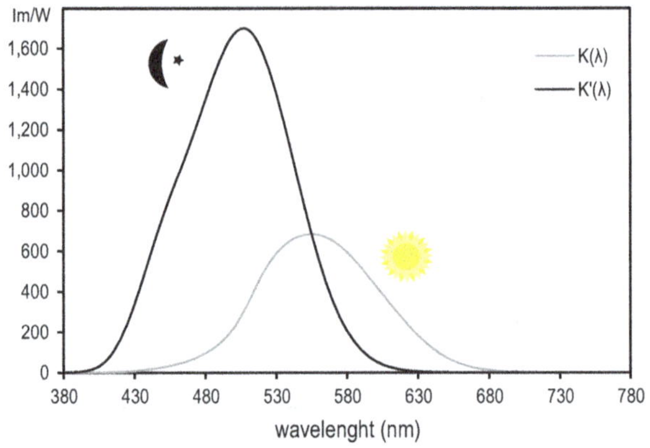


**Fig. 3.2** Diagram of the photopic and the scotopic light efficacy curves

and occurs when luminance levels are less than about 0.001 cd/m$^2$; and the mesopic vision, which occurs with intermediate luminance levels.

Based on experiments conducted by the CIE (Commission International de l'Éclairage), at the beginning of the twentieth century (Wyszecki and Stiles 2000), it has been discovered that the average human observer perceives as light only the electromagnetic radiations comprised approximately between the wavelengths of 380 and 780 nm. Moreover, the various wavelengths are not perceived with equal intensity, but in a variable way with a maximum sensitivity at 555 nm for the photopic vision. All this is expressed by the photopic spectral light efficacy function K(λ) or its normalised form, the photopic spectral light efficiency function V(λ). This is obtained by dividing K(λ) by its maximum value $K_{max} = K(555)$, which is at 555 nm (or $V(\lambda) = K(\lambda)/K_{max}$). Wavelength that is perceived as the yellow colour. Besides the photopic efficacy, there is also a scotopic spectral light efficacy curve K′(λ) and the corresponding scotopic spectral light efficiency curve is expressed by the function $V'(\lambda) = K'(\lambda)/K'_{max}$, since the maximum of K′(λ) or $K'_{max} = K'$ (507) occurs at the wavelength of 507 nm, which is perceived as the green colour. Figure 3.2 illustrates the curves of photopic spectral light efficacy K(λ) and of the scotopic one K′(λ), typical of the diurnal and nocturnal human vision.

From each spectral radiometric quantity it is therefore possible to determine a corresponding integrated photometric value (Table 3.2). The photometric quantities are distinguished from the radiometric ones due to the presence of the subscript *v* (visual) instead of the subscript *e* (energy). In the past, more often than not, in the lighting design the use of the subscript *v* was omitted since photometric quantities were assumed, but also because the design considered light to be achromatic. Today, however, thanks to the new LED sources, the situation has changed, because the design considers the spectral and chromatic aspects of light and this requires the use of both photometric and radiometric quantities; it is therefore appropriate to distinguish between them.

The luminous flux $\Phi_v$ describes how much light comes out of a light source. This is the correct quantity in order to establish how much light is produced, for example, by a lamp. Conversely, it is not correct to describe the quantity of light produced by the consumed electric power P (Watt) because, with the same electrical energy, the light sources can have a higher or lower light emission. Furthermore, the electric power P consumed by a lamp must not be confused with the radiant power $\Phi_e$; indeed, the former is a measure of electric power, while the second is a measure of power of the electromagnetic field. The ability of light sources to emit a luminous flux according to the consumed electrical power is described by a quantity known as luminous efficacy: $\eta = \Phi_v/P$, given by the ratio between the produced luminous flux and the electric power needed to produce it and is measured in lumens on Watt (lm/W). To relate the idea of luminous flux to our daily experience, it is good to know that, for example, the classic incandescent lamp with 100 W tungsten filament, now banned from the European market, emits a luminous flux of about 1,350 lm. It has therefore a very low efficacy $\eta = 13.5$ lm/W when compared to that of a compact 20 W fluorescent lamp that emits a flux of 1,200 lm and has an efficacy of $\eta = 60$ lm/W, and even lower if compared to a modern 11 W LED lamp that emits a flux of 1,420 lm and has an efficacy of $\eta = 129$ lm/W. From the environmental sustainability standpoint, it would therefore be better to use only the most efficient sources because they are the ones that, with the same emitted flux, consume less electricity. But often the increase in efficacy does not correspond to a completeness and richness of the spectrum. Before the advent of LEDs, the most efficient lamps, such as those with low pressure sodium vapour that reach an efficacy of 180 lm/W, are those that have a worse colour rendering index (CRI). High efficacy of traditional light sources does not usually provide a correct distinction of colours, given that they have an almost monochromatic yellow emission. With LED sources it is also used another performance parameter, the luminous efficacy of optical radiation (LER). This is obtained by the ratio between the luminous flux ($\Phi_v$) and the total radiant flux ($\Phi_e$) of a light source and is measured in lm/W.

The light intensity $I_v$ is used to describe the way in which light is distributed in the space around a source. This can be measured, through a goniophotometer, in all the spatial directions around the source. For real light sources, light intensity depends on the direction around the source. By convention, the direction is expressed with two polar angles $C, \gamma$ where the first identifies the horizontal rotation direction, while the second the elevation. The set of luminous intensity values can be represented by a table according to the two polar angles or, in 3D graphic format with the photometric solid. This is defined by the space around the centroid of the light emitted by the source and it is enclosed by a surface whose distance from the centre represents the intensity in that direction. Figure 3.3 illustrates the photometric solid of a luminaire with an indirect (upward) light emission component greater than the direct one (down), the emission centre is located at the origin of the axes.

The illuminance describes the amount of light that arrives on a surface. It is the most used quantity in light evaluation for lighting design. It can be easily measured by the illuminance meter. The illuminance decreases with the square of the distance from the light source and depends on the inclination with which the rays of light hit the surface: it is maximum when they arrive perpendicular to the surface and is nullified when the rays are parallel to the surface. Illuminance values can vary by

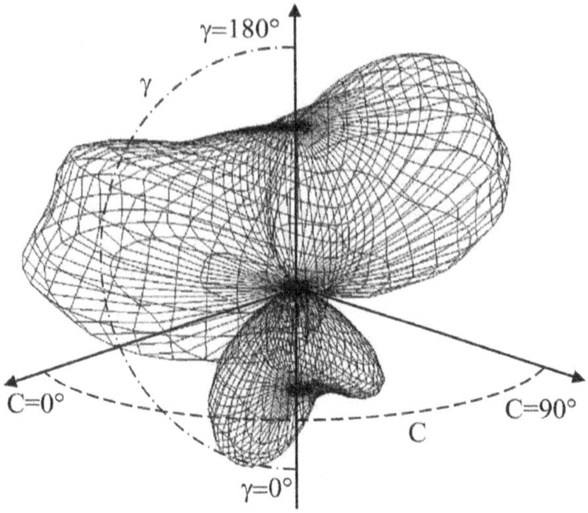


**Fig. 3.3** The photometric solid of a luminaire

**Table 3.3** Typical lighting values produced by natural and artificial light

| Context | Place | Lux |
|---|---|---|
| Natural lighting | Summer day with clear sky | 60,000–100,000 |
| | Summer day with overcast sky | 20,000 |
| | Winter day with overcast sky | 3,000 |
| | Moon night | 0.25 |
| | Starry night without the moon | 0.01 |
| Lighting according to EN 12464-1 standard in the interior workplaces | Office desk | 500 |
| | Drawing table | 750 |
| | Schoolroom | 300 |
| | Library reading area | 500 |
| | Corridors and circulation areas | 100 |
| | Stairs and elevators | 150 |

many orders of magnitude between natural and artificial light (Table 3.3); in fact, our visual system has considerable ability to adapt to these variations, more than it is simply described by the photopic and scotopic efficacy curves.

The *luminance* derives from the radiance and is the photometric quantity that makes it possible to evaluate the light that reaches the eyes of an observer. In particular, from the lighting design standpoint, it provides a quantification of the psychophysical perception of luminosity, of the perception phenomena of contrast and glare and therefore of the visual comfort aspects of the project. If a surface has a Lambertian behaviour, i.e. it reflects light uniformly in all directions, and is illuminated with the illuminance value $E_v$, then its luminance can be calculated with the simple formula $L_v = E_v \cdot \rho/\pi$, where $\rho$ is the reflectance of the surface with a value between 0 and 1 ($\rho = 0$ is black, while $\rho = 1$ is the ideal white). This relationship is very important because in everyday life most of the surfaces that are found in the living spaces of human beings have an almost Lambertian behaviour. For example, a very white sheet of paper, with a reflectance $\rho = 0.85$ placed on a desk illuminated with 500 lx, has a luminance of 135 $cd/m^2$. From the lighting design standpoint, it is important to understand a fundamental aspect: while illuminance depends on the quantity of light and the position and orientation of the surface with respect to the light source, the luminance of a reflecting surface depends on the illuminance that this receives and its reflectance, i.e. on how much light it reflects depending on its lightness. Furthermore, illuminance is an intrinsic property of the geometry of an environment with its light sources, while luminance also depends on the material characteristics of the surfaces and the position of the observer. A fundamental limitation of the human visual system is the phenomenon known as glare. This occurs when one looks directly, or indirectly through a specular reflection, at a light source. Glare also occurs, but to a lesser extent, in the presence of strong luminance contrasts, when for example we look at the sky through a window from inside a dark room. This phenomenon causes a reduction in the contrast of the scene in the darker areas and a reduction in the visibility of objects close to the light source or its reflection. The dazzling effect of a light source observed directly is inversely proportional to its size, in relation to the amount of light emitted and its distance. For this reason, the halogen sources and the LEDs used in interior spaces can have considerable glare problems, too often ignored.

### *3.1.3 White Light Tones*

To describe the colour tones of a white light source, lighting design uses the correlated colour temperature (CCT). This is defined as the temperature of the black body, in Kelvin (K) degrees, which produces a spectral emission with chromatic characteristics similar to the source itself. The concept of chromatic similarity is evaluated in terms of colour distance ($\Delta uv$) from the chromaticity of the black body, the curve of the Planckian locus, evaluated on the chromaticity diagram CIE UCS 1960. In order for the CCT of a light source to be definable, the value of this chromatic distance must be contained within a limit value: $\Delta uv \leq \pm 0.02$. If the absolute value of the chromatic distance $\Delta uv$ is greater than the limit value, the correlated colour temperature of a source cannot be defined and this indicates that the light source cannot be considered white but coloured.

## 3.2 LED Light and Health Risks

LEDs have been used since the seventies of the last century for signalling in instrumentation panels, but only after the recent spread of LED technology also in the field of interior lighting, the press has begun to take an interest in the issue of possible risks for health and a series of alarming news appeared in newspapers arguing that LED lighting would be harmful to human health (Cossar 2013; Ollove 2016; Bodkin 2018). Actually, these articles, written for sensationalistic purposes, take partial data from scientific publications, sometimes even questionable, reporting only some elements extracted from the research context and without evaluating the real application impact from the standpoint of general lighting for human beings. Given that one of the central themes of this text are LED light sources, this topic undoubtedly deserves an in-depth study.

At the European level, LED-based luminaires must comply with the general Directive 2001/95/EC that applies to all products placed on the market if their operating voltage is Extra-Low Voltage (ELV), i.e. less than 50 V AC or 75 V DC. If the operating voltage exceeds these values, the luminaires must comply with the Directive 2014/35/EU on the safety of low voltage electrical products.

A fundamental point on the topic of health risks, already introduced in the first chapter, is that the element to be analysed is not only the LED as a light source, but rather the multiple types of light that can be emitted by it and the way in which they are used, as it happens in the study of other effectively harmful light sources. Among these there is the sun, whose radiation, if taken inappropriately, can be extremely harmful for both the eyes and the skin. Even some types of discharge or halogen lamps produce a considerable amount of ultraviolet (UV) radiation, to the point that they need a filter to be used for general lighting.

### *3.2.1 Photobiology*

Photobiology is the area of biology that, based on the scientific method, studies the effect of non-ionising radiation on living organisms (Kohen et al. 1995). With the term photobiological hazard referring to humans we mean the assessment of health risks, mainly due to the exposure of the skin and eyes, which may be caused by these types of electromagnetic radiation. In order to assess whether a light radiation, UV or infrared (IR), can be harmful to human health, there are some fundamental characteristics that must be considered:

1. The real SPD of radiation, with particular attention to the blue region of the visual spectrum.
2. The amount of radiation that hits living tissues, which is evaluated in terms of irradiance, radiance or illuminance if only light is considered.
3. The duration of the radiation exposure.
4. The history of previous radiation exposure.

5. The real exposure to radiation, distinguishing that on the eyes from that on the skin. This depends on the amount of radiation by the exposure time, which is defined as a dose. For example irradiant exposure dose ($J/m^2$) or radiant exposure dose [$J/(sr\ m^2)$].
6. The state of health of the eyes and skin of the considered subject.
7. The type of visual function, distinguishing the direct fixed observation of the light source from the normal visual function in which the view moves continuously in the context.

### *3.2.2 Kind of Risks*

Often, when we talk about risks in relation to artificial lighting, we think about the safety of the luminaire. However, these aspects are already widely considered in the standards of the electrotechnical sector as for other products that use electricity. Instead, in the context of possible health risks in relation to light and other radiation such as UV and IR, we can list four main types of risks:

A. The risks that may arise as a consequence of a temporary direct or indirect vision of a source of intense light, which can cause a temporary visual problem, but dangerous if sight is important for the activity that is taking place, such as driving a vehicle or using a potentially dangerous work tool.
B. The photobiological hazards deriving from the direct exposure of the eyes or skin to a source of radiation. For example solar radiation. Without the need for special regulations, human beings know that they must protect their eyes from the sun and white people know that they must also protect their skin.
C. The risks deriving from the temporal characteristics of light emission by a light source. Such as the flicker and the effects that can be had with moving objects, for which a stroboscopic effect can lead to an incorrect evaluation of the actual rotary motion.
D. The risks that may arise from the circadian aspects of exposure to light or other effects related to personal well-being. It is in this context that it is necessary to evaluate whether artificial light can have a negative or positive influence on what should be the normal circadian cycle of a person.

This book is mainly devoted to the issues indicated in point D regarding current knowledge about circadian lighting. However, the evaluation of the other types of risks also contributes to a correct approach to human centric lighting and will therefore be introduced in the following paragraphs.

## 3.3 Glare and Health Risks

In order to illuminate, we need light sources, but these can be dazzling for the sight. In particular, a source is as dazzling as its angular size is smaller with respect to the observer and in relation to the amount of light emitted (ICNIRP 2013). We know from everyday experience the dazzling effect of the sun, but the same also happens with many types of traditional lamps and especially with LEDs. In fact, the modern LED sources emit a high luminous flux in relation to their dimensions, which are reduced, by technical nature, to the size of a few millimetres.

A fundamental limitation of the human visual system is the phenomenon known as glare (Boyce 2014; Mainster and Turner 2012). This occurs when observing directly, or indirectly through a reflection on a surface, a light source. However, glare also occurs, even if to a lesser extent, in the presence of strong luminance contrasts, when for example we look at the sky through a window from inside a scarcely illuminated room or in the presence of strong contrasts on light reflecting surfaces (IES 2011). Glare causes a reduction in the ability to perceive the contrast of the scene in the darker areas and at the same time a reduction in the visibility of objects close to the light source or its reflection.

### *3.3.1 Disabling Glare*

Regardless of whether glare originates from the direct vision of the light source or from reflection phenomena, its visual effect is twofold. Disabling glare occurs when the amount of light is so high that it diffuses within the vitreous body of the eye preventing a correct vision of the observed scene (Aslam et al. 2007). This phenomenon can persist for a limited period of time after the light source has been removed from the field of vision, due to the effect of the posthumous image on the retina.

An example of disabling glare occurs when driving a vehicle at night and another vehicle comes from the opposite direction with the headlights on. The headlights of vehicles can emit a maximum light intensity between 30,000 and 130,000 cd from a reflecting surface whose projection is generally less than a square decimetre, or even lower values in the case of LED headlamps, and this may result in a luminance of order of millions of cd/m$^2$ (Aslam et al. 2007). To make a comparison, let us consider that the sun at noon has a luminance that reaches up to 1,600,000,000 cd/m$^2$, while the display of a personal computer has a luminance of up to 300 cd/m$^2$ for the white point. The veil luminance, due to the light diffused within the eyes, effectively prevents vision with the real risk of not correctly perceiving an obstacle or a curve on the road. Human beings tend to instinctively avoid disabling glare and to react with actions aimed at avoiding the dangers due to the lack of temporary vision, provided they have time to do so (Gray and Regan 2007).

### 3.3.2 Discomfort Glare in Lighting Design

A more subtle form of glare, which can occur in lighting design, is instead the discomfort caused by excessive luminance contrasts. This does not prevent the vision but disturbs it, generating a consequent sensation of visual distress (CIE 117:1995).

In the context of lighting design, discomfort glare occurs in all situations in which the user, during a normal visual function, can simultaneously observe a source of light, direct or reflected, in his or her field of vision. Discomfort glare is a negative element in the ergonomics of vision, it disturbs the observer and, if persisting over time, can cause physical disorders such as headaches and migraines as well as a feeling of psychological distress (Huang and Menozzi 2014). In this sense, halogen lamps and the first generations of LED luminaires, used in interiors or even in street lighting, can present considerable glare problems, sometimes ignored in the lighting design. For many years, quality luminaires have been designed to avoid glare in the interior through the use of transverse blades or diffusers, which prevent the direct view of the light source inside the luminaire from the typical observation positions envisaged for the use of the product.

### 3.3.3 Glare Evaluation

To evaluate the visual comfort in interior spaces, the Söllner diagrams have been formerly used, which can be measured for each luminaire. These charts have now being abandoned due to a previous CIE standard being replaced by ISO 8995-1:2002/CIE S 008/E:2001, in which visual comfort is assessed using the Unified Glare Rating (UGR) (CIE 190:2010). The substantial difference between the two methods lies in the fact that the previous one is based on the evaluation of the luminance that defines glare only with respect to the characteristics of the luminaires, while the UGR method correctly considers the visual comfort of an entire lighting system, by taking into account:

- The position of the luminaires.
- The dimensions of the environment.
- The reflection indices of the background on which the luminaires are placed.
- The position of the observer in the environment.

In order to provide adequate and safe visual comfort for human activities, the standard EN 12464-1:2011 defines the UGR limit values for internal areas, depending on the many application areas of lighting. Lower values of UGR correspond to greater visual comfort in the position expected for the observer. An extract of the UGR limit values allowed by the EN 12464-1:2011 standard for some application areas in the workplace is shown in Table 3.4. In Fig. 3.4, it is presented an example of calculation of the UGR in various positions in the design of a reading area of a library.

The assessment of the UGR involves calculations whose complexity grows with the complexity of the environment. Today, this method has become widespread also

**Table 3.4** UGR limit values allowed by the EN 12464-1:2011 standard

| UGR limit value | Applications |
|---|---|
| ≤16 | Manual and CAD technical drawing, precision activity |
| ≤19 | Reading, writing, training, computer work, infirmaries |
| ≤22 | Canteens, gyms, accurate industrial processing |
| ≤25 | Stairs, elevators, toilets, warehouses, general industrial processing |
| ≤28 | Corridors, industrial processing with limited manual intervention |

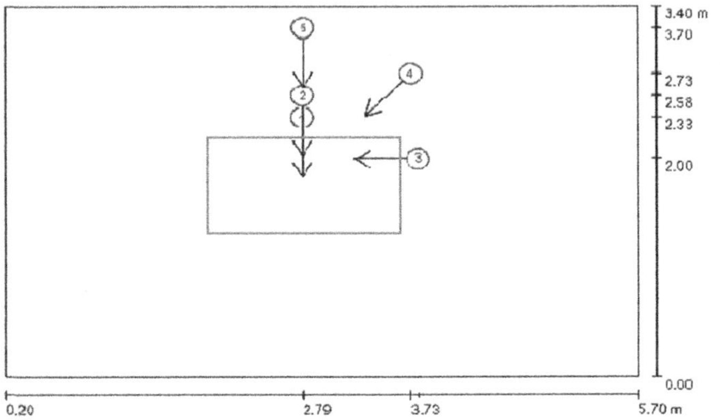


| No. | Reference | Position X | [m] Y | Z | Direction [°] | UGR |
|---|---|---|---|---|---|---|
| 1 | N2 | 2.794 | 2.377 | 1.250 | -90.0 | 18 |
| 2 | N3 | 2.794 | 2.578 | 1.620 | -90.0 | 15 |
| 3 | N4 | 3.794 | 1.997 | 1.620 | -180.0 | 17 |
| 4 | N5 | 3.731 | 2.777 | 1.620 | -136.0 | 17 |
| 5 | N7 | 2.800 | 3.200 | 1.620 | -90.0 | <10 |

**Fig. 3.4** Example of calculation of the UGR

thanks to the use of CAD software for lighting, which provides a fast and accurate calculation of the UGR in all the positions of interest for the environment considered by the project. The UGR makes it possible to evaluate and limit glare in the project and this also guarantees the fact that there is no disabling glare, for which the UGR index would certainly exceed the value of 28.

The UGR method is certainly more advanced than the old Söllner graphs, and can be used for indoor lighting applications. However, the luminance survey is one of the moments of verification and testing of the quality of the project that can now be facilitated through the use of image-based measuring instruments such as the video photometer. This tool provides a more complete set of luminance information than the

**Fig. 3.5** Image-based measurement of the luminance

classic luminance meter. The video photometer produces luminance measurements of entire shots with false-colour scales sometimes called pseudocolour image.

These false colour images make it possible to carry out an immediate analysis of the luminance levels of the lighting systems or of particular types of lamps for which a direct visual appreciation is expected rather than a single illuminating function. Some experiments carried out at the Laboratorio Luce of the Politecnico di Milano have shown that the video photometer method makes it possible to carry out a much faster but less precise luminance measurement, with an error in the order of $\pm 10\%$, compared to the measurements that can be carried out with a precision luminance meter. In Fig. 3.5, it is illustrated an image-based measurement with video photometer of the luminance of Room 31 of the Pinacoteca di Brera. The colour scale on the right shows the luminance values through false colours.

## 3.4 Photobiological Safety and Blue Light Hazard for LEDs

With reference to the safety of non-ionizing optical radiation, for some time there has been a recommendation in the USA, recently updated (ANSI/IES RP-27.1-15 2015), concerning the photobiological safety of artificial radiations, in the range from 200 to 3,000 nm, emitted by lamps and systems that integrate lamps, with reference to human applications. One of the basic principles of this recommendation is that attention is given not only to the SPD emitted by the light source, but also to the one actually received by a human being and that may differ from that emitted by the source, due to the modification of the radiation by the optical elements that are integrated into the product, such as reflectors and refractors. In this recommendation the dimension projected by the light source and the typical use distance by human beings are also taken into consideration. In particular, with reference to the distance from the source, the measurement is carried out at the distance at which there is an illuminance of 500 lx for the lamps used in general lighting, or at 20 cm for the other types (Sliney et al. 2016).

The contents of the first version of the US standard (ANSI/IESNA RP27.1.27 1996) were the basis of the international standard that, in 2002, was defined jointly by the International Electrotechnical Commission (IEC) and by the CIE as a world standard regarding the photobiological safety of lamps and lamp systems (IEC 62471/CIE S 009/E&F:2002). This standard is currently being reviewed by a CIE technical committee, the JTC-5, based on the new recommendations published by the International Commission on Non-Ionizing Radiation Protection (ICNIRP 2013).

In Fig. 3.6 is shown the spectral sensitivity to blue light B(λ) recommended by the ICNIRP 2013 for the human retina: the maximum sensitivity is defined for wavelengths between 435 and 440 nm. The curve A(λ) refers to aphakic subjects, i.e. without the crystalline lens (due to congenital anomaly or trauma), which is able to filter the short wavelengths in excess of 400 nm and therefore limit the amount that reaches the retina.

Subsequently, the IEC has published an international standard (IEC 62471:2006) which was adopted two years later by the European Committee for Electrotechnical Standardization (CENELEC) (EN 62471:2008) as a European standard, with some differences in exposure limits due to the different ICNIRP recommendations in force at the time of publication of the two standards. The types of photobiological damage set out in the regulation concern:

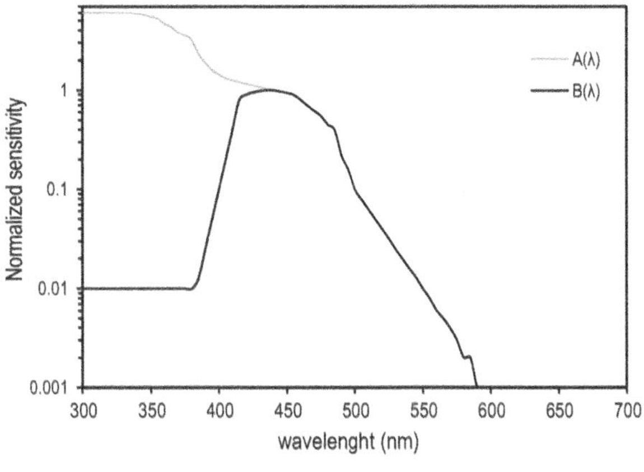


**Fig. 3.6** Spectral sensitivity to blue light recommended by the ICNIRP 2013

- The damage deriving from the exposure of the eyes (cornea) and the skin to the actinic UV with a wavelength between 200 and 310 nm.
- The damage deriving from the exposure of the eyes (cornea) to the UVA with a wavelength between 315 and 400 nm.
- The photochemical damage to the retina deriving from the exposure to light with short wavelengths, the blue light hazard (BLH), for wavelengths between 300 and 700 nm.
- The thermal damage to the retina resulting from exposure to strong light and IR, for wavelengths between 380 and 1,400 nm.
- Eye damage (cornea) resulting from IR in the wavelength range between 780 and 3,000 nm.
- Thermal damage to the skin resulting from radiation ranging from 380 to 3,000 nm.

The regulation is based on the relation between quantity of radiation and exposure time, the increase in one of the two factors must correspond to the decrease of the other; it also examines the angular dimension of the light source and finally classifies the sources into four groups, based on the risk level. The LEDs used for lighting do not have risks related to the emission of UV and IR radiation, but they could have them due to the damage to the retina related to the emission of blue light (Table 3.5).

Applying the standard EN 62471:2008, the spectral radiations have been measured, weighted according to the sensitivity B(λ) defined by the ICNIRP, of various light sources on the market, in common conditions of use, to evaluate the BLH, finding that:

- The linear fluorescent lamps belong to the RG0 group, favoured by their considerable size that guarantees a low radiance with respect to the observer. The same also applies to opal tubes containing LEDs.
- The RG1 group includes incandescent lamps, halogen lamps and high intensity discharge lamps with opal glass. This category also includes lamps that contain LEDs.

**Table 3.5** Summary of risk groups of light sources, defined in the standard IEC/EN 62471:2008, with reference to BLH

| Risk group | Description |
|---|---|
| RG0—exempt | The light source causes no hazard to the retina within 10,000 s of exposure |
| RG1—low risk | The source does not present a hazard under normal conditions of use. It exceeds the limits of RG0 but does not cause any hazard to the retina within 100 s of exposure |
| RG2—moderate risk | The source does not present a hazard as a consequence of a natural instinctive reaction that diverts the gaze |
| RG3—high risk | The source represents a hazard also as a consequence of a temporary exposure of the eye |

**Fig. 3.7** A power LED should never be observed directly when it is turned on

- On the limit between the group RG1 and RG2 are high intensity discharge lamps with transparent glass and white LEDs. However, this classification of the LEDs concerns only observed directly LED power sources (microchips), a situation that should never occur in properly designed luminaires (Fig. 3.7).
- The sun is classifiable between the groups RG2 and RG3.

The standard IEC 62471:2006 has a horizontal value regardless of the application and has been later integrated by the IEC with some technical reports. In IEC TR 62471-2:2009 guidelines were given for the production of lighting equipment for photobiological safety, with reference to the interpretation of the data provided by the manufacturers of lamps, the safety measures and the marking of the products. While in the IEC TR 62471-3:2015 guidelines for the safety of devices with intense pulsed light source (IPL) were defined, with reference to hazards for eyes and skin. In the standard IEC 62471:2006 the definition of lamps and luminaires for general lighting could lead to misunderstandings, so another technical report was published (IEC TR 62778:2014), of particular interest to the manufacturers of luminaires, aimed at defining the BLH for lamps and luminaires that emit light in the spectrum between 380 and 780 nm. The limits for the BLH have then been applied to other standards dedicated to the safety of LED modules (IEC 62031:2018) and the regulation on requirements and testing of luminaires (IEC 60598-1:2014 + AMD1:2017), which states that for luminaires that use light sources in the RG0 or RG1 groups, or have been evaluated as finished products included in the RG0 or RG1 groups, the requirements related to the blue light retinal hazard do not apply.

Regarding lighting design in the workplace, the European standard EN 62471:2008 has been adopted as a reference for non-coherent radiation from the European directive 2006/25/EC that defines the safety requirements for workers exposed to artificial optical radiation. The definition of this directive includes visible radiations, i.e. light, but also UV and IR, therefore all electromagnetic radiations that have a wavelength between 100 and 1 nm. This directive determines the exposure limits based on the ICNIRP guidelines (ICNIRP 1997, 2002), in force at the date of publication of the directive. Later these guidelines have been reviewed (ICNIRP

2013) to better consider new aspects emerging from research, such as the angular dimension of the light source, the revision of the function of thermal retinal hazard and the sensitivity spectrum of the cornea and the lens.

### *3.4.1 Blue Light Hazard: Myths*

As far as BLH is concerned, the differences between US and European standards currently concern product labelling and some factors that come into play in the calculations. The evolution of these rules has not been following a simple path and is still in progress (Sliney et al. 2016). What makes the situation even more complex is the fact that on the subject of safety there are often national rules and laws that have not yet been completely harmonised at European level and still different situations on other continents. On this topic, the press has also intervened with alarming and misleading information of little scientific value.

Examples of possible sources of information manipulation by journalists (Cossar 2013; Ollove 2016; Bodkin 2018) are some of the many recent scientific researches aimed at establishing a dose-harm ratio for health, due to the light produced by blue LEDs or white LEDs made with blue LEDs coated with yellow phosphors.

The research conducted by Krigel et al. (2016) on laboratory albino and brown rats presents the conclusion that the light produced by an LED source, compared to a CFL compact fluorescent one, produces damage to the retina of the rats after exposure to 500 lx of illuminance. Observing also that the damage is greater in albino rats than in brown ones. The research is presented rigorously from the physiological standpoint, but it contains a series of flaws from the standpoint of lighting for human beings, details below. Various light sources have been used in the research, but the two highlighted in the comparison are the LED with cold white light, with CCT of 6,300 K, and a generic CFL called "commercial for the home", for which no data has been provided. For a sensible comparison, it would have been appropriate to have the colour temperatures of both or, even better, the SPD. From a scientific standpoint, the comparison makes sense only if it is possible to characterise the spectral content of the two light sources with particular attention to the blue area of the spectrum. The rats were subjected for 24 h, with dilated pupils, to an illuminance on the eyes of 500 lx that is described as that prescribed for housing, but this does not correspond to the facts, since it is that prescribed for office desks (EN 12464-1:2011). In residential buildings the generally recommended illuminance is less than half (CIBSE 2013). Furthermore, vertical lighting on the eyes is always smaller than horizontal lighting on work surfaces. The 6,300 K LED light is presented as typical of homes, but such a cold light is hardly used for homes. The maximum values of CCT in residential areas are generally lower than 3,000 K, with a much lower content of blue light in the SPD. It is therefore very likely that the CFL used in this trial had a CCT of less than 3,000 K or even less. Finally, the laboratory rat is derived from the common rat (*Rattus norvegicus*), probably the most common mammal on the earth after the human being, but unlike us it is a nocturnal animal (Hanson and Berdoy 2010; Timm

2005). It is therefore natural that it is more sensitive to light and even more so in the case of an albino mutation. The research therefore has a scientific value for albino and brown rats but is not reliable for humans.

The research of Jaadane et al. (2015) and Shang et al. (2014) also concludes that LED light induces oxidative stress and retinal lesions, causes the necrosis of photoreceptors and the main cause of damage is the blue component of the spectrum. However, by reading in detail these studies, which were conducted on laboratory albino rats, we learn that after many hours, or days, of continuous exposure to the cold white and blue LED light, retinal damage was found. Moreover, in the first research the real data of the quantity of light received by the rats, in terms of lighting technology, is not clearly stated, while in the second one it is of 750 lx. The same authors admit that the retina model of albino rats is not recognised as scientifically proven to be applicable to humans and that such levels of artificial lighting are not the typical ones to which people are subjected.

### 3.4.2 *Blue Light Hazard: Truths*

To evaluate the meaning of research conducted in vitro or on mice, with respect to effects on human health, O'Hagan et al. (2016) have carried out a metrological research in which they compared the spectral irradiations produced, under typical conditions of use, by an incandescent lamp, an LED lamp with low CCT and the blue sky on a clear day in the United Kingdom. These were then evaluated with respect to the sensitivity curve of BLH, the B(λ) function defined by the ICNIRP. In the conclusions of this research it is observed that human beings have evolved under natural light and instinctively avoid looking directly at the sun, but not at the blue sky, which is not harmful to sight. However, the blue sky in the spectral range of the BLH produces a spectral irradiance (~1,000 $mW/m^2$) which is two orders of magnitude higher than that of a white LED lamp (~10 $mW/m^2$) or an incandescent lamp. Also the research conducted by James et al. (2017) and by Bullough et al. (2017) has drawn similar conclusions. The latter has also estimated that an LED source with 6,500 K CCT has a BLH which is two and a half times that of a 3,000 K one. However, the CCT alone cannot be taken into consideration; for assessing BLH between light sources based on different technologies, the SPD must always be considered. Possible BLH problems caused by LEDs may occur only for very high levels of illuminance on the retina. Figure 3.8 illustrates a comparison of the SPDs of the radiations produced by a white LED lamp, an incandescent lamp, both in normal conditions of use, and the blue sky. The graph also shows the sensitivity curve of BLH, B(λ), which is dimensionless.

Based on press reports, the European Commission has commissioned the Scientific Committee on Health, Environmental and Emerging Risks (SCHEER 2016) to investigate the potential risks of LEDs for human health. The committee has analysed the state of the latest medical research in the field and has concluded its work stating that although there has been research on animals or cells (Chamorro et al. 2013) that

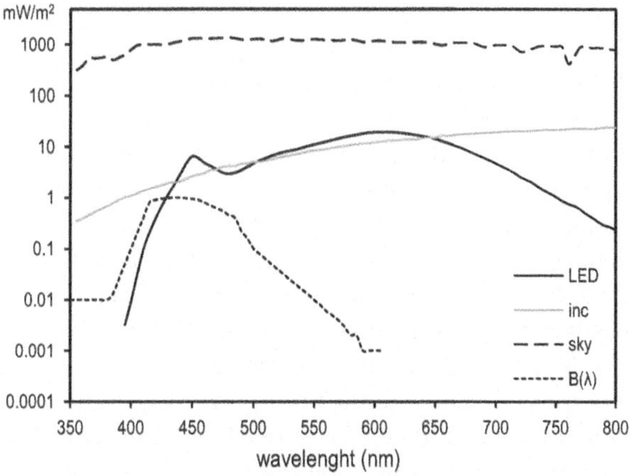


**Fig. 3.8** Comparison of different SPDs with BLH sensitivity curve

shows negative effects of the blue component of light, there is no clear evidence that general lighting using LED technology could pose a risk to human health. In particular, the emphasis is placed on the fact that the levels of light exposure used in studies carried out on animals or cells are not comparable to those typical of general lighting concerning human beings (SCHEER 2018). In the same document, however, it is recognised that the light used in the evening can influence the circadian rhythm. In addition, special attention should be given to children, due to the sensitivity of their retina to blue light, and also to the elderly, due to the degeneration of their visual system caused by age. It is known that in the elderly the aging of the lens and the structure of the eye cause a diffusion of light inside the eyeball, making it difficult to read blue lights and blue luminous words that can be considered a source of dazzle and discomfort.

In the case of children up to three years of age, the lens passes a greater amount of blue light and UVA that reach the retina. For this reason, the ICNIRP suggests evaluating the BLH of children with the curve A(λ) defined for people who are aphakic, i.e. those who do not have a lens. Because LEDs are increasingly used in children's toys too, Higlett et al. (2012) have proposed a method for determining the emission limits of LEDs in toys. At European level, since 2009 there has been a directive (Directive 2009/48/EC) that has defined the safety of toys and there is a specific standard (EN 62115:2005 + A12:2015) that was updated in 2015 and is further under review, to also cover the problem of BLH in toys (FprEN 62115:2016).

## 3.5 LEDs Temporal Light Modulation (Flicker)

The modulation of light over time (TLM) is defined as the rapid variation of the light emission of a source. From the perceptive standpoint, we use the term flicker to indicate the effect perceived by the human nervous system that can differ in the various subjects. The flicker can also not be seen directly but perceived subliminally by the human brain and affect visual (Veitch and McColl 1995) and cognitive (Knez 2014) performance. For many people, the flicker is perceived when the light source modulates the luminous flux with a frequency lower than a critical threshold, which also depends on the amplitude of the modulation. The frequency at which the image of a pulsating source merges into an apparently constant one is called the critical flicker fusion rate (CFF) (de Lange 1958). This varies from individual to individual and depends on numerous factors such as, for example, amplitude modulation, waveform, luminance adaptation, amplitude of the visual field subtended by the source, and it is generally between 60 and 100 Hz (Kelly 1961).

The photometric flicker is different from the electric flicker, which depends on the noise signals present on a power supply line that create a modulation of the luminous flux emitted on resistive loads, as for incandescent lamps that until recently have been used in our houses. In the latter case, the main factor responsible for the modulation of the luminous flux is the characteristics of the AC supply line, rather than the characteristics of the source or of its supply system. In the case of incandescent lamps, powered directly with AC, at mains frequency of 50 Hz (Europe, Asia, Australia, Africa and part of South America) or 60 Hz (North America and part of South America), the amplitude of modulation of the photometric flicker is generally contained within 10%, thanks to the thermal inertia of the tungsten filament.

### *3.5.1 Flicker Before LEDs*

The problem of flicker for traditional fluorescent lamps and discharge lamps has long been studied and debated when the diffusion of these lamps, with ferromagnetic ballasts, was very high. In these lamps there was typically a photometric flicker at twice the network frequency, i.e. 100 or 120 Hz. The flicker has been related to migraines (Shepherd 2010) and headaches (Wilkins et al. 1989), in addition to other health problems or neurological states of illness, including the onset of epileptic seizures in predisposed subjects (Wilkins et al. 2010; Harding and Harding 2010). Photosensitive epilepsy is estimated at 1.5/100,000 episodes per year, with a greater incidence in subjects aged 7–19 years (Quirk et al. 1995). The frequency range for which there is a hypothesis of 5% of epileptic seizures in patients is between 5 and 65 Hz, with a maximum sensitivity for 20 Hz, i.e. for the visible flicker (Harding and Jeavons 1994; Binnie et al. 2002). Approximately 1 in 10,000 people are affected by epileptic seizures due to light stimuli and 1 in 4,000 in the age group between 5 and 24. However, even in the non-epilepsy population an abnormal EEG response was demonstrated in 0.3–3% of the population as a reaction to light stimulations or images with strongly contrasted patterns (Fig. 3.9).

In Japan there was an incident in which, during the projection of a Pokemon cartoon, 685 children were taken to hospital for neurological reactions to flashes of coloured light coming from the television (Fisher et al. 2005). In particular, it is always better to avoid the contemporary occurrence of flicker and patterns. For

**Fig. 3.9** Example of a discomforting visual pattern due to its strong contrast

example, the worst patterns are those with light and dark stripes in which a pair of stripes subtends a visual angle of about 1/15 of a degree.

The progressive introduction of high frequency electronic power systems (20–40 kHz) for fluorescent lamps, also with the aim of improving the energy efficiency of the systems, has led to a decreased interest in this issue, also in terms of criticism made by users regarding the comfort and well-being guaranteed by these lighting systems.

### *3.5.2 Flicker of LEDs*

With the appearance of LED lighting, the main responsible for the flicker is the power system adopted. This is because the emitted flux is almost proportional to the current passing through the LED, so the flicker theme has once again returned to the attention of researchers. Indeed, the LEDs work in direct current (DC) and therefore require a power supply that converts AC into DC and may even allow the regulation of the emitted light flux. This conversion can be achieved by different methods that affect the costs of the power supply and the photometric flicker emitted by the LED sources. The control of the emitted flux can be implemented through the pulse with modulation (PWM) technique, which is widely used for the adjustment of the flux emitted by LED luminaires. The PWM, generating a square wave of variable width, causes the flicker; the perception of which is reduced in the case of high frequency of the PWM signal. For example, the flicker is not perceivable for a 1,000 Hz PWM while it can be at 70 Hz.

On the market there are LED sources in which quality power supply systems do not produce any flicker when the intensity is adjusted to the maximum or minimum,

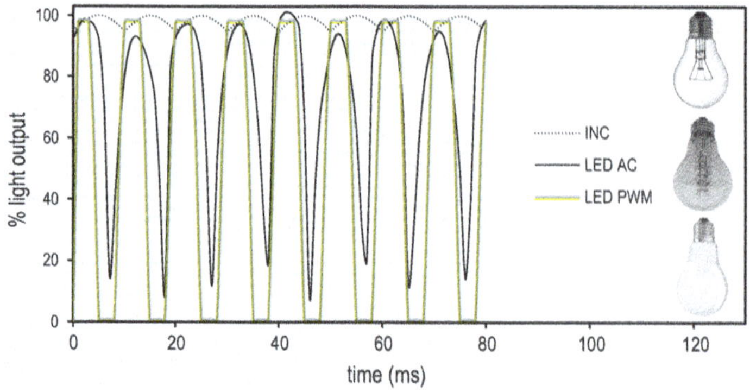


**Fig. 3.10** Comparison of the photometric flicker measured for three lamps

because the power system emits a steady direct current. In other low cost LEDs the flicker is perceivable by some subjects both in conditions of full light and minimal adjustment, and others in which the phenomenon only occurs when the light emission is adjusted to the minimum value. Finally, there are control systems that show the phenomenon during adjustment, but not during their stable operation at a pre-set level (Rand et al. 2007). Since the AC network frequency is 50 or 60 Hz, in the current white LED sources without flux adjustment, we can generally have a flicker with a dual frequency, i.e. 100 or 120 Hz. This occurs because many low-cost power supplies are based on a simple full wave diodes rectifier and sometimes this device is directly integrated with the LED sources.

In Fig. 3.10 is shown a comparison of the photometric flicker measured for three lamps. An incandescent lamp, which has a low amplitude modulation, with a flicker of ~5.3%; a low cost AC LED lamp, which features a flicker of ~86.5%; and an LED lamp, regulated in PWM to emit 50% of its luminous flux, which has a flicker of ~98%. In the three cases, the frequency of the photometric flicker derives from the AC mains voltage frequency supply in Europe.

### *3.5.3 Effects of Flicker on Human Beings*

With reference to human beings, the flicker phenomenon can be distinguished in different categories:

- Visual flicker: the visual system and the brain perceive the modulation of the light emitted by the source over time, which is detected and consciously perceived by users. It is perceptibly different by different subjects up to CFF at about 60–100 Hz.
- Invisible flicker: the modulation of light over time is present, but not perceived. It occurs for frequencies above 100 Hz.

- Stroboscopic flicker: the periodic modulation of light is due to the intrinsic characteristics of the power supply device and is noticeable only if there is a relative motion between the light source and the observer or the illuminated object. For example, when the head of the observer moves with respect to the sources or there are moving objects. As a consequence, a rotating object, like a machine tool, may appear to be still. In the research conducted by Bullough et al. (2011) stroboscopic effects have been perceived up to frequencies of 300 Hz.
- Phantom array effect: it occurs when the observer moves with respect to a source and because of the retinal persistence a series of multiple images of the light source are generated on the retina. Roberts and Wilkins (2013) have shown that this effect can be perceived up to frequencies of almost 2 kHz, especially during the saccadic movements of the eye (Hershberger and Jordan 1998).

In the case of LED sources, except for products of poor quality, there should be no problem with the visual flicker, as the sources that produce flicker do it at frequencies of 100 Hz or higher. In the research of Binnie et al. (1979), conducted on 20 patients with photosensitive epilepsy, it has been shown that, at the frequency of 100 Hz, there would be no possible negative effect with regard to the possibility of causing seizures.

However, the 100 Hz limit as an acceptable flicker level for new LED sources is a controversial topic. Indeed, it has been shown that in many subjects who are not able to perceive the flicker, for frequencies higher than CFF, physiological effects are measurable even above this frequency. For example, the electroretinogram measurement (ERG) shows that an invisible flicker up to 200 Hz is still transmitted through the retina (Berman et al. 1991; Burns et al. 1992). Moreover, for the 100 Hz flicker, which for many subjects is at the limit between the visual and the non-visual, it has been shown by Jaén et al. (2005), on a sample of 50 people, that with an increase in modulation amplitude, from 3 to 32%, there is an average decrease in visual performances. In another research, in a comparison between flicker frequencies of 50, 100 Hz and constant lighting, Kennedy and Murray (1991) have analysed the amplitude of the saccadic movements during a reading task on 60 people. In their investigation they have found that both the 50 and the 100 Hz flicker, normally considered more comfortable, because beyond the CFF, induce shorter and less accurate saccadic movements that would make reading more tiring. This research therefore poses questions regarding the LED sources with flicker considered safe at 100 Hz but also and especially on PC screens that generally have a refresh rate of 60 Hz.

Frequency is not the only parameter that must be considered to evaluate flicker. Indeed, at the same frequency there can be different waveforms and with different heights. To describe these differences, two other parameters have been historically used:

- Percent flicker: also called flicker modulation, is the percentage ratio between the difference of the maximum and minimum value of the signal, which represents the trend of the luminous flux over time, divided by the sum of the amplitudes of the maximum and minimum value of the signal.

- Flicker index: defined as the ratio between the area subtended between the maximum of the signal and its average level, divided by the sum of this area and the area that is subtended between the zero and the mean value of the signal.

The percent flicker is independent of the waveform, while the flicker index also depends on the waveform, so it provides more complete information (Musante and Rossi 2016) since the perceived effect of the flicker also depends on the waveform (de Lange 1961; Levinson 1960).

In the research conducted by Olsen et al. (2014) in an office environment the level of acceptance of lighting was evaluated on 18 subjects, after eight hours of work, under constant lighting and with a variation of the percent flicker from 29 to 100%, at a fixed frequency of 100 Hz. The results show that the subjects did not express negative opinions on the constant lighting and on that with a percent flicker of 29%, while they expressed a negative opinion on that with a percent flicker of 100%, with remarkably different judgments between the various people. Similar results have been found in the research carried out by Jaén et al. (2011), on twelve students between the ages of 20 and 22, which has compared lighting with a percent flicker of 3% to one of 32%, finding that in the second case there was a decrease in performance regarding discrimination activities and visual research.

The percent flicker and the flicker index are considered by the CIE not fully able to quantify the flicker perceived by humans (CIE TN 006:2016). Details on the calculation method of these indices are provided by IEEE 1789-2015, which was the first standard concerning the assessment of risks for human health due to LED flicker. This standard collects and composes the results of several previous studies and proposes a guideline and some frequency-based limits for the classification of LED products in two categories: low risk and risk free. The standard aims at assessing the potential health risks due to flicker and relates the frequency to the percentage of the admissible flicker. The principle applied is that the lower the frequency, the lower the percentage of the flicker must be. Figure 3.11 illustrates the relationship between frequency and flicker percentage proposed by the standard IEEE 1789-2015. Under the green curve there is no effect of the flicker. Under the yellow curve there is a moderate risk due to the flicker while above this curve there is a high risk. For low frequency values, lower than 100 Hz, the permitted modulation level is very low, less than 2% to avoid any risk and less than 5% to have a moderate risk. For frequencies above 100 Hz, such as a common incandescent lamp powered by 50 Hz AC and pulsating at a frequency of 100 Hz, the percentage values of the flicker permitted for both risk levels are higher.

A metric for the evaluation of the visual flicker has also been proposed as part of the ASSIST program (The Alliance for Solid-State Illumination Systems and Technologies), which has been running since 2002 at the Lighting Research Center (LRC) of the Rensselaer Polytechnic Institute with the participation of several companies (Bodington et al. 2016). In this document a perceptual modulation index $M_P$ has been proposed that takes into consideration both the frequency and the modulation of the light signal for each frequency component of the periodic signal. From the conceptual standpoint, this method is equivalent to converting the value of modulation for

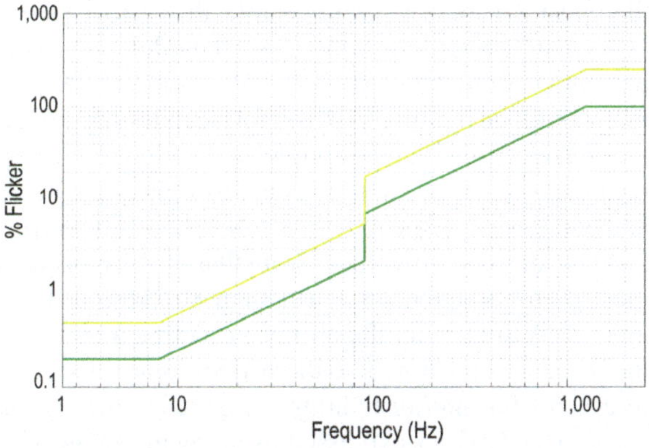


**Fig. 3.11** Relationship between frequency and flicker percentage

each frequency into a perceived modulation value. The perceived modulation values at the different frequencies are usually considered perceptually independent of each other and consequently the overall perception of the flux modulation $M_P$ is given by the combination of the various components. $M_P$ values lower than 1 involve a lower probability of perceiving the flicker, while values higher than 1 indicate a greater probability of perceiving it (ASSIST 2015).

On the electric side, the IEC has defined a series of standards with reference to the flicker. The IEC 61000-4-15:2010 provides guidelines on how to build an instrument to measure the electric flicker (flickermeter), defined by the index $P_{st}$ (short term flicker severity), for products supplied with AC voltages of 120 and 230 V and frequencies of 50 and 60 Hz. While the IEC 61000-3-3:2013 defines limits for voltage fluctuations and flicker on 230 V public power lines. In addition, the technical report IEC TR 61547-1:2017 explores in depth the flickermeter theme for the evaluation of the photometric flicker, independent of the power supply network and dependent instead on the supply systems of the lighting equipment and proposes an index $P_{st}^{LM}$ for the photometric flicker.

To date, there are no established methods to evaluate the phantom array and the stroboscopic effects. However, for the latter Perz et al. (2015) have proposed a stroboscopic effect visibility index (SVM). If the SVM value is higher than 1, there is a probability greater than 0.5 that the stroboscopic effect due to lighting is visible, while if the SVM is lower than 1 the stroboscopic effect can be visible with a probability that is lower than 0.5. The research by Bierman (2017) proposes a method for evaluating if the stroboscopic effect is acceptable to a human observer, by defining a stroboscopic acceptability metric (SAM). This method has also been incorporated into a document of the ASSIST 2017 program.

After its publication, the standard IEEE 1789-2015 has been deemed too strict by the United States National Electrical Manufacturers Association (NEMA), with respect to some particular applications: this would, in their opinion, lead to an unnecessary increase in the costs of production and purchase of power systems for LED luminaires and other sources (NEMA TLAs-2015). An example given to support their position concerns, for example, the case of the discharge sources used for many years in urban lighting, which have a modulation frequency of 100 or 120 Hz, with a percentage of flicker up to 84%. According to NEMA, the use of these products for many outdoor lighting systems for decades has not shown high health risks, hence the consideration that the simple application of the requirements set by the standard, regardless of the installation conditions, may not be the best solution to avoid waste and unnecessary expenses. Later, NEMA itself has proposed a standard (NEMA 77-2017) with some differences from that of IEEE because it uses two different metrics, the $P_{st}^{LM}$ for frequencies lower than 80 Hz and the SVM for stroboscopic effects above this limit, with the aim of providing a better description of the sensitivity of the human visual system to the stroboscopic effects. It also indicates different flicker thresholds for interior lighting compared to the external one.

Also the CIE has expressed itself regarding the assessment of the effect of flicker on humans, stating that in recent years there have been many bodies that have done research and expressed opinions, sometimes controversial, at the international level and advocating greater cooperation between these bodies (CIE TN 008:2017). In particular, a CIE technical committee, TC 1-38, is working to propose a predictive model of the flicker effects worldwide.

The problem of the correct evaluation of the flickering of lighting sources remains therefore a research topic and, in the coming years, further studies and more detailed guidelines will certainly be presented to combine comfort and safety with the economy of lighting systems. To date, SCHEER (2018) has nevertheless evaluated the standard IEEE 1789-2015 as the main one that, although in conservative way, makes an assessment of the risks of LED light sources with reference to the various flicker effects. It has also pointed out that there are LED light sources that do not have flicker effects, being powered in DC, as presented in the standard itself based on the measurements made by Lehman et al. (2011).

## 3.6 Fundamentals for the Evaluation of the NIF Effects of Light

Chapter 2 of this book has explored in depth the relationships between light and human physiology highlighted in forty years of scientific research. These relationships can be numerous, from brain electrical activity to cognitive performance, from effects on seasonal depression to senile dementia, from sleep disorders to the health of premature infants, from the effects of melatonin to the adjustment of the human circadian rhythm also with respect to modern life. In this section, we would rather focus on the quantitative evaluation of the effect that light can have on the melatonin level and therefore on the timing of the human circadian rhythm.

At the beginning of the third millennium, the need to define a circadian photometry started to surface, with the aim of evaluating the NIF effects of the light stimulus, with an action spectrum curve (Lipson 2012) different from that of human photopic spectral light efficiency V(λ). Indeed, within the scientific community (Rea et al. 2002b; van Bommel and van den Beld 2004) it has become evident that, from the standpoint of circadian stimulation, there are a number of lighting factors that need to be defined and measured using means other than photometry, in terms of SPD, duration, timing and exposure history. As also highlighted by the CIE TN 003:2015a the NIF effects that light can have on the human organism can fall within two macro categories, those of a neurophysiological nature and those regarding the melatonin level and the circadian rhythm.

Unfortunately, many researches that have been conducted to evaluate the NIF effects of light have quantified the latter in terms of illuminance $E_v$ and CCT. This is because the $E_v$ can be measured with low-cost instruments and the CCT is deduced from the technical characteristics declared by the producers of the light sources. The illuminance is certainly suitable for describing the light that hits surfaces, such as the worktop, but not to quantify the stimulus on the retina. A different evaluation can be made with regard to illuminances measured vertically, since they are more relevant to the light that can reach the eyes, but also in this case we know that the visual task can be variable and have different orientations. What is the relationship between the light that comes out of the source and the light that goes directly to the retina? Most of the light arrives indirectly to the retina because it is reflected by the surfaces, but the colour, or spectral reflectance, of these can significantly change the light radiation and therefore assuming the CCT of the sources as spectral information of the light that gets to the eyes is a procedure involving a high degree of approximation.

Today, there is a scientific consensus on what should be the pillars at the base of this research area:

- It is scientifically proven that light has non-visual effects on the human organism.
- There is a common consensus on the fact that the NIF effects cannot be described by a single response curve, as is the case with vision.
- The fifth photoreceptor discovered at the beginning of the third millennium, additional to the rods and the three types of cones, has been identified in the intrinsically photosensitive retinal ganglion cells (ipRGCs). Within these cells is the melanopsin photopigment, which plays a central role in the reception of light for NIF effects (Berson et al. 2002).
- The ipRGCs alone are not sufficient to explain the NIF effects of light, but it is believed that, since these also collect information from rods and cones, they also contribute to NIF effects (Panda et al. 2002; Hattar et al. 2003).
- The stimuli for the NIF effects generated and collected by the ipRGCs reach brain areas that are different from those responsible for vision, i.e. the hypothalamus and the suprachiasmatic nucleus (SCN), with different effects depending on the SPD and the temporal aspects of light stimulation.
- In order to compare the results of different studies, the light used in the experiments must be expressed in terms of spectral irradiance $E_{e,\lambda}$ on the eyes.

The CIE also defined the spectral sensitivity of the five known photoreceptors (Fig. 3.12), namely:

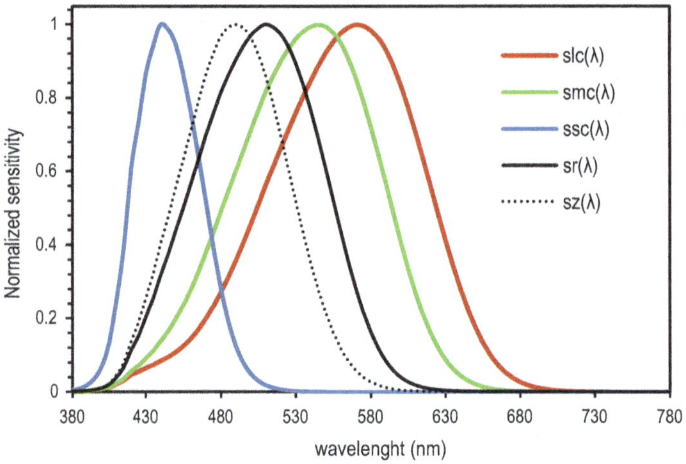


**Fig. 3.12** The spectral sensitivity of the five photoreceptors recommended by CIE TN 003:2015a

1. $s_{lc}(\lambda)$ is the spectral sensitivity of lc cones sensitive to long wavelengths, containing erythropsin, with maximum sensitivity at 570 nm.
2. $s_{mc}(\lambda)$ is the spectral sensitivity of mc cones sensitive to long wavelengths, containing chloropsin, with maximum sensitivity at 545 nm.
3. $s_{sc}(\lambda)$ is the spectral sensitivity of sc cones sensitive to short wavelengths, containing cyanopsin, with maximum sensitivity at 440 nm.
4. $s_{r}(\lambda)$ is the spectral sensitivity of rods, containing rhodopsin, with maximum sensitivity between 505 and 510 nm.
5. $s_{z}(\lambda)$ is the spectral sensitivity of the ipRGCs, containing melanopsin, with a maximum sensitivity at 490 nm.

These spectral sensitivities are defined at the external level of the cornea, considering the changes to the light spectrum resulting from the spectral transmittance of the transparent structures of the eye, such as the cornea, the lens and the retinal cells, before light reaches the photoreceptors. The spectral transmittance of the eye can vary with the different individuals and depends very much on age (Boettner and Wolter 1962). Over 65 years, there is a noticeable reduction in the short wavelengths that reach the retina (Lerman 1987; Barker and Brainard 1991; Artigas et al. 2012). The CIE definition, for the spectral sensitivity functions of the five photoreceptors, considers an average individual of 32 years of age. For research activities on the NIF effects of light, the CIE TN 003:2015a therefore suggests specifying light in terms of spectral irradiance on the cornea and evaluating the stimulus received by the five photoreceptors in terms of five total irradiations, weighted according to the five sensitivity curves, defined as α-opic irradiance: $E_{e,lc}$, $E_{e,mc}$, $E_{e,sc}$, $E_{e,r}$ and $E_{e,z}$. It also provides a software toolbox, consisting of an xls table, to calculate these values and to create graphs according to the spectral irradiance of light (CIE TN 003: b).

This research approach is aimed at evaluating the stimulus that reaches the five categories of photoreceptors, to determine the response curves, hopefully for all NIF effects of light, depending on the received stimuli and the time factor. Research activities in this sense are still ongoing, with the perspective of being able to define a global standard.

In parallel, in recent years, there has also been an application-oriented approach aimed at evaluating the response curve of melatonin suppression according to the light SPD in order to directly apply the results in the field of lighting design.

## 3.7 Action Spectrum for Melatonin Suppression

Assuming that the measurement of the amount of melatonin is the main marker in order to evaluate the status of the human circadian system (Rosenthal 1991; Lewy et al. 1999; Benloucif et al. 2005), many researchers have carried out studies aimed at evaluating the dose-effect relationships, which have taken into consideration the quantity of light (Brainard et al. 1988; Mclntyre et al. 1989; Boivin et al. 1996; Zeitzer et al. 2000) the CCT (Morita and Tokura 1996; Rea et al. 2001; Kraneburg et al. 2017) and also the spectrum (Brainard et al. 2001a; Thapan et al. 2001). In particular these last two researches have been the first ones trying to determine a spectral response curve of the melatonin suppression, due to a photopigment different from the cones and rods, depending on the wavelength. This type of investigation has appeared to be necessary when it has been discovered that light with higher CCT, i.e. tending towards cold white and blue, has a greater effect on the melatonin suppression, as well as monochromatic lights with a shorter wavelength (Brainard et al. 2001b). The question has then arisen of how the spectrum of light can influence melatonin suppression.

### *3.7.1 Sensitivity to Monochromatic Coloured Lights*

The research by Brainard et al. (2001a) has been conducted to evaluate the effect of monochromatic lights on the photoreceptor active in suppressing nocturnal melatonin in the blood, from 2:00 to 3:00, in 72 human subjects (37 women and 35 men) between 18 and 30 years of age. The participants have been first subjected to the Ishihara (Kindel 2005) and Farnsworth-Munsell D-100 (Farnsworth 1943) tests to include in the experimentation only subjects with a normal colour vision. Then they have been subjected to different monochromatic wavelengths, between 420 and 600 nm, allowing at least a week to elapse between two successive tests. The monochromatic lights have been used exposing the subjects through an integrating sphere with perception of the Ganzfeld type with wavelengths of 420, 440, 460, 480, 505, 530, 555, 575, and 600 nm. It is important to note that many of the circadian system stimulation experiments have been performed using light sources directed at the retina with perception of the Ganzfeld type. The term Ganzfeld (from the German full-field) is used to indicate those experiments in which the whole field of view is uniformly subjected to the same lighting conditions. The Ganzfeld condition, also used in parapsychology for sensory isolation experiments, is however a different condition from that which occurs in everyday reality, where light arrives on the retina after many reflections on objects, and what reaches the retina is more a complex image than a uniform light source throughout the visual field. In this research the value of corneal irradiance has been corrected, in order to evaluate the effect on photoreceptors, to consider the 45% reduction of the 420 nm light that arrives on the retina (Brainard et al. 1997), in relation to the age group of the subjects involved. Moreover, a response curve of an undefined opsin, present in the fifth photoreceptor, has been assumed to have a maximum at 464 nm. Finally, it has been verified that the level of melatonin suppression increased with increasing irradiance on the eyes. Later the experiment has been repeated by Brainard et al. (2008) to better characterise the melatonin suppression response at 420 nm.

The research by Thapan et al. (2001) has involved 22 healthy people (18 men and 4 women), with normal sleep-wake cycles, not using drugs, who have been first subjected to the Ishihara test to verify their normal reaction to colours. The exposure to monochromatic light has been performed using an integrating sphere, at night, for a duration of 30 min, in a time slot between 23:30 and 2:30 to assess the level of suppression of nocturnal melatonin. Six monochromatic lights with wavelengths of 424, 456, 472, 496, 520 and 548 nm have been used. In this case also, the indicated values have been corrected to consider the spectral transmittance of the lens and correctly evaluate the real effect on the retina. It has been observed that the level of melatonin suppression grows with the increase of light irradiance on the eyes and a maximum sensitivity of 459 nm is estimated for a hypothetical opsin different from that of the cones and rods.

The authors of these two researches therefore have concluded that there should be a fifth photoreceptor, different from the cones and rods, characterised by an opsin with a maximum sensitivity of ~460 nm, as sole responsible for receiving light for the suppression of melatonin. However, in another research, Rea et al. (2002a) have used two white lights, one warm and one cold, instead of monochromatic lights, confirming that the response curve for melatonin suppression shows a maximum for wavelengths between 420 and 520 nm, but they expressed doubts about the fact that the effect was due to a single photoreceptor. These researchers have rather hypothesised, also from the characteristic shape shown in Fig. 3.13, of the response curve deduced from previous researches by Brainard et al. (2001a) and Thapan et al. (2001), that in the effect of melatonin suppression there could also be a contribution of the rods and the cones sc.

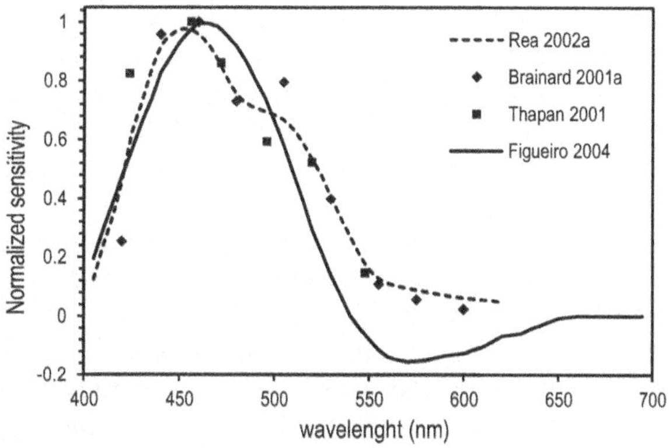


**Fig. 3.13** Different action spectrum for melatonin suppression

### *3.7.2 Sensitivity to White Lights*

In the research conducted by Figueiro et al. (2004) the fact that the melatonin suppression could not be due to a single photoreceptor was further highlighted. The same research has shown that 29 $\mu W/cm^2$ of light, produced by a blue LED, are more efficient in melatonin suppression than 170 $\mu W/cm^2$ of white light, produced by a mercury vapour lamp. The reason why in this comparison a mercury vapour lamp has been chosen, is motivated by the conformation of the SPD of the emitted light, characterised by two peaks in the yellow area and one in the blue area of the spectrum, with an energy greater than that of the blue LED. The experiment has been designed so that the mercury lamp would create a greater inhibition of melatonin, according to the invariance principle, which envisages that short wavelengths are more effective than the others. Indeed, the data obtained has shown that the blue monochromatic LED light, even if with a significantly lower irradiance, inhibits melatonin much more effectively than white light, with blue and yellow bands, coming from the mercury lamp. Therefore, in a balanced situation of the spectrum, with a broadband white light, there would be a lower influence on the circadian system. For this reason, a direct relationship cannot be established between the intensity of the light stimulus and the level of melatonin suppression according to the single wavelengths without considering also the whole SPD. Figure 3.13 illustrates the normalised punctual values of the response curves in the melatonin suppression determined by Brainard et al. (2001a) and by Thapan et al. (2001) compared with the interpolated curve calculated by Rea et al. (2002a) based on these series of values. The dotted curve has a trend that cannot be explained by the sensitivity of a single photoreceptor. The curve assumed by Figueiro et al. (2004) highlights the negative effect of wavelengths beyond ~550 nm in the suppression of melatonin.

These important results have been also confirmed by a subsequent study, in which only a mercury vapor lamp was used (Figueiro et al. 2005). In a first phase, the lamp was used to generate white light, while in a second phase the light of the lamp was spectrally filtered in such a way that only its blue component, around

436 nm, could reach the eye of the four people participating in the experiment. After the introduction of the filter, the melatonin suppression was higher, although the illuminance levels were much lower and the blue component of the light was identical or slightly lower. This shows that the effects of light on melatonin have a spectral opposing behaviour. A single blue light is more effective in suppressing melatonin than a more energetic white light consisting of a blue component plus a yellow component. The most important aspect of this principle is that the light sensitivity of the circadian system does not exhibit an additive behaviour with respect to the wavelengths considered individually, but it becomes more sensitive to blue light as the yellow-red light decreases. The hypothesis of a spectral opposition in circadian sensitivity, further verified in a subsequent study by the same researchers (Figueiro et al. 2008), envisions that the ipRGCs, receiving information also from the bipolar cells of the retina, are influenced by the signal of opponent colours (blue-yellow), according to the opponent colour theory formulated in 1892 by Hering (1964) and formalised from the colourimetry standpoint by Hurvich and Jameson (1957).

Indeed, other studies have shown the actual existence of a fifth photoreceptor based on vitamin A, the melanopsin contained in the ipRGCs, for which a maximum sensitivity of 480 nm on the retina was estimated (Berson 2003; Hankins and Lucas 2002; Dacey et al. 2005; Bailes and Lucas 2013). This value is corrected to 490 nm on the eyes to consider the spectral transmittance of the lens, which significantly reduces wavelengths shorter than 420 nm. The difference between the response curve of the ipRGCs and the response curve of the melatonin suppression, has made it possible to infer that this could also depend on the action of the other photoreceptors that transmit their information to the ipRGCs (Panda et al. 2003; Lucas et al. 2003; Hattar et al. 2003).

## 3.8 Circadian Photometry Hypothesis

Based on the previous paragraphs, it is now internationally accepted that the response curve in melatonin suppression is different from the photopic $V(\lambda)$ and scotopic $V'(\lambda)$ visual sensitivity. However, to date there is no officially recognised international standard to describe the circadian performance of light. Nonetheless, in recent years proposals have been made to try to define the bases of a circadian photometry.

### *3.8.1 Hypothesis Based on a Single Photoreceptor*

From an application standpoint, aimed at providing practical suggestions for designers, a research supported by Osram (Gall and Bieske 2004) has proposed the definition of circadian variables that can be determined by photometric ones, through the adoption of a circadian response curve $c(\lambda)$. This curve was obtained empirically as the average of the results of research conducted by Brainard et al. (2001a) and Thapan

**Table 3.6** The $a_{cv}$ factor values for some light sources

| Light source | CCT | $a_{cv}$ |
|---|---|---|
| Direct sun | 5,081 K | 0.76 |
| Blue sky | 19,963 K | 1.49 |
| Cloudy sky | 5,924 K | 0.88 |
| Incandescent lamp | 2,800 K | 0.35 |
| Neutral white metal halide | 3,640 K | 0.39 |
| High pressure sodium | 2,770 K | 0.28 |
| Generic warm white fluorescent | 2,827 K | 0.31 |
| Generic neutral white fluorescent | 3,678 K | 0.52 |
| Fluorescent daylight 765 | 6,750 K | 0.85 |
| Fluorescent daylight 865 | 6,400 K | 0.80 |
| Fluorescent daylight 965 | 6,500 K | 0.94 |
| Fluorescent daylight 880 | 8,000 K | 1.00 |
| Fluorescent truelite | 5,600 K | 0.76 |
| Blue LED $\lambda_p = 468$ nm | n.a. | 6.90 |
| White LEDs | n.d. | 1.05–2 |

et al. (2001). The circadian action factor $a_{cv}$ was then defined, as a multiplicative factor that makes it possible to calculate each circadian integral quantity from the equivalent photometric quantity. For example, the circadian illuminance is calculated as $E_c = a_{cv} \cdot E_v / K_m$, where $K_m$ is the maximum value (683 lm/W) of the photopic spectral light efficacy function $K(\lambda)$. The $a_{cv}$ factor increases almost linearly with the increase in CTT and this is in line with the first evidence from research carried out in previous years.

Based on these hypotheses, they have also made an assessment of the typical light sources present on the market at that time, to estimate the respective $a_{cv}$ factor, as described in Table 3.6. According to this method, we can deduce that the blue sky has a strong ability to affect our biological clock and even more does the blue LED.

From the metrological standpoint, the two researchers have stated that the circadian response curve $c(\lambda)$, necessary to determine the factor $a_{cv}$, can also be approximated with the colour matching function curve $\bar{z}(\lambda)$ of the CIE, presumably normalised to the unit value. They have also highlighted the important fact that the $a_{cv}$ factor cannot be evaluated solely based on the light source used, but must be evaluated based on the spectral irradiance actually present on the eyes of the observers, which depends on the light sources but also on the reflection factors of the surfaces in the environment. This definition also takes into account the amount of light needed to activate the circadian system, which should be in the range of total irradiations $E_e \approx$ 0.06–0.24 W/m$^2$. Based on this first hypothesis of circadian photometry, as regards lighting design, Diano (2004) has proposed an optimal combination between human activities, the circadian action factor and the colour temperature of the light sources (Table 3.7). To promote relaxation in a private environment, a low value of $a_{cv}$ is

**Table 3.7** The relationship between activity, factor $a_{cv}$ and CCT

| Situation/Activity | $a_{cv}$ | CCT |
|---|---|---|
| Moments of tranquillity, relaxation, private non-professional applications | <0.4 | Warm white < 3,300 K |
| Office, industry, street lighting, shops, markets, special lighting requirements | 0.3–0.8 | Neutral white 3,300–5,300 K |
| Optimal concentration at work, feeling of well-being combined with great visual efforts, therapies | >0.7 | Daylight > 5,300 K |

recommended, while a high value of $a_{cv}$ is recommended to facilitate intellectual work activities that require high concentration. The intermediate values of $a_{cv}$ are recommended for normal office work or manual tasks in industries, but also in public places such as shopping centres and street lighting.

In this first hypothesis of circadian photometry, however, there are many critical elements. The temporal factor of exposure to circadian light is lacking: the human circadian system has its own natural rhythm, with a slow response, and cannot be switched on and off at will. Rather, artificial light should facilitate the natural adaptation to the rhythms of natural light, consistent with the pattern of activities normally carried out by people. Furthermore, the proposed circadian response curve c(λ), based only on the response to the individual wavelengths, does not consider the effect of spectral opponency nor the contribution of the other photoreceptors to the ipRGCs in the stimulation of the circadian system. In particular, the c(λ) proposed by Gall and Bieske (2004) has a regular bell shape and ignores the irregular pattern between 470 and 510 nm that can be observed from the data previously measured by Brainard et al. (2001a) and Thapan et al. (2001). Because of this, it overestimates the circadian effect of white light sources with CCT greater than 4,000 K, as evidenced by Bellia and Seraceni (2014). Circadian photometry models that rely directly on the sensitivity curve of melanopsin do not offer a better perspective (Dacey et al. 2005; Bailes and Lucas 2013) like the one formulated by Enezi et al. (2011). In fact, also this last type of model, by not contemplating the contribution of the other photoreceptors, cones and rods, which carry information to the ipRGCs, presents a response curve that deviates from the values measured by Brainard et al. (2001a) and Thapan et al. (2001), committing a high error for all spectral contributions due to the shorter wavelengths of ~500 nm (Rea and Figueiro 2016).

### *3.8.2 Hypothesis Based on All Photoreceptors*

A more recent and different model of circadian photometry has been proposed by Rea et al. (2012). This circadian photometry is based on what is known about the neuroanatomy and neurophysiology of the human visual system, in addition to the characteristics observed by Brainard et al. (2001a) and Thapan et al. (2001) on the response to light of the human circadian system. The salient elements on which this proposal of circadian photometry is based can be summarised as follows:

- The calculation model is non-linear in order to better interpret the irregularities present between 470 and 510 nm in the values measured by Brainard et al. (2001a) and Thapan et al. (2001).
- All five known photoreceptors on the human retina, ipRGCs, cones and rods are considered.
- The phenomenon of spectral opponency is included in the calculations, considering the signal of the opposite blue-yellow (b-y) colours that affects the response of the ipRGCs.
- A threshold level (minimum effect) and saturation level (maximum effect) are defined for the suppression of melatonin, depending on the amount of light that reaches the retina.
- The spectral response curve for melatonin suppression is not only one, but it changes according to the SPD in the eyes since, from a physiological standpoint, the rods control the absolute sensitivity of the cones according to the amount of light they receive.

The spectral irradiance that reaches the eyes being known, the model calculates the equivalent circadian light $CL_A$ defined as radiant flux, weighed by the response curve defined by the model after one hour of exposure, divided by the area unit. This dimensionless quantity is not an illuminance defined in lux but could be compared to the illuminance concept. For example, in the presence of an eye illuminance of 1,000 lx with an SPD equal to the black body at 2,856 K, we have a $CL_A = 1{,}000$. Based on the $CL_A$, we can then calculate the circadian stimulus CS, which is a dimensionless value, between 0.1 and 0.7, proportional to the rate of suppression of nocturnal melatonin in humans. The value 0.1 corresponds to no circadian effect, while the increase of the value of CS increases the suppression effect of melatonin up to the maximum level, which is called saturation. From studies conducted on the elderly, young people and workers it has been observed that a light that produces a value of $CS = 0.3$, used for one hour in the first part of the day, is able to stimulate the circadian system, improving the quality of night-time sleep and people's mood.

Figure 3.14 shows the relative SPD of two white LEDs. A Lumileds Luxeon Rebel LXM3-PW81 2700K and a CREE XLamp XP-G 6500K. The diagram also shows the two respective action spectrum as defined by Rea et al. (2012) according to the SPD of the two LEDs considered individually, with an illuminance to the eye of 200 lx. The Lumileds obtains $CL_A = 162$ and $CS = 0.208$ while the CREE obtains $CL_A = 243$ and $CS = 0.277$.

For example, if we want to define the characteristics of an illuminance that does not affect the circadian system, we can propose a value of $CS < 0.05$, which involves a 5% melatonin suppression after one hour of exposure. For a light source with warm CCT (2,670 K) this corresponds to a limit of ~50 lx on the eyes. While for a cold CCT (6.520 K) this limit drops to ~26 lx on the eyes (Rea and Figueiro 2013). However, the authors of this study emphasise the fact that the CTT is not a valid method for

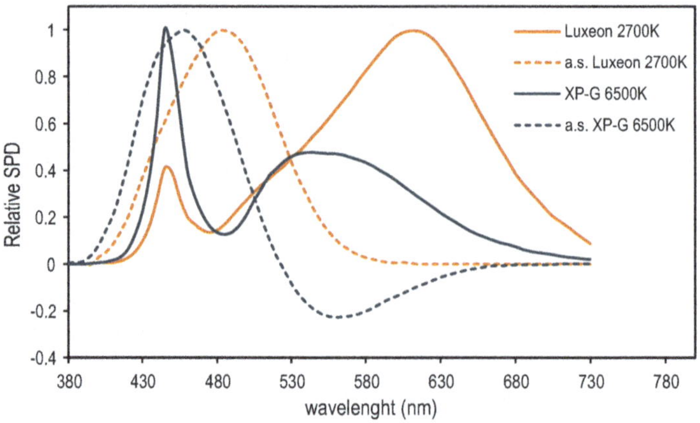


**Fig. 3.14** Comparison of action spectrum for circadian light of two LEDs

determining the CS factor with certainty, rather reminding the importance of being able to measure the true spectral irradiance on the eyes. The mathematical details of the calculation method of the two elements that underlie this circadian photometry proposal, $CL_A$ and CS, go beyond the scope of this book and are also difficult to apply in the daily practice of lighting design that today bases its calculations on CAD lighting systems. Fortunately, the authors of the research have made available two software tools that make it possible to calculate both $CL_A$ and CS, and also other parameters useful for experimentation as well as design practice. The first is an excel table (LRC 2017), while the latter is a calculation application directly accessible via a web page (LRC 2018). OSRAM SYLVANIA (2017) has also made available a useful program to perform colorimetric and photometric calculations based on the SPD that also has an option that allows the $CL_A$ and CS calculations. In addition to making calculations, these tools make it possible to create graphs. The recommended input quantity is the spectral irradiance to the eyes but, in the absence of the possibility of measuring this value, it is also possible to define the spectrum of the light sources according to the data supplied by the manufacturers and the contribution in terms of lux to the eyes brought by every source. The latter method, however, does not take into account the modifications of the light spectrum that reaches the eyes, due to the reflectances of the surfaces in the environment.

Bellia and Seraceni (2014) have also proposed a model to try to characterise the human circadian response curve, taking into account its non-linearity, but this is less complex from the calculation standpoint compared to the model of Rea et al. (2012). The aim has been to propose a calculation method closer to the methods of classical photometry. The approach of this model is more mathematical than neurophysiological. In this model, the circadian light $CL_A$ is calculated according to the illuminance to the eyes and two other parameters that are the circadian action factor $a_{cv}$, defined by Gall and Bieske (2004) and a parameter opp that takes into account the effect of spectral opponency. Based on the circadian light $CL_A$, the circadian CS stimulus can then be calculated as defined by Rea et al. (2012).

The group of LRC researchers who have proposed the circadian photometry based on the calculation of the circadian light $CL_A$ and of the CS circadian stimulating factor, have also carried out further field studies aimed at validating their method. A first research (Figueiro et al. 2006) has verified the model with respect to the suppression of nocturnal melatonin with two different SPDs and four levels of illuminance. The results have confirmed a good prediction from the calculation model although with a higher level of uncertainty for lower illuminance. In another study Figueiro et al. (2008) have demonstrated, by checking the suppression of nocturnal melatonin, that the subadditive response of the human circadian system to polychromatic lights is based on mechanisms present on the retina. In this case also, the results have confirmed a good prediction of the model proposed by them in agreement with other measures realised by Revell and Skene (2007) and Bullough et al. (2008) on lights both monochromatic and polychromatic.

At the end of this chapter, we observe that the CS metric, which is at the base of the proposal of Rea et al. (2012) for a circadian photometry, has a main limitation concerning the temporal aspects of lighting, such as the evaluation of duration, time and history of light exposure of people. Moreover, the "circadian" aspect of this photometry is based on the main marker for the evaluation of the state of the internal biological clock, i.e. melatonin, without considering the NIF effects of light on the alertness state of the central nervous system. Recent research has highlighted that this seems also influenced by red light, which has no effect on the adjustment of melatonin (Figueiro et al. 2016).

From a design standpoint, the assessments that can be made through the circadian photometry mean that the white light used in buildings could be less effective in suppressing melatonin than one might suppose, when compared to natural light, i.e. the blue sky with its strong content of short wavelengths. We must therefore ask ourselves whether, in what way and depending on which design parameters, artificial light should or should not stimulate the human circadian system in interior spaces. This will be the topic of the next chapters.

## References

ANSI/IES RP-27.1-15. (2015). *Recommended practice photobiological safety lamps lamp systems—General requirements*.

ANSI/IESNA RP27.1.27. (1996). *Photobiological safety of lamps and lighting systems*.

Artigas, J. M., et al. (2012). Spectral transmission of the human crystalline lens in adult and elderly persons: Color and total transmission of visible light. *Investigative Ophthalmology & Visual Science, 53*(7), 4076–4084. https://doi.org/10.1167/iovs.12-9471.

Aslam, T. M., Haider, D., & Murray, I. J. (2007). Principles of disability glare measurement: An ophthalmological perspective. *Acta Ophthalmologica Scandinavica, 85*(4), 354–360. https://doi.org/10.1111/j.1600-0420.2006.00860.x.

ASSIST. (2015). *Recommended metric for assessing the direct perception of light source flicker*. Available at: https://www.lrc.rpi.edu/programs/solidstate/assist/pdf/AR-FlickerMetric.pdf (Retrieved: July 16, 2018).

ASSIST. (2017). *Evaluating light source flicker for stroboscopic effects and general acceptability*. Available at: https://www.lrc.rpi.edu/programs/solidstate/assist/pdf/AR-FlickerStrobEval.pdf (Retrieved: July 16, 2018).

Bailes, H. J., & Lucas, R. J. (2013). Human melanopsin forms a pigment maximally sensitive to blue light ($\lambda$max $\approx$ 479 nm) supporting activation of Gq/11 and Gi/o signalling cascades. *Proceedings of the Royal Society B: Biological Sciences, 280*(1759). https://doi.org/10.1098/rspb.2012.2987.

Barker, F., & Brainard, G. (1991). *The direct spectral transmittance of the excised human lens as a function of age (FDA 785345 0090 RA)*. US Food and Drug Administration. Available at: https://www.chemie.uni-wuerzburg.de/fileadmin/08020000/user_upload/makula/transmittance.pdf (Retrieved: May 15, 2018).

Bellia, L., & Seraceni, M. (2014). A proposal for a simplified model to evaluate the circadian effects of light sources. *Lighting Research & Technology, 46*(5), 493–505. https://doi.org/10.1177/1477153513490715.

Benloucif, S., et al. (2005). Stability of melatonin and temperature as circadian phase markers and their relation to sleep times in humans. *Journal of Biological Rhythms, 20*(2), 178–188. https://doi.org/10.1177/0748730404273983.

Berman, S. M., et al. (1991). Human electroretinogram responses to video displays, fluorescent lighting, and other high frequency sources. *Optometry and Vision Science, 68*(8), 645.

Berson, D. M., Dunn, F. A., & Takao, M. (2002). Phototransduction by retinal ganglion cells that set the circadian clock. *Science, 295*(5557), 1070–1073. https://doi.org/10.1126/science.1067262.

Berson, D. M. (2003). Strange vision: Ganglion cells as circadian photoreceptors. *Trends in Neurosciences, 26*(6), 314–320. https://doi.org/10.1016/s0166-2236(03)00130-9.

Bierman, A. (2017). A model for predicting stroboscopic flicker. In *IES 2017 Annual Conference Proceedings*. New York: IES.

Binnie, C. D., de Korte, R. A., & Wisman, T. (1979). Fluorescent lighting and epilepsy. *Epilepsia, 20*(6), 725–727. https://doi.org/10.1111/j.1528-1157.1979.tb04856.x.

Binnie, C. D., et al. (2002). Characterizing the flashing television images that precipitate seizures. *SMPTE Journal, 111*(6–7), 323–329. https://doi.org/10.5594/j15330.

Bodington, D., Bierman, A., & Narendran, N. (2016). A flicker perception metric. *Lighting Research & Technology, 48*(5), 624–641. https://doi.org/10.1177/1477153515581006.

Bodkin, H. (2018). New LED streetlights may double cancer risk, new research warns. *The Telegraph*, April 26, 2018.

Boettner, E. A., & Wolter, J. R. (1962). Transmission of the ocular media. *Investigative Ophthalmology & Visual Science, 1*(6), 776–783.

Boivin, D. B., et al. (1996). Dose-response relationships for resetting of human circadian clock by light. *Nature, 379*(6565), 540–542. https://doi.org/10.1038/379540a0.

Boyce, P. (2014). *Human factors in lighting* (3rd ed.). Boca Raton: CRC Press.

Brainard, G. C., et al. (1988). Dose-response relationship between light irradiance and the suppression of plasma melatonin in human volunteers. *Brain Research, 454*(1), 212–218. https://doi.org/10.1016/0006-8993(88)90820-7.

Brainard, G. C., Rollag, M. D., & Hanifin, J. P. (1997). Photic regulation of melatonin in humans: Ocular and neural signal transduction. *Journal of Biological Rhythms, 12*(6), 537–546. https://doi.org/10.1177/074873049701200608.

Brainard, G. C., et al. (2001a). Action spectrum for melatonin regulation in humans: Evidence for a novel circadian photoreceptor. *Journal of Neuroscience, 21*(16), 6405–6412. https://doi.org/10.1523/jneurosci.21-16-06405.2001.

Brainard, G. C., et al. (2001b). Human melatonin regulation is not mediated by the three cone photopic visual system. *Journal of Clinical Endocrinology and Metabolism, 86*(1), 433–436. https://doi.org/10.1210/jcem.86.1.7277.

Brainard, G. C., et al. (2008). Sensitivity of the human circadian system to short-wavelength (420-nm) light. *Journal of Biological Rhythms, 23*(5), 379–386. https://doi.org/10.1177/0748730408323089.

Bullough, J. D., et al. (2008). On melatonin suppression from polychromatic and narrowband light. *Chronobiology International, 25*(4), 653–656. https://doi.org/10.1080/07420520802247472.

Bullough, J., et al. (2011). Effects of flicker characteristics from solid-state lighting on detection, acceptability and comfort. *Lighting Research & Technology, 43*(3), 337–348. https://doi.org/10.1177/1477153511401983.

Bullough, J. D., Bierman, A., & Rea, M. S. (2017). Evaluating the blue-light hazard from solid state lighting. *International Journal of Occupational Safety and Ergonomics: JOSE*, 1–10. https://doi.org/10.1080/10803548.2017.1375172.

Burns, S. A., Elsner, A. E., & Kreitz, M. R. (1992). Analysis of nonlinearities in the flicker ERG. *Optometry and Vision Science, 69*(2), 95–105. (Official Publication of the American Academy of Optometry).

Chamorro, E., et al. (2013). Effects of light-emitting diode radiations on human retinal pigment epithelial cells in vitro. *Photochemistry and Photobiology, 89*(2), 468–473. https://doi.org/10.1111/j.1751-1097.2012.01237.x.

CIBSE. (2013). *Lighting guide 9: Lighting for communal residential building LG9*. Available at: https://www.cibse.org/knowledge/knowledge-items/detail?id=a0q20000008I7kJAAS (Retrieved: July 8, 2018).

CIE 117:1995. *Discomfort glare in interior lighting*.

CIE 190:2010. *Calculation and presentation of united glare rating tables for indoor lighting luminaires*.

CIE TN 003:2015a. *Report on the first international workshop on circadian and neurophysiological photometry, 2013*. Available at: http://www.cie.co.at/publications/report-first-international-workshop-circadian-and-neurophysiological-photometry-2013 (Retrieved: April 21, 2018).

CIE TN 003:2015b. *Toolbox*. Available at: http://files.cie.co.at/784_TN003_Toolbox.xls (Retrieved: May 15, 2018).

CIE TN 006:2016. *Visual aspects of time-modulated lighting systems—Definitions and measurement models*.

CIE TN 008:2017. *Final report CIE stakeholder workshop for temporal light modulation standards for lighting systems*.

Cossar, V.-M. (2013). LED lights: Should we worry about damage to our eyes? *Metro*, December 9, 2013.

Dacey, D. M., et al. (2005). Melanopsin-expressing ganglion cells in primate retina signal colour and irradiance and project to the LGN. *Nature, 433*(7027), 749–754. https://doi.org/10.1038/nature03387.

de Lange, H. (1958). Research into the dynamic nature of the human fovea → cortex systems with intermittent and modulated light. I. Attenuation characteristics with white and colored light. *JOSA, 48*(11), 777–784. https://doi.org/10.1364/josa.48.000777.

de Lange, H. (1961). Eye's response at flicker fusion to square-wave modulation of a test field surrounded by a large steady field of equal mean luminance. *JOSA, 51*(4), 415–421. https://doi.org/10.1364/josa.51.000415.

Diano, J. (2004). *Freschezza di idee: la luce come regolatore del nostro orologio biologico, in Convegno Nazionale Associazione Italiana Di Illuminazione*. Genova: AIDI.

Directive 2001/95/EC of the European Parliament and of the Council of 3 December 2001 on general product safety (Text with EEA relevance).

Directive 2006/25/EC of the European Parliament and of the Council of 5 April 2006 on the minimum health and safety requirements regarding the exposure of workers to risks arising from physical agents (artificial optical radiation) (19th individual Directive within the meaning of Article 16(1) of Directive 89/391/EEC).

Directive 2009/48/EC of the European Parliament and of the Council of 18 June 2009 on the safety of toys (Text with EEA relevance).

Directive 2014/35/EU of the European Parliament and of the Council of 26 February 2014 on the harmonisation of the laws of the Member States relating to the making available on the market of electrical equipment designed for use within certain voltage limits Text with EEA relevance.
EN 62471:2008. *Photobiological safety of lamps and lamp systems*.
EN 12464-1:2011. *Light and lighting—Lighting of work places—Part 1: Indoor work places*.
EN 62115:2005 + A12:2015. *Electric toys. Safety*.
Enezi, J., et al. (2011). A "melanopic" spectral efficiency function predicts the sensitivity of melanopsin photoreceptors to polychromatic lights. *Journal of Biological Rhythms, 26*(4), 314–323. https://doi.org/10.1177/0748730411409719.
Farnsworth, D. (1943). The Farnsworth-Munsell 100-hue and dichotomous tests for color vision. *JOSA, 33*(10), 568–578. https://doi.org/10.1364/josa.33.000568.
Figueiro, M. G., et al. (2004). Preliminary evidence for spectral opponency in the suppression of melatonin by light in humans. *NeuroReport, 15*(2), 313.
Figueiro, M. G., et al. (2005). Demonstration of additivity failure in human circadian phototransduction. *Neuro Endocrinology Letters, 26*(5), 493–498.
Figueiro, M. G., Rea, M. S., & Bullough, J. D. (2006). Circadian effectiveness of two polychromatic lights in suppressing human nocturnal melatonin. *Neuroscience Letters, 406*(3), 293–297. https://doi.org/10.1016/j.neulet.2006.07.069.
Figueiro, M. G., Bierman, A., & Rea, M. S. (2008). Retinal mechanisms determine the subadditive response to polychromatic light by the human circadian system. *Neuroscience Letters, 438*(2), 242–245. https://doi.org/10.1016/j.neulet.2008.04.055.
Figueiro, M. G., et al. (2016). Light at night and measures of alertness and performance: Implications for shift workers. *Biological Research for Nursing, 18*(1), 90–100. https://doi.org/10.1177/1099800415572873.
Fisher, R. S., et al. (2005). Photic- and pattern-induced seizures: A review for the Epilepsy Foundation of America working group. *Epilepsia, 46*(9), 1426–1441. https://doi.org/10.1111/j.1528-1167.2005.31405.x.
FprEN 62115:2016. *Electric toys—Safety*. Available at: https://www.cenelec.eu/dyn/www/f?p=104:110:311166124775401::::FSP_ORG_ID,FSP_PROJECT,FSP_LANG_ID:1257159,61861,25 (Retrieved: August 13, 2018).
Gall, D., & Bieske, K. (2004). Definition and measurement of circadian radiometric quantities. In *Proceedings of the CIE Symposium '04 on Light and Health* (pp. 129–132).
Gray, R., & Regan, D. (2007). Glare susceptibility test results correlate with temporal safety margin when executing turns across approaching vehicles in simulated low-sun conditions. *Ophthalmic and Physiological Optics, 27*(5), 440–450. https://doi.org/10.1111/j.1475-1313.2007.00503.x.
Hankins, M. W., & Lucas, R. J. (2002). The primary visual pathway in humans is regulated according to long-term light exposure through the action of a nonclassical photopigment. *Current Biology, 12*(3), 191–198. https://doi.org/10.1016/s0960-9822(02)00659-0.
Hanson, A. F., & Berdoy, M. (2010). Rats. In V. V. Tynes (Ed.), *Behavior of exotic pets* (pp. 104–116). Oxford: Wiley-Blackwell.
Harding, G. F. A., & Jeavons, P. M. (1994). *Photosensitive epilepsy* (New ed.). London: Mac Keith Press.
Harding, G. F. A., & Harding, P. F. (2010). Photosensitive epilepsy and image safety. *Applied Ergonomics, 41*(4), 504–508. https://doi.org/10.1016/j.apergo.2008.08.005.
Hattar, S., et al. (2003). Melanopsin and rod-cone photoreceptive systems account for all major accessory visual functions in mice. *Nature, 424*(6944), 76–81. https://doi.org/10.1038/nature01761.
Hering, E. (1964). *Outlines of a theory of the light sense* (1st ed.) (L. M. Hurvich & D. Jameson, Trans.). Harvard University Press.
Hershberger, W. A., & Jordan, J. S. (1998). The phantom array: A perisaccadic illusion of visual direction. *The Psychological Record, 48*(1), 21–32. https://doi.org/10.1007/bf03395256.
Higlett, M. P., O'Hagan, J. B., & Khazova, M. (2012). Safety of light emitting diodes in toys. *Journal of Radiological Protection, 32*(1), 51. https://doi.org/10.1088/0952-4746/32/1/51.

Huang, Y.-Y., & Menozzi, M. (2014). Effects of discomfort glare on performance in attending peripheral visual information in displays. *Displays, 35*(5), 240–246. https://doi.org/10.1016/j.displa.2014.08.001.

Hurvich, L. M., & Jameson, D. (1957). An opponent-process theory of colour vision. *Psychological Review, 64*(6, Pt.1), 384–404. https://doi.org/10.1037/h0041403.

ICNIRP. (1997). Guidelines on limits of exposure to broad-band incoherent optical radiation (0.38 to 3 microM). International commission on non-ionizing radiation protection. *Health Physics, 73*(3), 539–554.

ICNIRP. (2002). General approach to protection against non-ionizing radiation. *Health Physics, 82*(4), 540–548.

ICNIRP. (2013). ICNIRP guidelines on limits of exposure to incoherent visible and infrared radiation. *Health Physics, 105*(1), 74. https://doi.org/10.1097/hp.0b013e318289a611.

IEC 60598-1:2014 + AMD1:2017. *CSV luminaires—Part 1: General requirements and tests.*

IEC 61000-3-3:2013. *Electromagnetic compatibility (EMC)—Part 3-3: Limits—Limitation of voltage changes, voltage fluctuations and flicker in public low-voltage supply systems, for equipment with rated current ≤16 A per phase and not subject to conditional connection.*

IEC 61000-4-15:2010. *Electromagnetic compatibility (EMC)—Part 4-15: Testing and measurement techniques—Flickermeter—Functional and design specifications.*

IEC 62031:2018. *LED modules for general lighting—Safety specifications.*

IEC 62471/CIE S 009/E&F:2002. *Photobiological safety of lamps and lamp systems.*

IEC 62471:2006. *Photobiological safety of lamps and lamp systems.*

IEC TR 61547-1:2017. *Equipment for general lighting purposes—EMC immunity requirements—Part 1: An objective light flickermeter and voltage fluctuation immunity test method.*

IEC TR 62471-2:2009. *Photobiological safety of lamps and lamp systems—Part 2: Guidance on manufacturing requirements relating to non-laser optical radiation safety.*

IEC TR 62471-3:2015. *Photobiological safety of lamps and lamp systems—Part 3: Guidelines for the safe use of intense pulsed light source equipment on humans.*

IEC TR 62778:2014. *Application of IEC 62471 for the assessment of blue light hazard to light sources and luminaires.*

IEEE 1789-2015. *IEEE recommended practices for modulating current in high-brightness LEDs for mitigating health risks to viewers.*

IES. (2011). *The lighting handbook* (10th ed.). New York: Illuminating Engineering Society.

ISO 8995-1:2002/CIE S 008/E:2001. *Lighting of work places—Part 1: Indoor.*

Jaadane, I., et al. (2015). Retinal damage induced by commercial light emitting diodes (LEDs). *Free Radical Biology and Medicine, 84,* 373–384. https://doi.org/10.1016/j.freeradbiomed.2015.03.034.

Jaén, E. M., et al. (2005). Office workers visual performance and temporal modulation of fluorescent lighting. *LEUKOS, 1*(4), 27–46. https://doi.org/10.1582/leukos.01.04.002.

Jaén, E. M., Colombo, E. M., & Kirschbaum, C. F. (2011). A simple visual task to assess flicker effects on visual performance. *Lighting Research & Technology, 43*(4), 457–471. https://doi.org/10.1177/1477153511405409.

James, R. H., et al. (2017). Evaluation of the potential optical radiation hazards with led lamps intended for home use. *Health Physics, 112*(1), 11–17. https://doi.org/10.1097/hp.0000000000000580.

Kelly, D. H. (1961). Visual responses to time-dependent stimuli.* I. Amplitude sensitivity measurements. *JOSA, 51*(4), 422–429. https://doi.org/10.1364/josa.51.000422.

Kennedy, A., & Murray, W. S. (1991). The effects of flicker on eye movement control. *The Quarterly Journal of Experimental Psychology Section A, 43*(1), 79–99. https://doi.org/10.1080/14640749108401000.

Kindel, E. (2005). Ishihara. Nine decades on, a Japanese army doctor's invention is still being used to test colour vision. *Eye Magazine, 14*(56).

Knez, I. (2014). Affective and cognitive reactions to subliminal flicker from fluorescent lighting. *Consciousness and Cognition, 26,* 97–104. https://doi.org/10.1016/j.concog.2014.02.006.

Kohen, E., Santus, R., & Hirschberg, J. G. (1995). *Photobiology* (1st ed.). San Diego: Academic Press.

Kraneburg, A., et al. (2017). Effect of colour temperature on melatonin production for illumination of working environments. *Applied Ergonomics, 58,* 446–453. https://doi.org/10.1016/j.apergo.2016.08.006.

Krigel, A., et al. (2016). Light-induced retinal damage using different light sources, protocols and rat strains reveals LED phototoxicity. *Neuroscience, 339,* 296–307. https://doi.org/10.1016/j.neuroscience.2016.10.015.

Lehman, B. et al. (2011). Proposing measures of flicker in the low frequencies for lighting applications. In *2011 IEEE Energy Conversion Congress and Exposition* (pp. 2865–2872). https://doi.org/10.1109/ecce.2011.6064154.

Lerman, S. (1987). Chemical and physical properties of the normal and aging lens: Spectroscopic (UV, fluorescence, phosphorescence, and NMR) analyses. *American Journal of Optometry and Physiological Optics, 64*(1), 11–22.

Levinson, J. (1960). Fusion of complex flicker II. *Science, 131*(3411), 1438–1440. https://doi.org/10.1126/science.131.3411.1438.

Lewy, A. J., Cutler, N. L., & Sack, R. L. (1999). The endogenous melatonin profile as a marker for circadian phase position. *Journal of Biological Rhythms, 14*(3), 227–236. https://doi.org/10.1177/074873099129000641.

Lipson, E. (2012). Action spectroscopy: General problems. In A. Griesbeck, M. Oelgemöller & F. Ghetti (Eds.), *CRC handbook of organic photochemistry and photobiology* (3rd ed.). Boca Raton: CRC Press.

LRC. (2017). *Circadian stimulus calculator*. Lighting Research Center. Available at: http://www.lrc.rpi.edu/resources/CSCalculator_2017_10_03_Mac.xlsm (Retrieved: July 20, 2018).

LRC. (2018). *Web CS calculator*. Lighting Research Center. Available at: https://www.lrc.rpi.edu/cscalculator/ (Retrieved: July 20, 2018).

Lucas, R. J., et al. (2003). Diminished pupillary light reflex at high irradiances in melanopsin-knockout mice. *Science (New York, N.Y.), 299*(5604), 245–247. https://doi.org/10.1126/science.1077293.

Mainster, M. A., & Turner, P. L. (2012). Glare's causes, consequences, and clinical challenges after a century of ophthalmic study. *American Journal of Ophthalmology, 153*(4), 587–593. https://doi.org/10.1016/j.ajo.2012.01.008.

Mclntyre, I. M., et al. (1989). Human melatonin suppression by light is intensity dependent. *Journal of Pineal Research, 6*(2), 149–156. https://doi.org/10.1111/j.1600-079x.1989.tb00412.x.

Mecherikunnel, A. T., & Richmond, J. C. (1908). *Spectral distribution of solar radiation*. NASA. Available at: https://ntrs.nasa.gov/archive/nasa/casi.ntrs.nasa.gov/19810016493.pdf (Retrieved: February 14, 2018).

Morita, T., & Tokura, H. (1996). Effects of lights of different color temperature on the nocturnal changes in core temperature and melatonin in humans. *Applied Human Science, 15*(5), 243–246. https://doi.org/10.2114/jpa.15.243.

Musante, F., & Rossi, M. (2016). The evaluation of flicker in LED lighting systems. *LUCE, 54*(318), 101–106.

NEMA TLAs-2015. *Temporal light artifacts (flicker and stroboscopic effects).*

NEMA 77-2017. *Temporal light artifacts: Test methods and guidance for acceptance criteria.*

O'Hagan, J. B., Khazova, M., & Price, L. L. A. (2016). Low-energy light bulbs, computers, tablets and the blue light hazard. *Eye, 30*(2), 230–233. https://doi.org/10.1038/eye.2015.261.

Ollove, M. (2016). Some cities are taking another look at LED lighting after AMA warning. *Washington Post*, September 25, 2016.

Olsen, J., et al. (2014). Human factors study on light modulation in indirect office lighting. In *Proceedings of the Human Factors and Ergonomics Society Annual Meeting* (pp. 1104–1108). https://doi.org/10.1177/1541931214581231.

OSRAM SYLVANIA. (2017). *LED color calculator*. Available at: https://www.osram.us/cb/tools-and-resources/applications/led-colourcalculator/index.jsp (Retrieved: May 10, 2018).

Panda, S., et al. (2002). Melanopsin (Opn4) requirement for normal light-induced circadian phase shifting. *Science (New York, N.Y.), 298*(5601), 2213–2216. https://doi.org/10.1126/science.1076848.

Panda, S., et al. (2003). Melanopsin is required for non-image-forming photic responses in blind mice. *Science (New York, N.Y.), 301*(5632), 525–527. https://doi.org/10.1126/science.1086179.

Perz, M., et al. (2015). Modeling the visibility of the stroboscopic effect occurring in temporally modulated light systems. *Lighting Research & Technology, 47*(3), 281–300. https://doi.org/10.1177/1477153514534945.

Quirk, J. A., et al. (1995). First seizures associated with playing electronic screen games: A community-based study in Great Britain. *Annals of Neurology, 37*(6), 733–737. https://doi.org/10.1002/ana.410370606.

Rand, D., Lehman, B., & Shteynberg, A. (2007) Issues, models and solutions for triac modulated phase dimming of LED lamps. In *2007 IEEE Power Electronics Specialists Conference* (pp. 1398–1404). https://doi.org/10.1109/pesc.2007.4342199.

Rea, M. S., Bullough, J. D., & Figueiro, M. G. (2001). Human melatonin suppression by light: A case for scotopic efficiency. *Neuroscience Letters, 299*(1), 45–48. https://doi.org/10.1016/s0304-3940(01)01512-9.

Rea, M. S., Bullough, J. D., & Figueiro, M. G. (2002a). Phototransduction for human melatonin suppression. *Journal of Pineal Research, 32*(4), 209–213. https://doi.org/10.1034/j.1600-079x.2002.01881.x.

Rea, M. S., Figueiro, M. G., & Bullough, J. D. (2002b). Circadian photobiology: An emerging framework for lighting practice and research. *Lighting Research & Technology, 34*(3), 177–187. https://doi.org/10.1191/1365782802lt057oa.

Rea, M. S., et al. (2012). Modelling the spectral sensitivity of the human circadian system. *Lighting Research & Technology, 44*(4), 386–396. https://doi.org/10.1177/1477153511430474.

Rea, M. S., & Figueiro, M. G. (2013). A working threshold for acute nocturnal melatonin suppression from white light sources used in architectural applications. *Journal of Carcinogenesis & Mutagenesis, 4*(3), 1–6. https://doi.org/10.4172/2157-2518.1000150.

Rea, M. S., & Figueiro, M. G. (2016). Light as a circadian stimulus for architectural lighting. *Lighting Research & Technology, 50*(4), 497–510. https://doi.org/10.1177/1477153516682368.

Revell, V. L., & Skene, D. J. (2007). Light-induced melatonin suppression in humans with polychromatic and monochromatic light. *Chronobiology International, 24*(6), 1125–1137. https://doi.org/10.1080/07420520701800652.

Roberts, J., & Wilkins, A. (2013). Flicker can be perceived during saccades at frequencies in excess of 1 kHz. *Lighting Research & Technology, 45*(1), 124–132. https://doi.org/10.1177/1477153512436367.

Rosenthal, N. E. (1991). Plasma melatonin as a measure of the human clock. *The Journal of Clinical Endocrinology & Metabolism, 73*(2), 225–226. https://doi.org/10.1210/jcem-73-2-225.

SCHEER. (2016). *Scientific committee on health, environmental and emerging risks—Public health—European commission*. Available at: https://ec.europa.eu/health/scientific_committees/scheer_en (Retrieved: July 7, 2018).

SCHEER. (2018). *Opinion on potential risks to human health of light emitting diodes (LEDs)*. European Commission. Available at: https://ec.europa.eu/health/sites/health/files/scientific_committees/scheer/docs/scheer_o_011.pdf (Retrieved: August 6, 2018).

Shang, Y.-M., et al. (2014). White light-emitting diodes (LEDs) at domestic lighting levels and retinal injury in a rat model. *Environmental Health Perspectives, 122*(3), 269–276. https://doi.org/10.1289/ehp.1307294.

Shepherd, A. J. (2010). Visual stimuli, light and lighting are common triggers of migraine and headache. *Journal of Light & Visual Environment, 34*(2), 94–100. https://doi.org/10.2150/jlve.34.94.

Sliney, D. H., Bergman, R., & O'Hagan, J. (2016). Photobiological risk classification of lamps and lamp systems—History and rationale. *LEUKOS, 12*(4), 213–234. https://doi.org/10.1080/15502724.2016.1145551.

Thapan, K., Arendt, J., & Skene, D. J. (2001). An action spectrum for melatonin suppression: Evidence for a novel non-rod, non-cone photoreceptor system in humans. *Journal of Physiology, 535*(1), 261–267. https://doi.org/10.1111/j.1469-7793.2001.t01-1-00261.x.

Timm, R. M. (2005). *Norway rats, Internet center for wildlife damage management*. Available at: http://icwdm.org/handbook/rodents/NorwayRats.asp (Retrieved: August 8, 2018).

van Bommel, W., & van den Beld, G. (2004). Lighting for work: A review of visual and biological effects. *Lighting Research & Technology, 36*(4), 255–266. https://doi.org/10.1191/1365782804li122oa.

Veitch, J. A., & McColl, S. L. (1995). Modulation of fluorescent light: Flicker rate and light source effects on visual performance and visual comfort. *International Journal of Lighting Research and Technology, 27*(4), 243–256. https://doi.org/10.1177/14771535950270040301.

Wilkins, A. J., et al. (1989). Fluorescent lighting, headaches and eyestrain. *Lighting Research & Technology, 21*(1), 11–18. https://doi.org/10.1177/096032718902100102.

Wilkins, A. J., Veitch, J., & Lehman, B. (2010). LED lighting flicker and potential health concerns: IEEE standard PAR1789 update. In *2010 IEEE Energy Conversion Congress and Exposition* (pp. 171–178). https://doi.org/10.1109/ecce.2010.5618050.

Wyszecki, G., & Stiles, W. S. (2000). *Color science: Concepts and methods, quantitative data and formulae* (2nd ed.). Hoboken: Wiley.

Zeitzer, J. M., et al. (2000). Sensitivity of the human circadian pacemaker to nocturnal light: Melatonin phase resetting and suppression. *The Journal of Physiology, 526*(Pt 3), 695–702. https://doi.org/10.1111/j.1469-7793.2000.00695.x.

# Chapter 4
# Case Studies: Natural Light in Interior Spaces

## 4.1 The Natural Light

It is intuitive to say that the best light for human beings is natural light. It has been so since the dawn of the human race. However, modern rhythms and lifestyles are both the consequence and the reason for a broader assessment of the lighting to which people in industrialised societies are subject today.

Unlike artificial light, natural light is characterised by considerable variability in terms of quantity, direction and spectral power distribution (SPD). This variability is primarily due to the time of day, but also to meteorological conditions, latitude, orographic and urban context. If on the one hand natural light is strongly desired for the interior lighting of buildings, on the other hand it can involve problems. In fact, direct sunlight can disturb because it is too intense. It is also a carrier of radiation such as ultraviolet (UV), which, if taken in excess, can be harmful to health, and infrared (IR), which introduce heat, causing high energy consumption for air conditioning buildings in hot weather. Nonetheless, natural light is very welcome in buildings, as demonstrated also in a study of Shao et al. (1998) on the use of lightpipes to bring light into the interior and also confirmed by other research on the use of natural light in interior spaces (Begemann et al. 1997; Heschong et al. 2002; Borisuit et al. 2015). Figure 4.1 shows two examples of natural light. Noon in Laguna Miñiques 4,115 m of altitude, Andes, Atacama Desert, one of the driest sites in the world. Sunset over the rooftops of Milan Italy, with a considerable amount of water vapour in the atmosphere.

Depending on the parameters shown above, the variability of natural light in outdoor areas can range from a minimum illuminance to the ground of 1,000 lx in winter, with a heavily overcast sky, to a maximum of 150,000 lx in summer, with a clear sky. Compared to artificial lighting, which is in the order of tens of lux outside and hundreds of lux inside, natural light is therefore up to three orders of magnitude greater. ISO and CIE have published an international standard that defines the luminous properties of natural light so that it can be considered in lighting design calculations (ISO 15469:2004(E)/CIE S 011/E:2003). This standard defines a



M. Rossi, *Circadian Lighting Design in the LED Era*, Research for Development,
https://doi.org/10.1007/978-3-030-11087-1_4

Fig. 4.1 Two examples of natural light: Andes and Europe

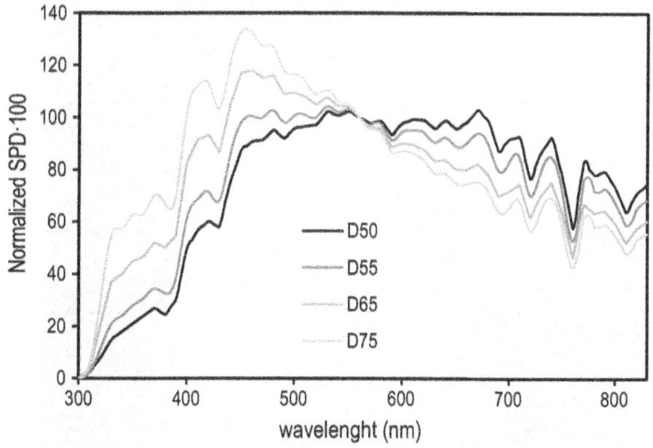


**Fig. 4.2** SPD of natural light, from 5,000 to 7,500 K

mathematical model that provides the luminance distribution of the various positions of the sky observed from the ground. The calculations take into account the time of day and the climatic conditions for the sky, which vary from clear to overcast. Modern CAD software for lighting incorporates calculation models such as the CIE or equivalent, to make it possible to calculate the contribution of natural light in interior spaces.

The SPD of natural light changes a lot during the day, from dawn to dusk and also depending on the cloud cover. The correlated colour temperature (CCT) can vary considerably, from 4,000 K on an average cloudy day up to 40,000 K, measurable when the blue sky is completely free from clouds and haze. To describe the chromaticity of natural light, CIE recommends the use of standard D-series illuminants (daylight, CIE 015:2004), from D50 to D75, which have CCT from 5,000 to 7,500 K respectively (Fig. 4.2).

In addition, there are two other fundamental parameters that influence the amount of natural light in the interiors and these depend on the construction characteristics of the building: the size of the windows and the orientation of the openings with respect to the cardinal points. This topic, of an architectural nature, goes beyond the scope of this book. However, it is interesting to note that even the lighting sector has presented guidelines on the design of buildings, in order to optimise the use of natural light, such as the guide by CIBSE (2014). This guide deals with aesthetic topics, but also with physical aspects, in the architectural and engineering design of the building, from lighting bases to modern automatic systems for the management and filtering of natural light, both for new buildings and for those undergoing renovation. There is also a trend, which presumably will become more generalised in the future, towards a new and increasingly integrated design methodology, in which the lighting designer intervenes from the early design stages of a building, in collaboration with architects and engineers, for the definition of all aspects concerning lighting, both artificial and natural. In Fig. 4.2 we can see two buildings that have completely different windows for the entry of natural light. The crystal tower Gran Torre Santiago in Chile on the left and the Chrysler Building in New York in the USA on the right (Fig. 4.3).

**Fig. 4.3** Two buildings that have completely different windows for the entry of natural light

As we have seen in Chap. 1, since the end of the nineteenth century two processes have had a fundamental impact on the societies of the industrialised countries and on people's exposure to light. On the one hand, there has been a progressive shift of workers from the exterior of the countryside to the interiors of factories in the industry and offices in the service sector. On the other hand, artificial light has started to replace natural light in the interiors, at night but also during the day, when it is not available due to urban, architectural, geographical, orographic, or climatic reasons, or for the setting of the sun, in a society that never sleeps. In fact, human beings no longer follow the cycle of natural light and often perform work or social activities well beyond the setting of the sun.

Because of this radical change in the lifestyles of human beings, who today spend 80–90% of their time in interior spaces (NRC 1981; Mccurdy and Graham 2003), in this chapter we analyse the quality of circadian light in some types of interiors, with particular attention to the contribution of natural light.

## 4.2 Evaluation of Light in an Office

The active population in the working world spends most of day time away from home. In modern society, most of the works in internal spaces take place in offices, which may have different conformations, but are typically characterised by the fact that they have desks for individual work and tables for meetings, visual display unit (VDU) workstations and spaces of other nature. A topic of fundamental importance for the lighting design of the future is being able to assess what are the circadian characteristics of light that manages to penetrate these interior spaces.

### *4.2.1 The Case Study*

In 2011, the Italian national agency for new technologies, energy and sustainable economic development (ENEA) commissioned Laboratorio Luce of the Politecnico di Milano to evaluate, from a physiological standpoint and a visual comfort perspective, the lighting present during the daytime hours in a workplace. ENEA has considered a number of alternatives related to public interiors and among these has chosen a library as the test environment on which to carry out the checks. This to be able to evaluate the impact of lighting in interiors on workers, in this case librarians, who spend the entire working day in the building. The place chosen has been the main hall of the municipal library of Marcallo con Casone in the province of Milan (Italy) located at a latitude of ~45°.

The environment consists of an internal space with a rectangular plan, 5.24 m wide, 17 m long and 3.57 m high, featuring a series of windows on the two longer sides (Fig. 4.4). The three windows of the NNW-oriented side are placed at a height of 2.30 m, and have a size of 1.1 × 1.1 m and are without any natural light control system (Fig. 4.4). The openings on the opposite side, SSE-oriented, are of two types: there are two French windows 0.9 × 2.60 m and two windows 0.9 × 1.60 m (Fig. 4.5). All the openings on the SSE side are equipped with Venetian blinds on the inside and shutters on the outside to control the flow of natural light. During the inspections, it has been observed that the Venetian blinds of the two French windows are almost always kept closed by the occupants to avoid disturbing glare due to natural light. Two main activities are carried out in the premises: book consultation and office activities for library management.

In the library there is an artificial lighting system consisting of 11 luminaires 1 × 58 W, with anti-glare transverse blades, containing 58 W linear fluorescent lamps,

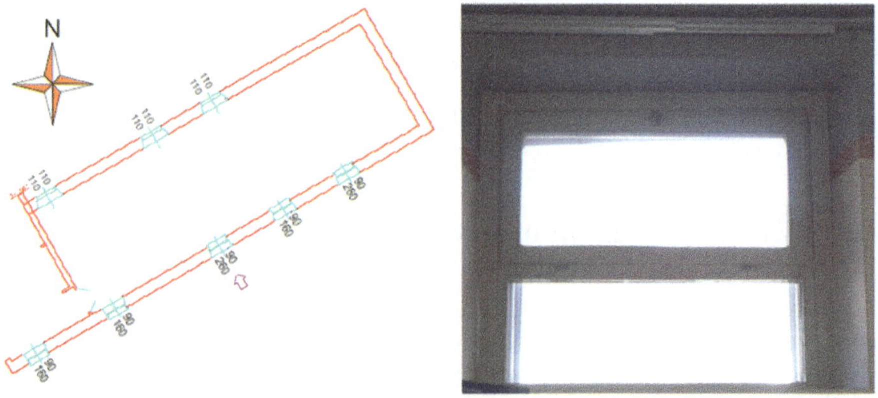


**Fig. 4.4** Marcallo con Casone Library. The plan and the windows oriented to NNW

**Fig. 4.5** The SSE-oriented French windows and the windows equipped with Venetian blinds

installed at a height of 3.15 m above the floor (Fig. 4.6). Each lamp emits a luminous flux of 5,200 lm with a CCT of 4,000 K and a colour rendering index $R_a = 85$. The luminaires are arranged in a modular system, in the centre of the room and in such a way as to ensure the illumination of both the reading area and the office area. The power supply of the luminaires is independent for each zone; however, no dimmer switch system has been provided.

The luminaires used have an efficiency of about 51% with an exclusively direct emission that does not bring any indirect light contribution towards the ceiling, whose daylighting is mainly provided by the contribution of natural light that enters from the windows, in a greater quantity from the SSE-oriented ones. In Fig. 4.7 is illustrated the photometric solid of the luminaires that shows an emission exclusively directed downwards.

**Fig. 4.6** The artificial lighting system

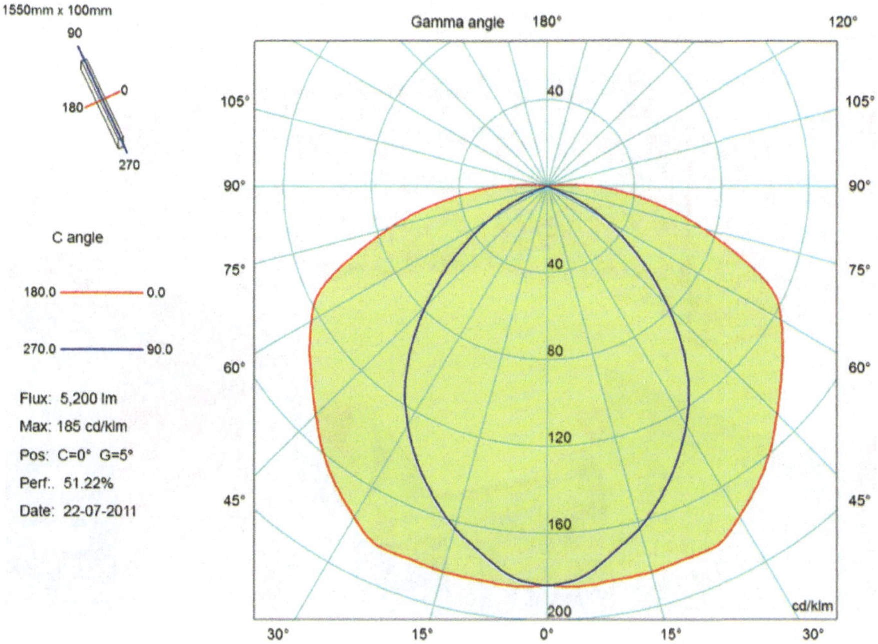


**Fig. 4.7** The photometric solid of the luminaires

### 4.2.2 Evaluation of Visual Comfort with the UGR Method

In order to evaluate visual comfort through the calculation of the Unified Glare Rating (UGR) index (CIE 190:2010), a software simulation has been carried out with a lighting CAD program, using a simplified geometry of the environment. In order to calculate reliable values, the reflectances of the masonry structures and of the furnishings have been measured and assigned in such a way as to try to obtain the same luminance values, in the comparison between the CAD simulation and the real environment, within the visual field of the observers. Luminance measurements have been made in the absence of natural light contributions at 21:00, to be able to validate the calculations made by the lighting CAD, with a margin of difference between real and simulated values contained within ±5%. Based on this CAD model, the value of UGR has been calculated for different observing positions along the sides of the reading table and in the office area. Figure 4.8 shows one of the image based luminance measurements of the reading area, at 21:00, in the presence of artificial lighting only. Luminance measurements have been performed with a Konica Minolta CA2000 and with a TechnoTeam-Rollei LMK 2000 video cameras.

In the reading area of the library, UGR values between 15 and 27 have been determined, depending on the position of the observer with respect to the consultation table, and near the table the maximum value is 23 (Fig. 4.9, Table 4.1). Not all evaluation positions are possible in real space due to the obstruction of carts or

**Fig. 4.8** Marcallo con Casone Library. Image based luminance measurements

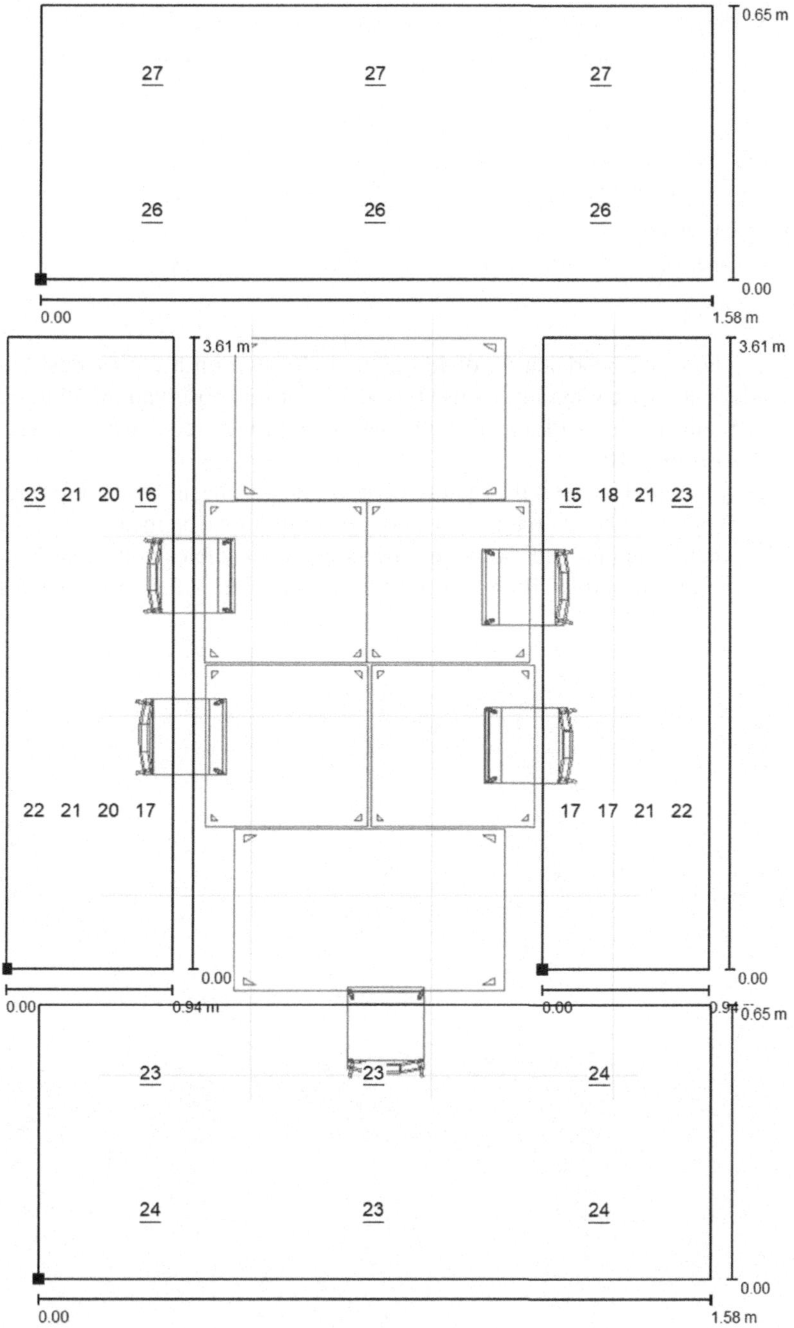


Fig. 4.9 The UGR index in various positions around the reading area

**Table 4.1** Photometric measures in the Marcallo con Casone Library

| | Measured values | | | Values according to EN 12464-1:2011 | |
|---|---|---|---|---|---|
| Area | $E_m$ (lx) | UGR | Uniformity $U_0$ | $E_m$ (lx) | UGR |
| Office desk | 321 | 18 | 0.62 | 300 | 19 |
| Office VDU | 173 | 18 | n.a. | n.a. | 19 |
| Reading room tables | 295 | 15–17 | 0.52 | 500 | 19 |
| Reading room floor | 119 | n.a. | 0.11 | n.a. | n.a. |

shelving. The office workstation of librarians is located on a corner desk on the NNW side, with a view mainly focused to SSW. For this observation point a UGR = 18 has been calculated (Table 4.1). On the same desk there is also a workstation with a VDU (Fig. 4.10).

To verify compliance with the requirements of the European standard for lighting indoor workplaces (EN 12,464-1:2011), horizontal illuminance measurements have been also carried out on the desk in the office area, on the tables in the reading area, on the floor, plus the vertical illuminance on the office VDU. All measurements have

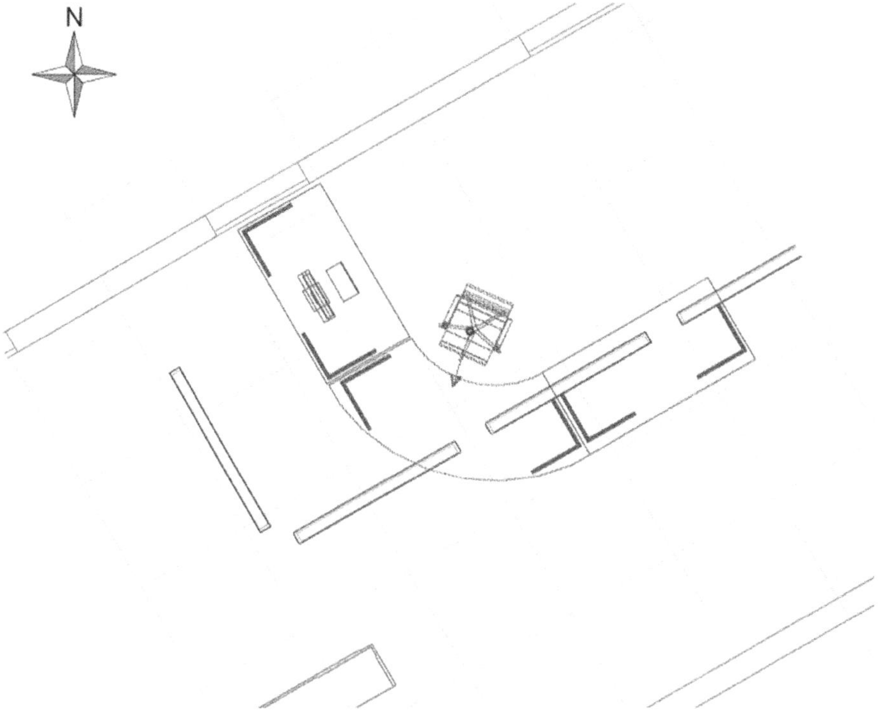


**Fig. 4.10** Marcallo con Casone Library. The office work area of librarians

been made after leaving the luminaires switched on for at least one hour in order to stabilise them.

The measures of average illuminance, uniformity values, calculated UGR values and comparison with the provisions of the European standard for the lighting of workplaces are presented in Table 4.1. The illuminance measurements have been carried out with an LMT Pocket Lux-2 lux meter.

The measurements made with artificial lighting alone have shown that the lighting system complies with the limit criteria set for the UGR in order to guarantee good visual comfort for the planned activities. In particular, on the reading tables the UGR is 4–2 points lower than the limit set by the standard, even if it has been found that the level of illuminance on the reading tables is 41% lower than what is prescribed by the standard for library reading tables. It should be noted that the poor uniformity of the walking surface depends essentially on the presence of the furnishings, especially in areas where the space between chairs and tables is very limited.

### *4.2.3 Discussion on Evaluation of Circadian Lighting*

To measure the circadian light, a mix of artificial and natural lighting has been used to estimate the effect over half a day. In particular, only the librarian's work station has been considered, since the library's employees have a permanent presence in the space. For these workers, the question has arisen of being able to assess whether the illuminance they receive in the eyes is able to influence the circadian system, as naturally occurs outside. The evaluation of the circadian effect has adopted the two methods proposed by Gall and Bieske (2004) for the calculation of the circadian action factor ($a_{cv}$) and by Rea et al. (2010) for the calculation of the circadian stimulus (CS), presented in Chap. 3 of this book. To this end, the study has considered the visual space typically framed by the workers, containing both the worktop, when they perform visual tasks on the desk, and part of the surrounding environment, when they look in front of them (Fig. 4.11).

To determine the $a_{cv}$ factor, a method has been used, developed by Lab. Luce of the Politecnico di Milano, which is based on the use of a TechnoTeam-Rollei LMK 2000 videophotometer for the acquisition of high dynamic images. This videophotometer is in fact a camera equipped with a common low-dynamic charge-coupled device (CCD) sensor. The device has a resolution of 650 × 515 points and acquires a series of five photographs, in rapid succession, using different exposure times; these five low dynamic frames can then be combined to obtain a single high dynamic image (Debevec and Malik 1997). For the calculation of the luminance values, the instrument uses a weighted average of the values for the red, green and blue (RGB) channels, obtained from the colour CCD matrix of the camera. On the CCD matrix, each point consists of a square macro-cell that contains two green, one red and one blue pixels. The CCD sensor, by its nature, is not selectively sensitive to the incident light spectrum. In order for a particular pixel to be sensitive to the red, green or blue band, colour filters are placed in front of the pixels. The green pixels are greater in

**Fig. 4.11** The visual space typically framed by the workers in the library

number because the human eye is more sensitive to the details of the scene in the central band of the visible spectrum. The spectral adaptation to the human photopic efficiency V (λ) is not achieved by means of a filter, but by a weighted average of the values of the three colours, set by the manufacturer based on the SPD of the standard illuminant A defined by the CIE (CIE 015:2004), which corresponds approximately to the SPD of incandescent lamps. This method involves the fact that the measurement of scenes illuminated with sources having different SPDs can lead to making errors, even significant ones; for this reason, the manufacturer provides a correction factor, to be entered in the measurement management and processing program, to take into account the possible different spectral distributions.

During this research, a calibration method has been implemented to estimate the spectral response of the TechnoTeam-Rollei LMK 2000 videophotometer, starting from the knowledge of the input spectral distribution and the response of the CCD sensor (Barnard and Funt 2002; Solli et al. 2005). The procedure makes it possible to take into account also the non-linearity of the camera. In fact, a linearisation method of the response has been applied, concurrently with the spectral sensitivity estimation procedure of each of the three chromatic channels RGB. The final aim has been to obtain a pseudocolour image, with a graduated chromatic scale, able to represent the spatial distribution of the $a_{cv}$ factor within the framed scene. This system obviously provides an approximation, with respect to the measurement of the SPD with a spectrophotometer and the error may be ±15%, depending on the SPD of the lighting present in the room. The measured SPD of the radiance from the fluorescent lamps of the lighting system installed in the library is shown in Fig. 4.12. It has been found that the actual CCT has been 3,824 K with a $R_a = 82$.

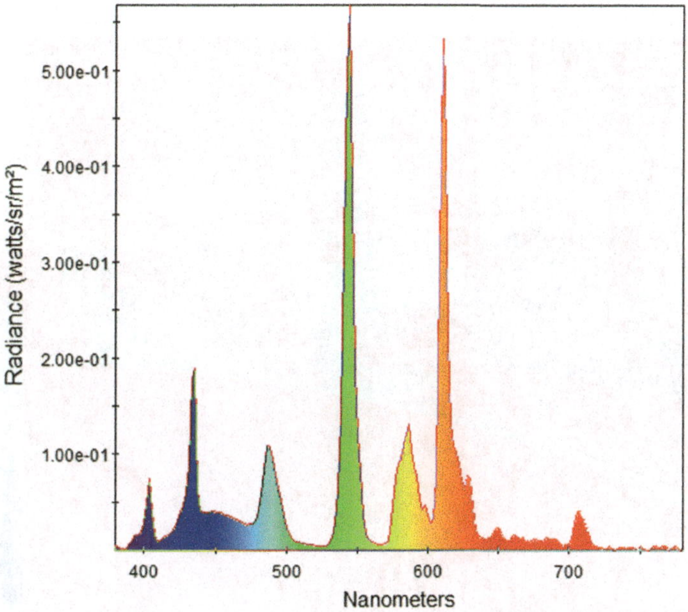


**Fig. 4.12** The measured SPD of the lighting system installed in the library

In the case of the fluorescent lamps, the SPD has been measured with a Spectrascan PR701s spectroradiometer, and an error of ±3% has been estimated with respect to the circadian response curve c(λ) defined by Gall and Bieske (2004). This error is to be considered acceptable, given that the sensitivity curve of the TechnoTeam-Rollei LMK 2000 videophotometer has been determined by the linear combination of the individual channel responses.

To evaluate the $a_{cv}$ factor, the measurements have been conducted from 15:00 to 21:00 on Monday, July 18, 2011, at time intervals of one hour, coinciding with the closure of the library to the public. The weather conditions of the day have been optimal, with an almost clear sky with rare passages of cloudy veils. These climatic conditions, during the summer, are an optimal test bench for the evaluation of natural light, because they configure a context of maximum availability of natural light. Figure 4.13a shows the measured $a_{cv}$ factor, from the typical observation position of the workplace, at 15:00 while Fig. 4.13b shows the $a_{cv}$ factor at 19:00.

The pseudocolour images representing the $a_{cv}$ factor have led to deduce that the values of circadian stimulation, which the light generates in the worker's visual field, would not be able to correctly stimulate the circadian system during the daytime hours of the afternoon, as it happens outside. In fact, remembering that the sensitivity to the circadian system spatially integrates the light stimulation received from the photoreceptors, Fig. 4.13 shows that the values of $a_{cv}$ are generally slightly higher than 0.45. The spatial values of $a_{cv}$ that, in some limited areas of the visual field, are less or greater than the range 0.45–0.7, are due to the presence of less illuminated

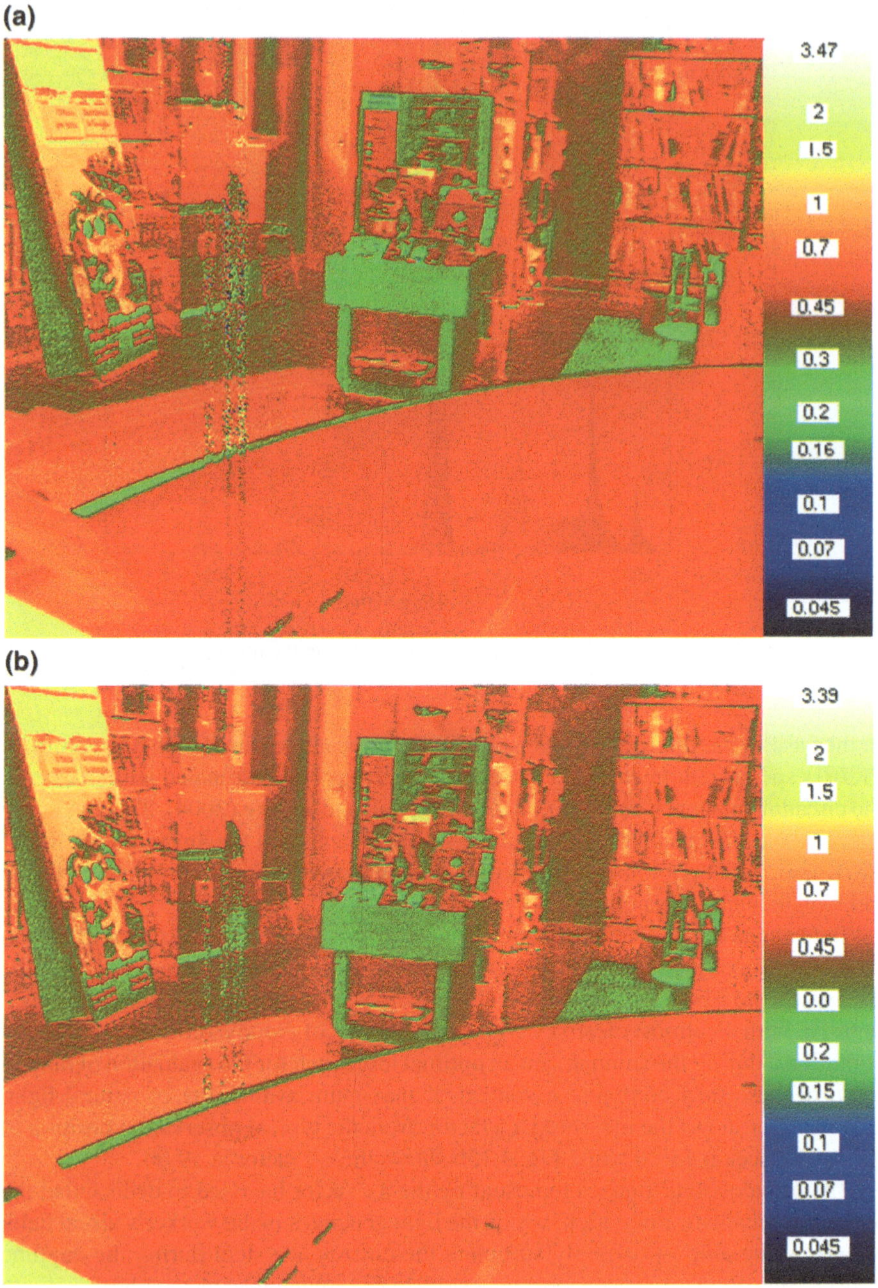


**Fig. 4.13** The measured $a_{cv}$ factor at 15:00 (**a**) and at 19:00 (**b**)

surfaces, which therefore reflect less light towards the eyes of the observer, or to the effect of the spectral reflectance of the surfaces (Bellia et al. 2017).

The limited circadian stimulation in the range of hours considered has two fundamental reasons. The first reason is that despite the library having two sides with windows, these are limited in number and size, with the consequence that the contribution of natural light to the luminance of the visual field of the observer, in the workplace, is very low. The second reason is that the artificial lighting system offers no possibility of controlling the emission of SPD, which has a reduced content of short wavelengths, and is therefore not very effective in stimulating the circadian system.

These evaluations have been also confirmed with a SpectraScan PR701s Spectroradiometer, equipped with a CR 55 adapter for measuring the spectral irradiance $E_{e,\lambda}$, with which the irradiations have been measured on the eyes of the workers, in the library office area. The instrument has also been equipped with appropriate shielding to limit its acceptance angle, so as to reproduce that of human vision. In fact, the SPD shape of the irradiations measured from 15:00 to 21:00, at 1-h intervals, shown in Fig. 4.14, is practically similar to the SPD shape of the measured radiance of the artificial light sources, shown in Fig. 4.12. Also the calculation of the CCT and the $R_a$ derived from the SPD, in Table 4.2, confirms that the illumination comes almost exclusively from the luminaires.

In the measurement of irradiations, very limited variations in the quantity of light have been observed in the afternoon, due to the reduced amount of natural light that enters through the openings. The minimum level of natural light is at 21:00, when the sun is almost set on July 18; while the maximum level is at 19:00, as it has been logical to expect, since at that time the sunlight can enter from the windows on the NNW

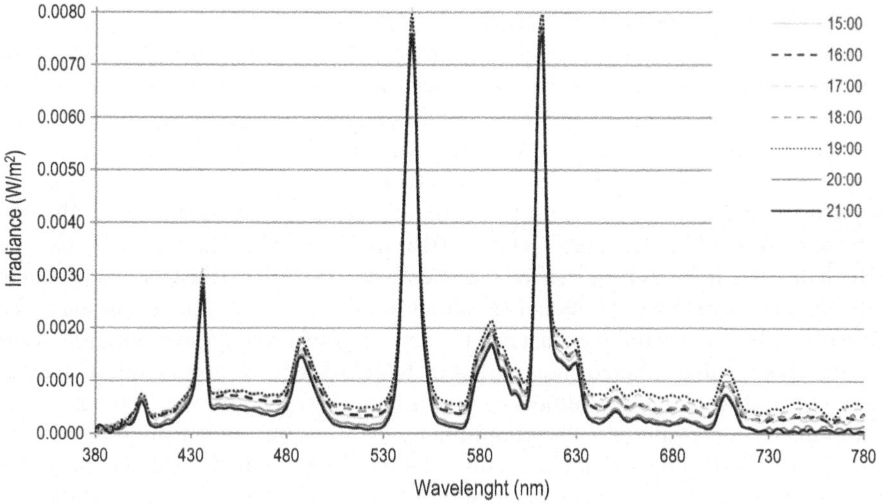


**Fig. 4.14** Spectral radiations measured at eye level of workers from 15:00 to 21:00

**Table 4.2** CCT and $R_a$ values at the eyes of workers from 15:00 to 21:00

| Hour | CCT (K) | $R_a$ |
|---|---|---|
| 15:00 | 3,741 | 86 |
| 16:00 | 3,768 | 86 |
| 17:00 | 3,796 | 87 |
| 18:00 | 3,908 | 87 |
| 19:00 | 3,763 | 88 |
| 20:00 | 3,691 | 84 |
| 21:00 | 3,603 | 83 |

**Table 4.3** CS factor and illuminance values at the eyes of workers

| Hour | CS | $E_{v,eye}$ (lx) |
|---|---|---|
| 15:00 | 0.09 | 72 |
| 16:00 | 0.12 | 77 |
| 17:00 | 0.15 | 85 |
| 18:00 | 0.14 | 79 |
| 19:00 | 0.15 | 86 |
| 20:00 | 0.06 | 66 |
| 21:00 | 0.03 | 64 |

side of the building, which are not equipped with any system limiting the natural light flux. It is also significant to observe the abrupt variation in the contribution of natural light from 19:00 to 20:00, since in this last hour the sun, although still present on the horizon, is masked by the adjacent buildings. Also at 15:00 the contribution of natural light is on average limited and below that of 19:00, since the sun is on the side of the library that has no external openings.

It should be remembered that the $a_{cv}$ factor, not considering the subadditivity of the human circadian system to the various wavelengths of light (Figueiro et al. 2008), overestimates the effect of short wavelengths and underestimates the counter-effect of medium wavelengths in the circadian stimulation (Rea and Figueiro 2016). The circadian effect of light is thus overestimated by the $a_{cv}$ factor. For this reason, we wanted to explore more in depth the issue of circadian stimulation using also the CS factor defined by Rea and Figueiro (2016) in the version available at the date on which this research has been carried out (Rea et al. 2005). Also the calculation of the CS factor has not shown significant variations of the index, which has been calculated from the value of spectral irradiance to the cornea of workers in the office area of the library. The results obtained are reported in Table 4.3, which shows modest changes in the CS index, in the central hours of the afternoon from 16:00 to 19:00, however lower than 0.3, considered as the minimum necessary value, after an exposure of 1 h, to have an effect on the circadian system. These changes in the CS value in the range 0.03–0.15 are mostly due to changes in the amount of light at eye level, rather than to the spectral composition of the incident radiation.

## 4.3 Evaluation of Natural Light in a Residence

As part of a research aimed at exploring in depth the topics of lighting design in the sustainable home of the future, in 2016 ENEA instructed Laboratorio Luce of the Politecnico di Milano to investigate the question of whether the amount of natural light present within a home is able to positively impact the human circadian system. To this end, a very common model of living space has been selected: a classic three-room flat located in a block of flats of a large city. The plan of the flat chosen for the study of circadian stimulation due to natural light in a residential interior is shown in Fig. 4.15.

In order to evaluate the circadian stimulation, in a first phase the natural lighting conditions have been simulated using a three-dimensional model, created with a lighting CAD system. This also in order to try to develop a working methodology useful for the design. In a second phase, radiometric measurements have been performed to determine the circadian stimulation in two different periods of the year.

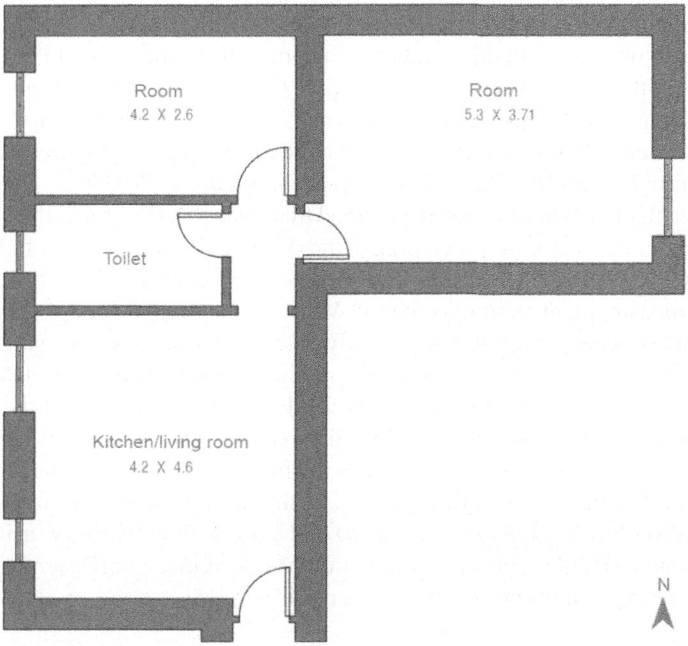


**Fig. 4.15** The plan of the flat chosen for the study

### *4.3.1 Simulations of Natural Light for the Assessment of Circadian Light*

The simulation model of natural light has been applied to a real-world case study, choosing a residential flat in which different lighting scenarios have been tested using the Dialux Evo lighting CAD software. The geometric survey has been done on site and entered in the CAD. It is a three-room flat of 60 $m^2$ configured to accommodate one to four people, located on the third floor of a residential building with double exposure to the east and west, in the municipality of Milan, at a latitude of 45.5°. From the urban standpoint, the flat has a very favourable position, as it does not suffer occlusions from other buildings. It faces a large internal courtyard westwards and a park/garden eastwards. The flat has been geometrically modelled to obtain a virtual environment identical to the real one, in order to have a corresponding place on which to carry out the radiometric measurements. The simulation of natural light has been carried out over a year, taking into account four dates corresponding to the astronomical beginning of the seasons: the spring equinox, the summer solstice, the autumn equinox and the winter solstice. All simulations have been performed during the daytime hours with hourly intervals (Table 4.4).

In order to proceed with the simulation of natural light, the topic of the variability of the light of the sky has been studied in depth. This depends on the latitude, the time and the weather conditions. At sunset, in positions close to the sun, the CCT value may be less than 3,000 K, a partially overcast sky may have a colour temperature of 5,000 K or higher, while a clear sky can reach values above 20,000 K. The research by Chain et al. (1999) has proposed a method that relates the CCT and the luminance distribution, in which, for the processing of the sky model, a series of spectral radiance measurements of the celestial vault have been carried out from April to September 1999 in a measurement station located in Valux-en-Velin near Lion, France. Based on these measurements, a link has been proposed between these two parameters: a high luminance value corresponding to a moderate value of the correlated colour temperature and vice versa. Other models for determining the CCT of the sky have been developed in the computer graphics field (Hosek and Wilkie 2012, 2013), for the photo-realistic representation of scenes in outdoor environments during daytime hours. Applying these calculation models, at the latitude of the flat in question, a sky simulation has been made on June 20 at 12:00, with a clear sky, using a high dynamic image (HDR). The same simulation has been made on December 21 with two different sky conditions: clear and overcast.

**Table 4.4** Days and times selected for the simulation of natural light

| Date | Season | Start time | End time |
|---|---|---|---|
| March 20th | Spring | 7:00 | 18:00 |
| June 21st | Summer | 6:00 | 21:00 |
| September 22nd | Autumn | 8:00 | 19:00 |
| December 21st | Winter | 9:00 | 16:00 |

**Table 4.5** Sky conditions assessed with and without curtain

| Simulations without curtain | | | Simulations with curtain | |
|---|---|---|---|---|
| Clear | Intermediate | Overcast | Clear | Intermediate |

Based on the studies performed by Hernández-Andrés et al. (2001), Andersen et al. (2012), Mardaljevic et al. (2014) the study then moved on to the CCT estimate in a day with different sky conditions. In particular, by applying some simplifications, it has been determined that an intermediate sky corresponds to an illuminant D55 (5,500 K), an overcast sky with diffused radiation corresponds to an illuminant D65 (6,500 K) and a clear sky corresponds to an illuminant D75 (7,500 K). This distinction is fundamental in determining the influence of natural light on the circadian system since, for example, the illuminant D75, containing a greater quantity of short wavelengths, has greater circadian efficacy than D65 and D55.

A further study has been carried out considering also the presence of a light diffusing curtain on the French window, in the only conditions of natural lighting with a clear and intermediate sky, as it happens in the real environment, in order to shield direct light that is potentially dazzling for users (Table 4.5). The simulation of the curtain within the calculation program has been carried out by replacing the window and the related distribution of the luminances, determined in turn by the spatial distribution of the luminances of the celestial vault, the geographical location, the time and the geometric characteristics of the installation, with a luminaire geometrically extended with dimensions equal to those of the window (Kobav and Bizjak 2005) and providing an adequate photometric distribution (Spitzglas 1984). For the calculation of the natural lighting mediated by the interposition of the curtain, an approximation has been made, having set a transmission factor of the fabric equal to 0.3 and having considered the behaviour of natural light that filters through the fibres, as a diffuser in Lambertian mode, i.e. without components concentrated in any direction (Ng et al. 2001).

## *4.3.2 Statistics of Meteorological Data*

In order to verify the impact of natural light on the NIF effects of light in the various periods of the year, the frequency of the different sky conditions has been evaluated, using meteorological statistical data referring to the period between 1996 and 2000. These data are derived from the images of the Meteosat satellite and makes it possible to know the solar radiation and the total light availability actually present in Europe in the past (Satel-Light 2018). Based on these data, it has been possible to determine the average percentage during the year in Milan, Italy, where the flat is located. The seasonal percentage, total percentage, number of days and conditions of the sky in Milan over a year are summarized in Table 4.6.

The data have made possible to determine that for 191 days (52% average) the sky is clear, for 99 days (27% average) the sky is intermediate and for 75 days (21%

**Table 4.6** Sky conditions in Milan over a year

| | | Days | Days % | Clear % | Clear days | Intermediate % | Intermediate days | Overcast % | Overcast days |
|---|---|---|---|---|---|---|---|---|---|
| Winter | Dec | 11 | 35 | 39 | 4.29 | 33 | 3.63 | 28 | 3.08 |
| | Jan | 31 | 100 | 44 | 13.64 | 30 | 9.3 | 26 | 8.06 |
| | Feb | 28 | 100 | 62 | 17.36 | 24 | 6.72 | 14 | 3.92 |
| | Mar | 19 | 61 | 62 | 11.78 | 21 | 3.99 | 17 | 3.23 |
| | *Tot* | *89* | | *52.9* | *47* | *26.6* | *24* | *20.6* | *18* |
| Spring | Mar | 12 | 38.7 | 62 | 7.44 | 21 | 2.52 | 17 | 2.04 |
| | Apr | 30 | 100 | 47 | 14.1 | 26 | 7.8 | 27 | 8.1 |
| | May | 31 | 100 | 50 | 15.5 | 31 | 9.61 | 19 | 5.89 |
| | Jun | 20 | 66.7 | 57 | 11.4 | 27 | 5.4 | 16 | 3.2 |
| | *Tot* | *93* | | *52.1* | *48.44* | *27.2* | *25.33* | *20.7* | *19.23* |
| Summer | Jun | 10 | 33.4 | 57 | 5.7 | 27 | 2.7 | 16 | 1.6 |
| | Jul | 31 | 100 | 63 | 19.53 | 24 | 7.44 | 13 | 4.03 |
| | Aug | 30 | 100 | 63 | 18.9 | 25 | 7.5 | 12 | 3.6 |
| | Sep | 21 | 70 | 58 | 12.18 | 26 | 5.46 | 16 | 3.36 |
| | *Tot* | *92* | | *61.2* | *56.31* | *25.1* | *23.1* | *13.7* | *12.59* |
| Autumn | Sep | 10 | 33.4 | 58 | 5.8 | 26 | 2.6 | 16 | 1.6 |
| | Oct | 31 | 100 | 42 | 13.02 | 30 | 9.3 | 28 | 8.68 |
| | Nov | 30 | 100 | 42 | 12.6 | 27 | 8.1 | 31 | 9.3 |
| | Dec | 20 | 64.5 | 39 | 7.8 | 33 | 6.6 | 28 | 5.6 |
| | *Tot* | *91* | | *43.1* | *39.22* | *29.2* | *26.6* | *27.7* | *25.18* |

average) the sky is overcast. This analysis suggests that, for about a fifth of the year and mainly in autumn and winter, the sky is overcast. Therefore, for those periods of the year we can estimate a decrease in the amount of light within the flat having, as a consequence, a decrease on the NIF effects of natural light.

For the flat considered, the following spaces have been assessed: the living room/kitchen and the two rooms, one used as a study and the other as a bedroom, considered as the most important spaces as users spend most of their time there. On the other hand, the areas not being considered are the bathroom, since it is a place with a limited access and the transit areas such as the hallway, for the reasons explained above and because of their modest size in the specific housing sample. The main furnishings in the rooms have been inserted in order to correctly position the viewpoints of the observer for the different activities considered and plausible in a residential context.

### *4.3.3 The 3D Model of the Flat and the Observer Positions*

With reference to the interior of the flat, it has been decided to use a virtual simulation model with materials, textures and colours different from those of the real physical environment. An achromatic grey material has been therefore used in which only the reflection factors change. These have been chosen from the average of the ranges proposed by the standard EN 12464-1:2011: for walls (0.5), ceiling (0.7) and floor (0.2). For furnishings, the study has considered an achromatic grey with a reflection coefficient between 0.5 and 0.7. The use of grey achromatic materials is a convenient simplification compared to studies in which materials with different colours have been used and more complex approximations of the spectral distribution of light calculated on nine bands of the spectrum (Wandachowicz 2006).

After having created the 3D model of the flat (Fig. 4.16), attention has been paid to the definition of the use of its spaces.

Assuming that people who carry out a working activity are generally at home for a limited percentage of their daytime in a year, we have focused on people who are more likely to stay at home for many diurnal hours, i.e. the elderly. In order to evaluate the effects of lighting on the individual in terms of adjustment of the circadian system, it has been therefore necessary to focus on the visual system of an aging population, considering as elderly users a heterogeneous group of subjects who are over 60. It has been taken into account this age limit because it does not necessarily reflect a change in the social activities and roles of the individual, such as retirement, but because it relates to the physiological changes in the human visual system. Indeed, at the age of 60 a series of physical changes occur that affect vision with a number of problems such as presbyopia and the possibility of lens problems. It is an established fact that the human lens tends to turn yellow with age, considerably modifying the total spectral transmission of visible light (Barker and Brainard 1991; Artigas et al. 2012). Although the lens of adults and older people shows great variability in spectral

**Table 4.7** Definition of the activities of the observer model

| Room | Function | Activities | Time | Duration | Posture | Height of the eyes m | Look | Place |
|---|---|---|---|---|---|---|---|---|
| Bed room | Rest and relax | Watching TV | 21:00 | 3 h | Sitting | 1.20 | Front | 1.3 |
| | | Reading a book | 21:00 | 3 h | Sitting | 1.20 | Inclined 30° | 5.6 |
| | | Reading from a tablet | 21:00 | 3 h | Sitting | 1.20 | Inclined 30° | 5.6 |
| | | Sleeping | 23:00 | 8 h | Laying down | 0.7 | Upward | 2.4 |
| | | Falling asleep | 23:00 | 10′ | Laying down | 0.7 | Upward | 2.4 |
| | | Waking up | 07:00 | 10′ | Laying down | 0.7 | Upward | 2.4 |
| Living room/kitchen | Food preparation | Cooking | 7:00<br>12:00<br>19:00 | 10′<br>20′<br>20′ | Standing | 1.70 | Inclined 30° | 17 |
| | | Cutting Cleaning Handling | 7:00<br>12:00<br>19:00 | 10′<br>20′<br>20′ | Standing | 1.70 | Inclined 30° | 12.18 |
| | | Washing | 7:00<br>12:00<br>19:00 | 10′<br>10′<br>10′ | Standing | 1.70 | Inclined 30° | 12 |
| | Food consumption | Eating | 7:00<br>12:00<br>19:00 | 20′<br>30′<br>30′ | Sitting | 1.20 | Front, inclined 30° | 1-4<br>11<br>13-15 |
| | Working | Manipulating objects | 9:00<br>13:00 | 3 h<br>6 h | Sitting | 1.20 | Inclined 30° | 11<br>13-15 |
| | Rest and relax | Reading a book | 9:00<br>13:00 | 3 h<br>6 h | Sitting | 1.20 | Inclined 30° | 9,10<br>16 |
| | | conversing | 9:00<br>13:00 | 3 h<br>6 h | Sitting | 1.20 | Front | 5-7 |
| Study | Studying working | Reading from a monitor | 9:00<br>13:00 | 3 h<br>6 h | Sitting | 1.20 | Front | 1 |
| | | Manipulating objects | 9:00<br>13:00 | 3 h<br>6 h | Sitting | 1.20 | Inclined 30° | 4 |
| | Reading | Reading a book | 9:00<br>13:00 | 3 h<br>6 h | Sitting | 1.20 | Front | 2.3 |
| | | Reading a book | 9:00<br>13:00 | 3 h<br>6 h | Sitting | 1.20 | Inclined 30° | 5.6 |

**Fig. 4.16** The top view of the 3D model of the flat subject of the research

transmissions, it filters the UV and short wavelengths of the spectrum, and it causes a noticeable decrease in visible light, especially after the age of 70.

For the three rooms taken into consideration by the research, the prevailing points of view have been determined according to the presumed functions and activities within the various spaces. Consequently, the preferential positions of head, height and gaze of the user have been deducted and consequently the points of view in which to calculate the vertical illuminance on the eye (Table 4.7). In the bedroom specifically, the points of view of two observers on the bed have been analysed from various observation positions in order to consider different activities such as watching TV, reading a book, resting when falling asleep, sleep and awakening. Also in the living room/kitchen the points of view of different observers have been considered in various situations. The activities have been divided between static activities on the table, sofa, armchair and dynamic ones on the kitchen worktops. The various observation positions have been chosen in order to consider different activities such as food preparation with washing, handling and cooking, food consumption, reading a book, work and concentration activities at the table and even relaxation and conversation activities while seated. In the study room the viewpoints of observers have been analysed in two different situations, at the table and on the sofa in various observation positions, to consider actions such as reading a book, work activities and concentration at the table. Figure 4.17 shows a schematic representation of the three rooms with the observation positions of the activities defined in Table 4.7.

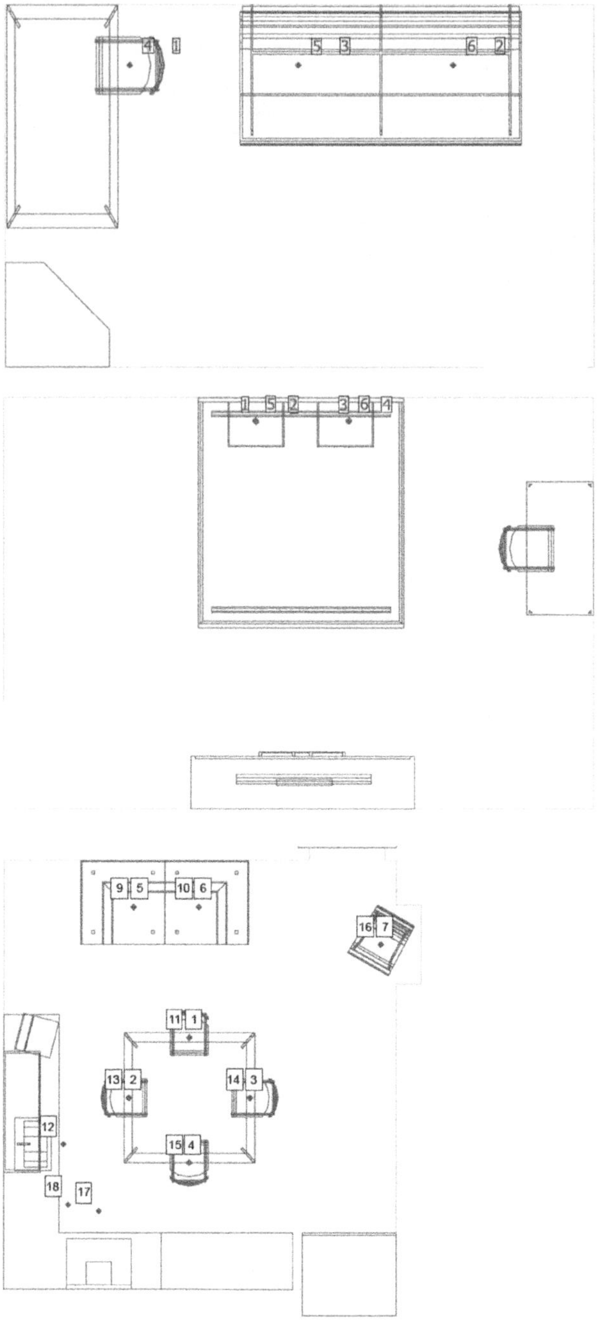


**Fig. 4.17** Schematic representation of the three rooms with the observation positions

### 4.3.4 The Time Factor and the Activities

Since the circadian cycle is defined according to the time factor, the latter has also been considered in relation to the flat and its typical occupants: the elderly. The day has been divided into three distinct periods during which there are different NIF lighting effects:

- Morning (6:00–10:00). It is the period when illuminance derived from natural light should synchronise the circadian biological clock and activate concentration and productivity in most people.
- Mid-morning and afternoon (10:00–18:00). In this period, high levels of lighting can help increase alertness and concentration.
- Late afternoon and night (18.00–6:00). In these phases of the day, exposure to circadian light should be avoided, as it may decrease melatonin production and keep the cortisol high. As a consequence, there would be an alteration of the normal human circadian cycle, i.e. the desynchronisation of the organism with respect to the solar day.

Based on these time periods, a full-day scheme has been created that considers the three moments in which exposure to light takes place and in which NIF effects are most desired with respect to the synchronisation of the circadian system. These moments are localised in different rooms of the flat in relation to the activities most carried out within the same room, defined by the rhythms of sleep-wake and breakfast, lunch and dinner, which contribute to create the daily rhythm (Fig. 4.18). This scheme is useful for defining the place where, most likely, users will find themselves during the day, in relation to the amount of light required for the circadian system at that time (Mardaljevic et al. 2014). Obviously, this scheme depends on the type of activity that must be carried out and has been structured according to the lifestyle of an

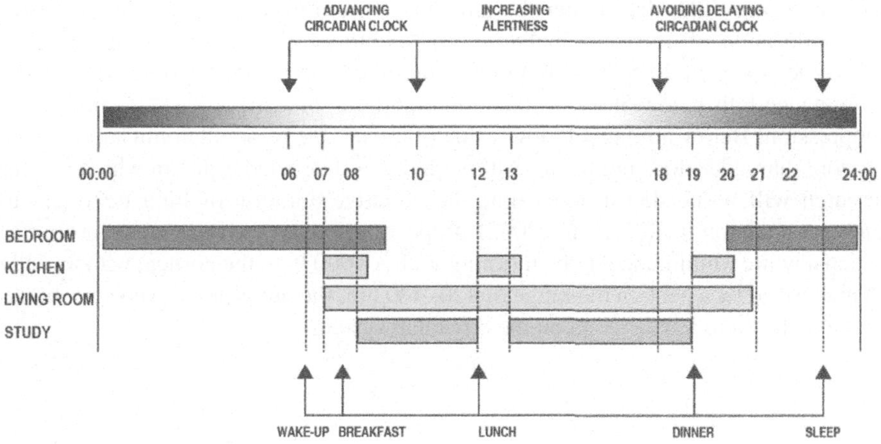


**Fig. 4.18** Plausible exposure to light in the three rooms of the flat

elderly person, who often has a reduced activity in external environments and a high permanence in the home environment.

From an analysis of the interviews on the behavioural habits of some elderly subjects, considering the activities of the people, we can outline for example the following sequence of activities over a typical day:

- Waking up in the morning in the darkness of the bedroom, with blinds generally closed.
- Subsequent awakening phase, more or less prolonged, in a sitting position on the bed, to reactivate the circulation of the limbs.
- First passage in the bathroom.
- Activities in the kitchen where breakfast is prepared and consumed on the table.
- Personal hygiene in the bathroom, generally for a short period of time.
- Dressing in the bedroom, often using the bed to sit during this activity.
- The following hours are distributed in the other rooms of the house and depend on functional activities based on the abilities and limitations of the elderly person, for example, the preparation and consumption of meals, the household chores, relaxation activities, such as reading a book, watching TV and engaging in hobbies like sewing.
- At the end of the day, after the preparation and consumption of dinner, the activities take place mainly in the living room, or in the bedroom, where most of the subjects spend their time watching TV or reading a book.

Also with reference to the time factor, another element to take into consideration is the duration of exposure to light in the various phases of the day. In fact, in order to have an effect on the circadian system, it is necessary to have a longer exposure to light than that required for vision. The human circadian system works at a much slower pace, mainly because it is based on the secretion of hormones in the blood, such as melatonin and cortisol, and less on neurological processes to synchronise the various organs of the body. It has also been known for some time (Mclntyre et al. 1989) that, with the same SPD, the greater the amount of illuminance on the eyes the faster the suppression of nocturnal melatonin in the blood. 25% melatonin suppression can be achieved in 20 min with an eye illuminance of 1,000 lx; otherwise, if the illuminance falls below 500 lx, it takes about 60 min to achieve the same melatonin suppression. Below a relatively low value of about 200 lx, a melatonin suppression of more than 25% does not occur. Although this is a debated topic, in which further research will be needed to determine the effective duration of light treatment to suppress melatonin, Rea et al. (2002) propose a first approximation according to which a white illuminance in the morning with ~1,000 lx to the cornea, with an SPD characterised by a peak in the range of 420–480 nm, the duration of exposure should be about 1–2 h to have effects on the circadian system.

### 4.3.5 The Circadian Photometry Model Used

The circadian photometry proposal of Rea et al. (2012), Rea and Figueiro (2013) has been used to evaluate the circadian stimulation of natural light. Based on this method, it has been possible to determine at what photopic illuminance value a source is equivalent, regarding the NIF effects of the light, to another with a different SPD. Considering the standard illuminants D55, D65 and D75, as appropriate representations of natural light due to the intermediate, overcast and clear sky respectively, the illuminance levels corresponding to the minimum threshold and the maximum value (Andersen et al. 2012) of circadian effect have been determined for the SPD of these three types of sky (Fig. 4.2).

Taking into account the known uncertainties, the evaluation of the NIF effects of light has used a simple ramp function as a reasonable model to represent the probability that the vertical illuminance at one point, extended for a sufficient time and according to a determined direction of observation, is sufficient to influence the circadian system. For each SPD this can be done with a low probability (0%) at the lower stimulation limit, and with a high probability (100%) at the upper stimulation limit, with a linear interpolation between these two values. The illuminance value at which the minimal stimulation of the circadian system is obtained has been determined for $CS = 0.29$, while the illuminance value at which the maximum stimulation or saturation of the circadian system is obtained has been determined in correspondence with $CS = 0.56$. Also the average aging of the lens has been taken into consideration for people over 70 (Artigas et al. 2012). In this way, it has been possible to define two potential medium circadian observatories: those included in an age group between 23 and 69 and those aged 70 and over. The limit values used in the evaluation of natural light in the home, for an elderly observer, are 372–1,584 lx with an intermediate sky, 298–1,303 lx with an overcast sky and 257–1,138 lx with a clear sky. For a user in the younger age group, 23–69, the illuminance values necessary for the circadian stimulation are significantly lower because in these subjects a greater amount of blue light reaches the retina. Figure 4.19 shows the minimum and maximum levels of circadian effect depending on the amount of illuminance in the eyes, SPD and age: 23–69, or $\geq 70$.

### 4.3.6 The Lighting Obtained with the CAD

Once all the parameters, defined in the previous paragraphs, have been set, simulations have been carried out during the day, for four days, in the four seasons, as reported in Table 4.4, with respect to the three possible sky conditions: overcast, clear and intermediate, as listed in Table 4.5.

With respect to the case study considered, the *overcast condition* occurs for a significant portion of the year, about 21%, and mainly in autumn and winter. The natural light simulations carried out highlight the limited amount of illuminance to

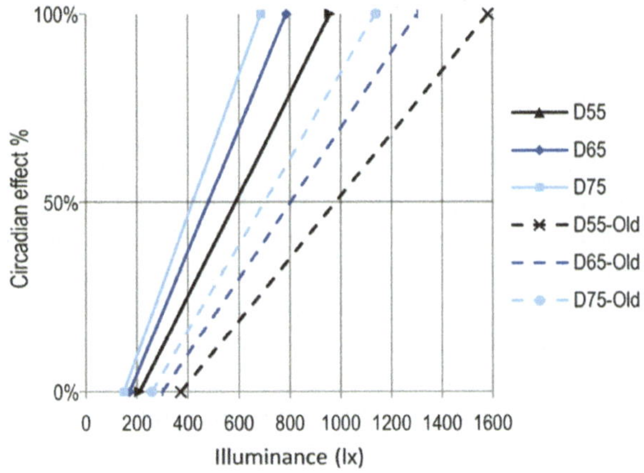


**Fig. 4.19** Minimum and maximum levels of circadian effect. The minimum level of circadian effect is equivalent to CS = 0.29 and the maximum level to CS = 0.56

the cornea, which is not sufficient to activate the adjustment of the circadian system, especially in winter and autumn.

Figure 4.20 illustrates the lighting levels during the day for the *bedroom*, in comparison with the minimum illuminance levels required for the circadian stimulation of young and old users, in overcast sky conditions. The grey vertical bands indicate times when there is a low probability of using this room. In the bedroom, analysing the results obtained, starting from the morning awakening and considering the period between 6:00 and 10:00, the level of illuminance that reached the eye in all seasons is much lower than the threshold for circadian activation, both for a young observer, with a threshold of 171 lx on the eyes, and for an elderly observer, with a threshold of 298 lx. Despite eastern exposure of the room, there is not a good amount of light in the morning under overcast conditions to stimulate the biological clock of individuals in the awakening phase. The subjects have been considered in a relaxed position (p. 2 and 4) and sitting on the bed (p. 1 and 3), assuming the different awakening phases.

The lighting levels during the day for the *kitchen/living room*, in comparison with the minimum illuminance levels required for the circadian stimulation of young and old users, in overcast sky conditions, are shown in Fig. 4.21. The grey vertical bands indicate times when there is a low probability of using this room. Analysing the period between 10:00 and 18:00, it has been assumed the position of the user in the study or the kitchen/living room in a continuous way, spending time in one or the other to carry out different activities. In these two rooms, high levels of eye illuminance are needed in the early morning to stimulate the awakening and timing of the circadian cycle, but also in the early afternoon to increase alertness, concentration and avoid post-prandial sleep. In the 10:00–18:00 time frame, given the exposure to the west, the kitchen/living room in the four seasons is not particularly bright and activating at the circadian level for those who are in the following situations: sitting in the armchair, sitting on the sofa and sitting at the table, as shown in Fig. 4.21. The only observation position that is fully activating, but only during the summer, between

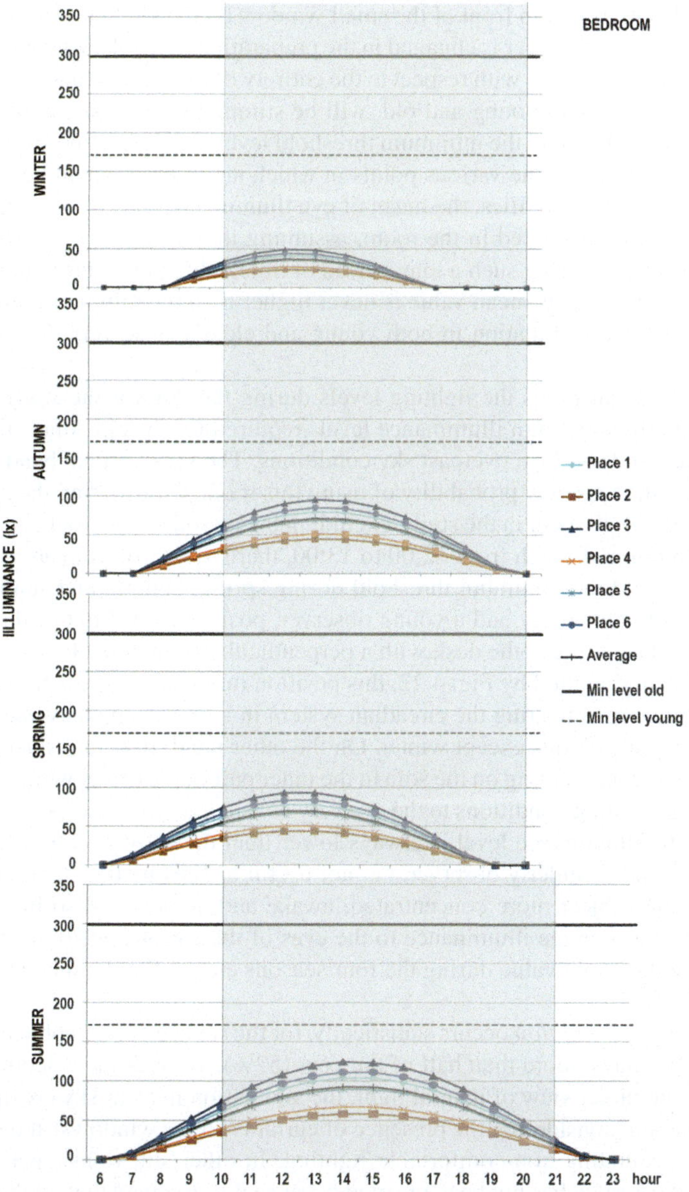


**Fig. 4.20** Lighting levels in the bedroom in overcast sky conditions

11:30 and 17:00, is that in front of the small window on the kitchen worktop (p. 18). In this situation an observer is engaged in the preparation of meals, which in any case is a limited period of time with respect to the entirety of the day. Actually, only in this position, the user, both young and old, will be stimulated because the illuminance on the eyes is higher than the minimum threshold level. Analysing the kitchen/living room with reference to the various points in which an observer could be positioned to perform different activities, the mean of eye illuminances has been calculated for all the positions considered in the room, assuming that a subject tends to move in different positions during such a long period of time, which includes both static and dynamic activities. This mean value is never higher than the minimum threshold to obtain a circadian stimulation in both young and elderly observers throughout the year.

Figure 4.22 illustrates the lighting levels during the day for the *study*, in comparison with the minimum illuminance levels required for the circadian stimulation of young and old users, in overcast sky conditions. The grey vertical bands indicate times when there is a low probability of using this room. Considering the same time frame 10:00–18:00 spent in the study, excluding the period related to the preparation and consumption of lunch from 12:00 to 13:00, there is clearly a circadian stimulation well beyond the minimum threshold during spring, autumn and summer. This applies both to an elderly and a young observer, positioned in front of the window (p. 1 and 4), i.e. sitting at the desk with a perpendicular or inclined look towards the worktop. As highlighted by Fig. 4.22, this position turns out to be the best one from the perspective of activating the circadian system in a time frame between 9:30 and 15:30 during all seasons except winter. On the other hand, if we consider the other positions of the user, sitting on the sofa in the inner part of the room, while reading or relaxing, the lighting conditions to the eye vary considerably with respect to the desk position. The illuminance level is always lower than the activation threshold, both for a young and an elderly observer, i.e. not useful to activate the circadian system and make the subject more concentrated, awake and attentive. Also in this room, considering the average illuminance to the eyes of the various positions, this never exceeds the threshold value during the four seasons except for summer and only for the young user.

The *clear sky condition* occurs statistically, for the selected geographical position, for about 191 days, more than half of the year (52%). To limit the discomfort glare caused by the direct view of natural light, the simulation in clear sky conditions has been carried out considering the presence of curtains on the windows, at times when the sun is generating non-comfortable lighting. In other situations, such as in the bedroom, the curtain has not been inserted because it is assumed that, in that specific time, the observers are in another room and engaged in a different activity than sitting or lying in bed.

Figure 4.23 shows the lighting levels during the day for the *bedroom*, in comparison with the minimum illuminance levels required for the circadian stimulation of young and old users, in clear sky conditions. The grey vertical bands indicate times when there is a low probability of using this room. Analysing the results obtained in the bedroom, starting from the morning awakening, between 6:00 and 10:00, the

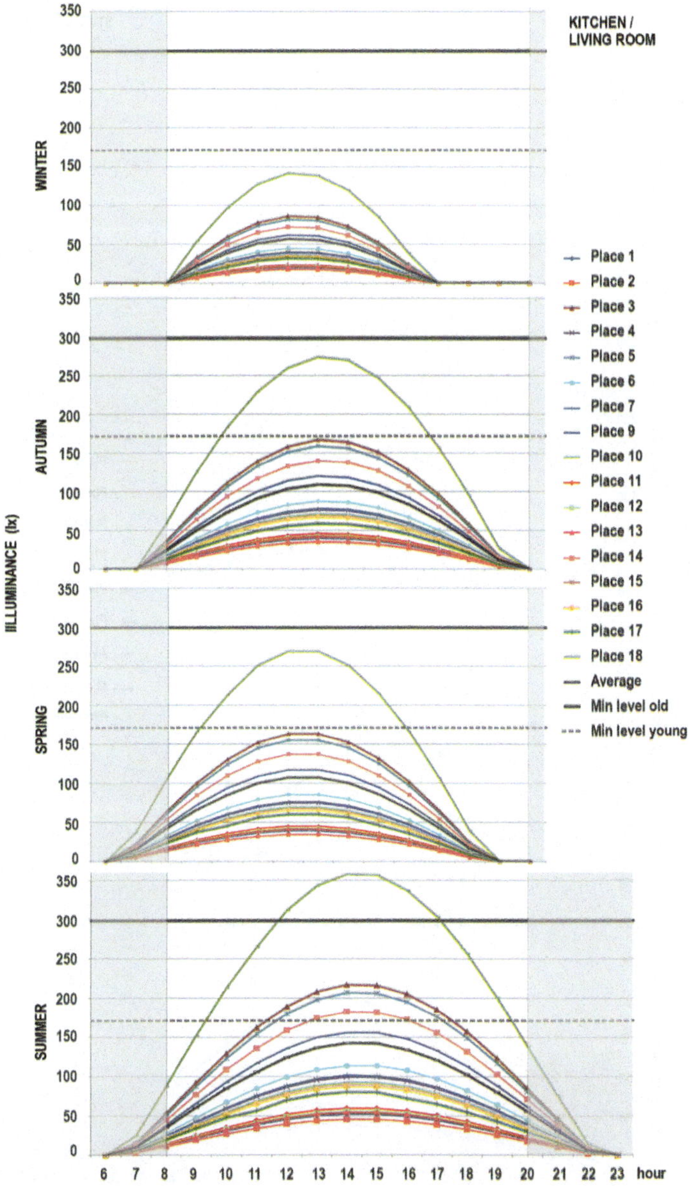


**Fig. 4.21** Lighting levels in the kitchen/living room in overcast sky conditions

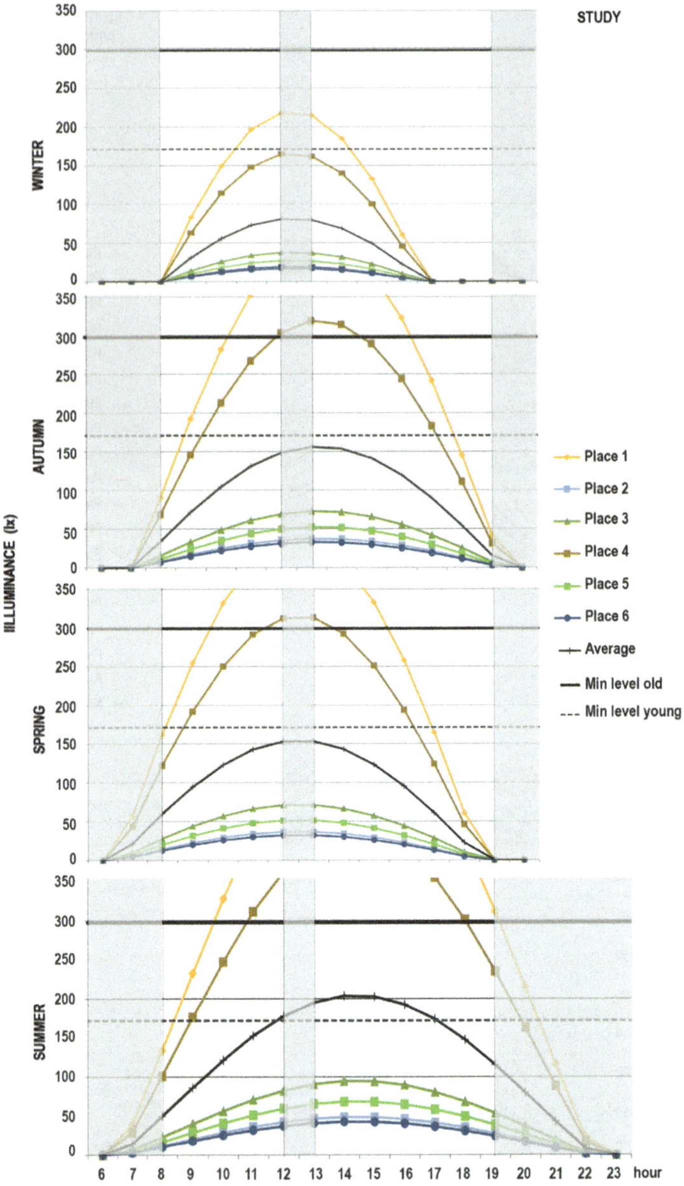


**Fig. 4.22** Lighting levels in the study in overcast sky conditions

level of illuminance reaching the eyes is fully able to satisfy the circadian stimulation in spring and autumn, both for a young and a senior user, with illuminance values that reach the maximum level of circadian stimulation. In particular, there is a high level of illuminance reaching the eyes from 8:00 to 10:00 both sitting (p. 1 and 3) and laying down on the bed (p. 2 and 4). In winter, however, the illuminance level reaching the eyes does not get to the minimum levels necessary for circadian stimulation except for a limited period of time, from 9:00 to 10:00, for sitting and lying observers. In winter, an awakening before this time frame does not receive enough light to activate the circadian system. Finally, in the summer season the minimum threshold level is activated from 8:00 to 10:00 only for the sitting position of the observers, while it is not sufficient for the laying down position.

The lighting levels during the day for the *kitchen/living room*, in comparison with the minimum illuminance levels required for the circadian stimulation of young and old users, in clear sky conditions, are illustrated in Fig. 4.24. The grey vertical bands indicate times when there is a low probability of using this room. The green vertical bands indicate times in which the use of a curtain has been considered. An in-depth analysis has been made for the time frame between 10:00 and 18:00 spent in the *kitchen/living room*. It has been assumed that the user may use the room in different observation points, engaged in various activities that lead him/her to move continuously between different areas. For this reason, the research has considered both the illuminance results obtained with respect to the individual observation positions and those that take into account the average illuminance, calculated with respect to all the observation points. In clear sky conditions, given the western exposure, during the morning, i.e. in the time frame 10:00–13:00, natural light is generally ineffective with regard to circadian stimulation for subjects positioned in the following places: sitting in an armchair (p. 7 and 16), sitting on the sofa (p. 5, 6, 9 and 10) and sitting at the table (p. 1–4, 11 and 13–15). As we can see in Fig. 4.24, similar to the overcast sky condition, the only position of an elderly observer where there is a minimum level of circadian stimulation in the morning (during spring, summer and autumn), is standing in front of the kitchen worktop looking towards the window. However, this observation point is assumed to be held too briefly, generally less than 30 min, to effectively activate the circadian system. In the winter season no circadian stimulation is recorded at any time of the day, taking into account an average of all the potentially assumed positions, both by an elderly and a young observer. Even in clear sky conditions, therefore, the winter season determines lighting conditions inside the house that do not trigger the circadian synchronisation. During the early and late afternoon, the light conditions of the room improve, encouraging circadian stimulation in several positions taken by an elderly observer and especially considering the average of the positions potentially taken in the room. In particular, during the autumn season, there is a circadian stimulation of the observer positioned in front of the kitchen worktop and of those sitting at the table looking towards the larger window (p. 3 and 14) in the time frame 15:00–18:00. If we consider the average of the possible positions of an elderly observer, the circadian activation period is very small, but still reasonably sufficient to determine the circadian stimulation, in the 16:00–17:30 time frame. Similar results are also obtained in spring, when there is

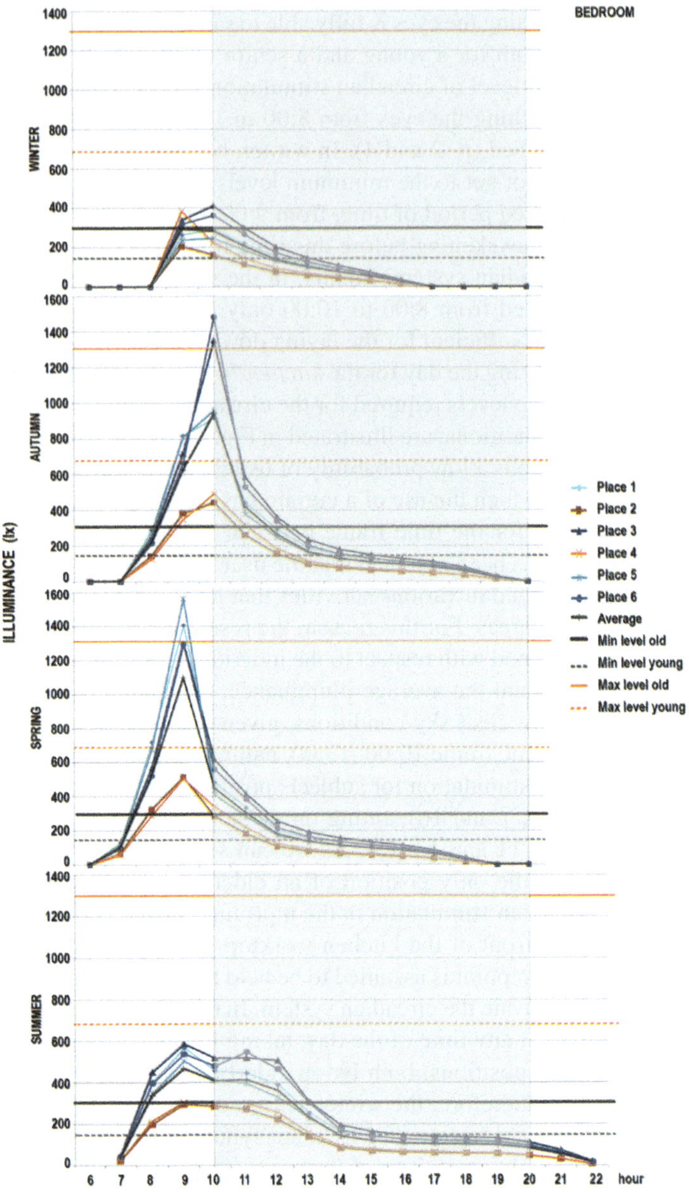


**Fig. 4.23** Lighting levels in the bedroom in clear sky conditions

an increase of both the times of possible exposure to circadian stimulation and the positions of the observer that make this possible. In addition to the positions of the observer sitting at the table and at the front of the kitchen worktop, the activation also extends to the sofa and the armchair furthest from the windows. The period of possible stimulation expands from 15:30 to 18:30 and the amount of light generates a circadian stimulation up to and beyond the maximum levels of sensitivity, but for limited time intervals. This trend increases during the summer, when the natural lighting that enters the room facilitates the circadian stimulation in most of the observation points facing the windows. From 16:30 to 20:00, being inside the kitchen/living room in various considered points, as can be seen from the average, determines the circadian stimulation even for an elderly observer.

Figure 4.25 shows the lighting levels during the day for the *study*, in comparison with the minimum illuminance levels required for the circadian stimulation of young and old users, in clear sky conditions. The grey vertical bands indicate times when there is a low probability of using this room. The green vertical bands indicate times in which the use of a curtain has been considered. As for the time frames 10:00–12.00 and 13.00–18:00 spent in the study, there is a strong circadian stimulation in conditions of clear sky during spring and autumn in the time frames 10:00–12:00 and 15:00–18:00, and in summer in the time frames 10:00–12:00 and 13:00–18:00. As we can see in Fig. 4.25, this is true both for an elderly and a young observer, but only for position in front of the window (p. 1 and 4). Indeed, this position is the best one from the standpoint of the activation of the circadian system. During winter, the circadian activation period in this position decreases to one hour in the morning (11:00–12:00) and half an hour after lunch (13:00–13:30). On the other hand, for all other observer positions in the study during all seasons, with the exception of a short summer period in the afternoon hours, natural lighting is never effective in stimulating the circadian system. This happens because the only natural lighting that enters from the window cannot reach the observation points furthest away from the opening, in which observers could stop for relaxation and reading activities.

Due to the geographic position chosen, the *intermediate sky condition* occurs for about 99 days, i.e. over a quarter of the year (27%). To reduce glare in situations where the observer receives direct natural light, limiting the visual ability due to lens scattering, the simulation in intermediate sky conditions has been also performed by inserting a curtain on the windows, at the times when the sun generates non-comfortable light conditions. As it has been logical to expect, the intermediate sky is significantly less effective than the clear sky in all seasons, both because it generates less illumination in the interior, and because it is the one with the lower CCT, 5,500 K, and is therefore less effective for the human circadian system. For this reason, it is only partially more effective than overcast sky for circadian stimulation in some particular conditions, which are described in detail below.

Analysing the results, starting from the morning awakening in the *bedroom* and considering the period between 6:00 and 10:00, the level of illuminance reaching the eye stimulates the circadian activation only in spring and autumn. In particular, there is an illuminance on the eyes in the time frame 9:00–10.00 in autumn and in the time frame 8:00–10:00 in spring. The results of the simulations describe a circadian

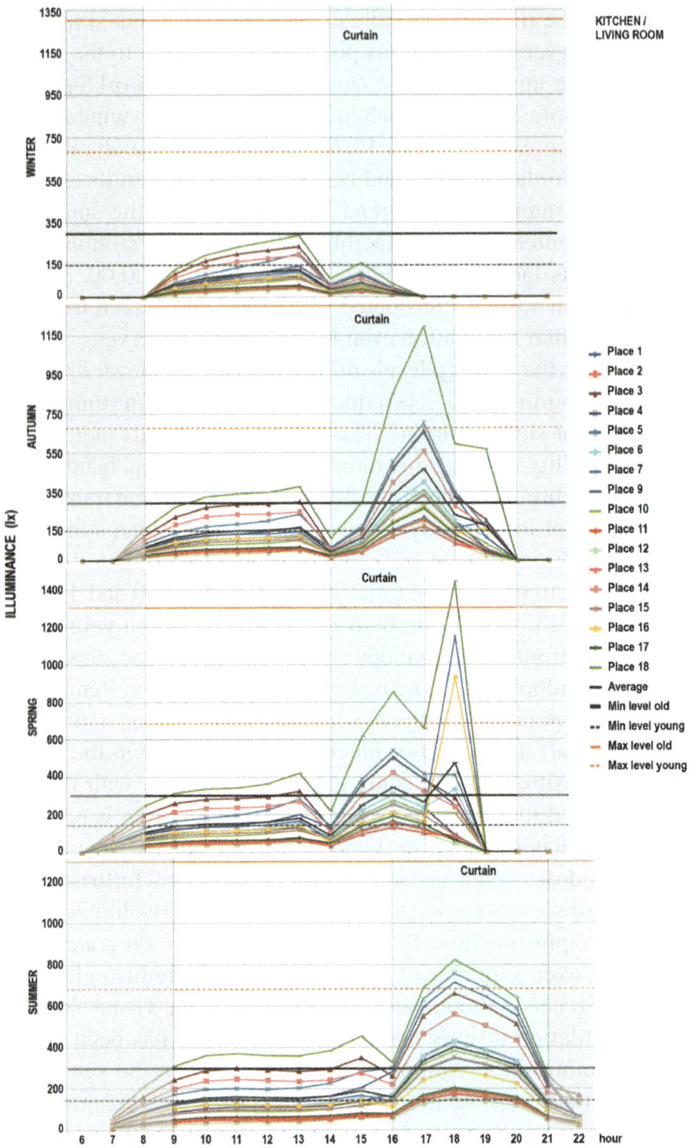


**Fig. 4.24** Lighting levels in the kitchen/living room in clear sky conditions

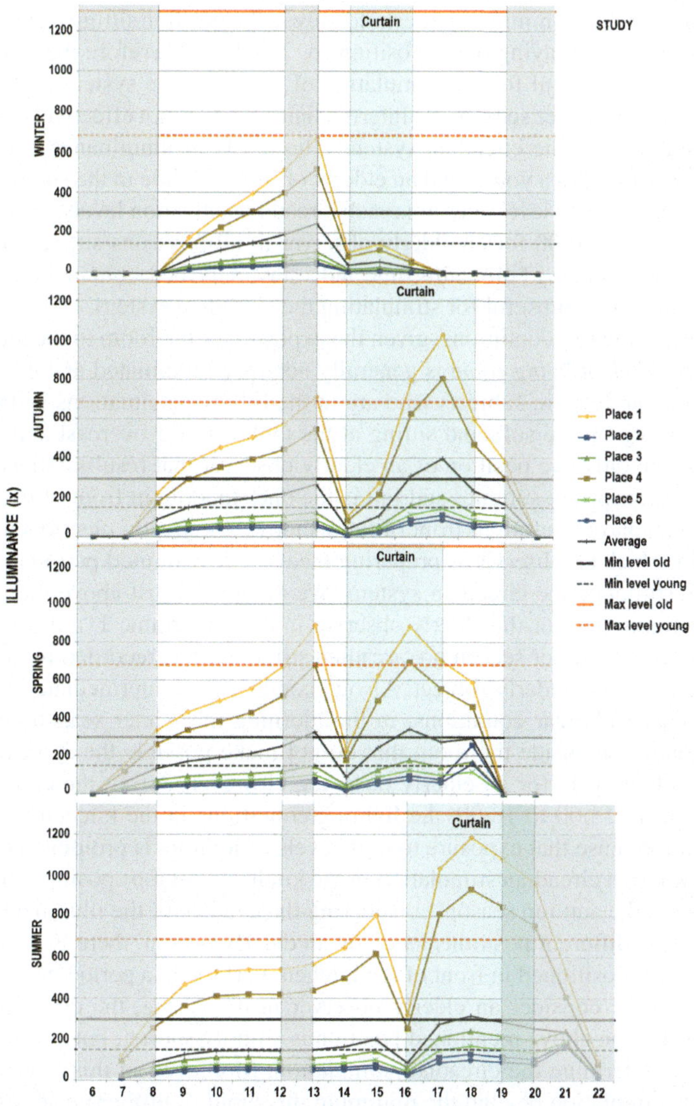


**Fig. 4.25** Lighting levels in the study in clear sky conditions

stimulation beyond the minimum threshold only in the position sitting on the bed (p. 1 and 3), but not in the laying down position (p. 2 and 4). Therefore the illumination is not particularly useful for the stimulation of the circadian system at the time of awakening. In the winter season, the intermediate sky does not effectively contribute to the stimulation of the circadian system. The levels of illuminance reaching the eyes are lower for both a young and an elderly observer. While in the summer season the cornea illuminance level does not reach circadian activation levels, except for the limited time 9:30–10:00 for an elderly observer in all positions sitting on the bed. Therefore, an awakening before this time and a stay in the bedroom in the early hours of the morning are not useful for stimulating the circadian system.

In intermediate sky conditions, given the exposure of the room to the west, in the morning the *kitchen/living room* is generally not very illuminated and not effective for the circadian level in subjects who are in the following situations: sitting in the armchair, sitting on the sofa and sitting at the table. As for overcast and clear sky conditions, the only eye position of an elderly observer that results activated in the morning during spring, summer and fall is the one standing in front of the worktop of the kitchen and therefore in front of the window. As already mentioned, this is a situation in which an observer is preparing meals, a very limited period that cannot effectively stimulate the circadian system. We get a level just above the minimum circadian stimulation for the elderly observer in the time frame 11:30–13:30 in fall and spring. In the winter season no circadian stimulation is recorded for any observation position of an elderly subject. Also in conditions of intermediate sky, as well as in overcast and clear conditions, in the morning the winter season determines lighting conditions inside the room that do not make possible the synchronisation of the circadian cycle for an elderly observer. However, during the early and late afternoon, from 13:00 to 18:00, the lighting conditions in the kitchen/living room improve in the sense that exposure to high levels of lighting is prolonged over time. This determines a circadian stimulation in the kitchen worktops position. In particular, as regards the autumn season, considering the average of the illuminance values detected in the different positions taken by an elderly person, there is a stimulation of the observer positioned in front of the kitchen worktop in a period between 15:30 and 17:30. If we consider an elderly observer in the spring, the circadian activation period for the above mentioned position is slightly shorter, but still reasonably sufficient to determine the circadian stimulation. Moreover, in this season there is a circadian stimulation beyond the minimum threshold in further positions, such as the position of an observer sitting on the sofa and the armchair furthest from the windows. Conversely, if we consider the average illuminance of the eyes, detected in all the possible positions of an elderly observer in the spring period, in the living area, the minimum level of circadian stimulation is almost never reached.

The morning time and afternoon time spent in the *study* result in effectively activating the circadian system in intermediate sky conditions:

- in spring in the time frames 9:00–12:00 and 13:15–16:30;
- in autumn in the time frames 10:00–12:00 and 13:00–16:30;
- in summer during the time frames 10:00–12:00, 13:00–16:00 and 17:00–20:00.

This applies to both an elderly and a young observer, but only in a position facing the window (p. 1 and 4). This position, taken when studying, reading and working at the table, turns out to be the best from a standpoint of activation of the circadian system during these seasons. During the winter season, however, the period of circadian stimulation, even in the studying position at the desk in front of the window, decreases to one hour in total, half an hour before and half an hour after the 12:00–13:00 lunch break. Finally, for all the other positions of the observer in the study, natural lighting during all four seasons is never effective in stimulating the circadian system, both for an elderly and a young observer.

### 4.3.7 Discussion of the CAD Simulations

At the conclusion of the study simulated with a lighting CAD, related to a residential context at the ~45th parallel, it is observed that the natural lighting available on average in various positions of the house is able to stimulate, to a level just above the minimum, the human circadian system, only in favourable weather conditions and at certain times of the year. A complete stimulation, as happens outside, is obtained only in very particular conditions, positions and in limited periods of time over the year. This assessment concerns the elderly subject, for whom the study estimates a longer stay at home during the day. However, even for the young person there are many periods, weather conditions and positions inside the house that are not satisfactory for a correct circadian stimulation during the day.

Keeping in mind the limitations of the study, which has been conducted on a type of housing with a specific orientation, positioning and location, we can summarise the various results in the following way. Regarding the lighting useful to the awakening in the bedroom, considering a period between 6:00 and 9:00, the overall results show that, in the four seasons, in overcast and intermediate sky conditions, the levels of illuminances reaching the eyes are not sufficient to properly stimulate the circadian system in the morning. These conditions occur in 48% of the annual period. Obviously, the situation improves considerably with clear sky conditions.

A correct circadian stimulation period in the first half of the morning (6:00–10:00) is not guaranteed in the kitchen/living room with overcast and intermediate conditions, during the whole year, i.e. 48% of the year. However, even in clear sky conditions, the internal lighting situation in this room is not able to stimulate the circadian system effectively due to the orientation and depth of the room. The results suggest that it would be helpful to increase the amount of light within this room in the morning throughout the year and with all sky conditions. This would increase the exposure of elderly and young subjects to an effective quantity and quality of light for the NIF effects.

### 4.3.8 Measurement of the Circadian Light in Winter

In order to validate the results of the simulation made with the CAD, in a subsequent phase, measurements have been also carried out of the real natural light present in the most significant environment of the flat chosen as a case study. These checks have been made in winter and spring, in the kitchen/living room, which is the most inhabited during the day. In order to determine the contribution of natural light in terms of NIF effects, the measures have concerned the $E_{v,eye}$ illuminance at eye level and, in the same position, also the SPD of the irradiance $E_{e,\lambda}$, the chromaticity and the CCT. The equipment used has been a portable Konica Minolta CL-500 Spectra-radiometer, which makes it possible to measure the spectrum in the 360–780 nm range, with 1 nm steps. Luminance measurements have been also performed using a Konica Minolta LS-100 luminance meter and a Canon Eos 550D camera with a standard lens (18–55 mm), together with the Luminance HDR (Kaneider et al. 2017) and HdrScope (Kumaragurubaran and Inanici 2013) software for the calibration of HDR images and the calculation of luminance images.

In the *winter period*, the measurements have been made on December 21, 2016 at 1-h intervals from 9:00 am to 16:00; not before, since the sun rises at 8:00 and no later, since the sun sets at 16:43. Four observation positions have been considered in the room, depending on the main activities normally taking place. With reference to Fig. 4.17 and Table 4.7, three positions have been chosen around the table and one position in the chair furthest from the window, i.e. positions 1, 3, 4 and 7, in which the characteristics of the light that reaches the observer's eyes have been measured. On the day chosen for the measurements the sky has been overcast until 12:00 and clear in the afternoon.

Table 4.8 shows the CCT values measured in the kitchen/living room from 9:00 to 16:00 in the four locations considered on December 21, 2016. The daily averages for each individual position and the averages at each hour on all positions are also reported. By analysing the CCT of natural light, present in the four selected positions, it has been found that this depends on the weather, the sky conditions, the season, the time and also the position. The measured values of the CCT in the observation positions vary from a minimum of 4,229 K to a maximum of 5,643 K, with an average value of 4,969 K. The measured variations are a consequence of the time, the different weather conditions in the morning, with respect to those in the afternoon, and the influence of the interior surfaces of the room, such as furnishings and plasters on the walls. Indeed, the values measured in positions that do not face the window (p. 1 and 4) have a CCT that is on average lower than those facing the window (p. 3 and 7).

Measured values of the chromatic coordinates x, y, with respect to the Planckian place curve, in the four selected observation points are shown in Fig. 4.26. In each diagram the points correspond to different measurement times. The upper curve is the place of chromaticity of natural light as the CCT changes. The chromatic measurements have made it possible to evaluate the influence of the inter-reflections of light, among the internal elements of the room, on the SPD of natural light. The chromatic coordinates x, y CIE 1931 (CIE 015:2004) show that in position 7, the

**Table 4.8** The CCT values measured in the kitchen/living in winter

| Hour | Sky | P. 1 (K) | P. 3 (K) | P. 4 (K) | P. 7 (K) | Average |
|---|---|---|---|---|---|---|
| 09:00 | Overcast | 4,891 | 5,643 (max) | 4,937 | 5,190 | 5,165 |
| 10:00 | Overcast | 4,941 | 5,546 | 4,999 | 5,190 | 5,169 |
| 11:00 | Overcast | 4,809 | 5,245 | 4,786 | 5,049 | 4,972 |
| 12:00 | Overcast | 4,834 | 5,302 | 4,768 | 5,070 | 4,994 |
| 13:00 | Clear | 4,667 | 5,131 | 4,463 | 4,989 | 4,812 |
| 14:00 | Clear | 4,496 | 5,130 | 4,229 (min) | 4,551 | 4,601 |
| 15:00 | Clear | 5,155 | 5,537 | 5,130 | 5,324 | 5,287 |
| 16:00 | Clear | 4,733 | 4,758 | 4,602 | 4,912 | 4,751 |
| Average | n.a. | 4,816 | 5,287 | 4,739 | 5,035 | 4,969 |

one farthest from the windows, the colour of the light differs the most from that of natural light due to the greater number of inter-reflections that modify the SPD and therefore the colour too. This has been also demonstrated in the recent research by Bellia et al. (2017). In other positions, in particular that in which the windows are directly observed (p. 3), the change in the colour of natural light is more contained, also because the internal surfaces are predominantly white. In the afternoon, however, the colour of natural light inside the room is strongly influenced by the direct sunlight filtered by the curtain on the main window.

To evaluate the NIF effects of light, the research has used the same model of circadian photometry of Rea et al. (2012) and Rea and Figueiro (2013) in CAD simulations, but with two substantial differences. The first concerns the CCT of daylight that, in the case of CAD simulation, has been estimated with that of the standard illuminants D55, D65 and D75 (Fig. 4.2). Instead, in the metrological verifications the real CCT has been measured in the kitchen/living room. This has been measured at the eye level of the four positions considered and the SPD of an illuminant D, of the daylight series, with a CCT equal to the measured one has been used. The second difference concerns the illuminance $E_{v,eye}$ that have been also actually measured rather than calculated.

In the evaluation using circadian photometry, the elderly observer has been considered as a model of the user of the house during the day. The spectral transmission curve of the artificial lens has been then considered to determine the true SPD of light reaching the retina of the elderly (Artigas et al. 2012). In this way, it has been possible to compare the level of illuminance measured in the eyes, with the minimum and maximum levels of circadian stimulation, according to the SPD of the real light that reaches the eyes.

Figure 4.27 shows the measured values of eye illuminance in winter, in the four selected positions, compared with the minimum and maximum levels of circadian stimulation for an elderly observer, which are in turn expressed according to the CCT measured at each hour. As illustrated in the figure, the values of illuminance to the eyes, recorded in the room during a winter day, are practically never sufficient to

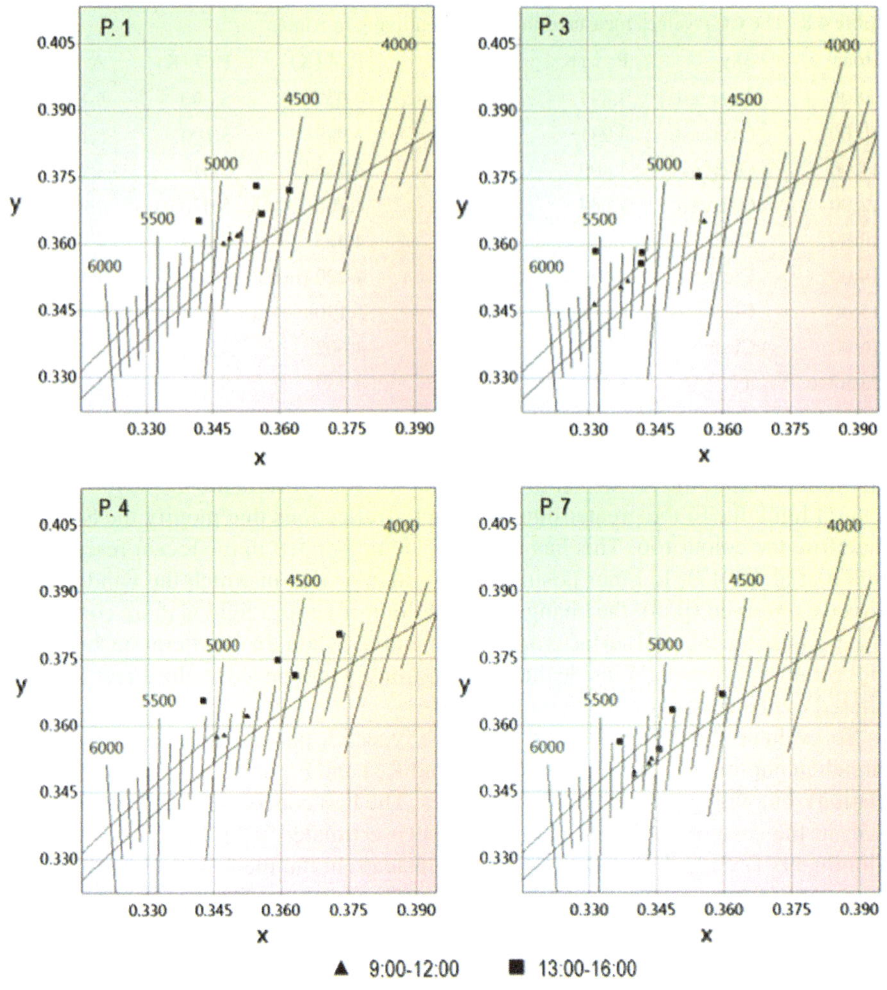


**Fig. 4.26** Chromatic measurements in the kitchen/living room in winter

stimulate the circadian system of the elderly, not even at a minimum level. In particular, this negative situation also appears in position 3, which should be theoretically favorable, since it is that facing the two windows.

In position 3, a minimum level of circadian stimulation is reached only at 14:00, in a limited time interval and in any case in a different period of time from the fundamental one in the morning. Indeed, an adequate stimulation in the morning is the one that makes it possible to properly phase the circadian cycle of the elderly. The fact that position 3, even if located in front of the two main windows, is not able to receive a suitable amount of light for correct circadian stimulation, can be attributed to two main factors. The first factor is the winter season with the presence of overcast

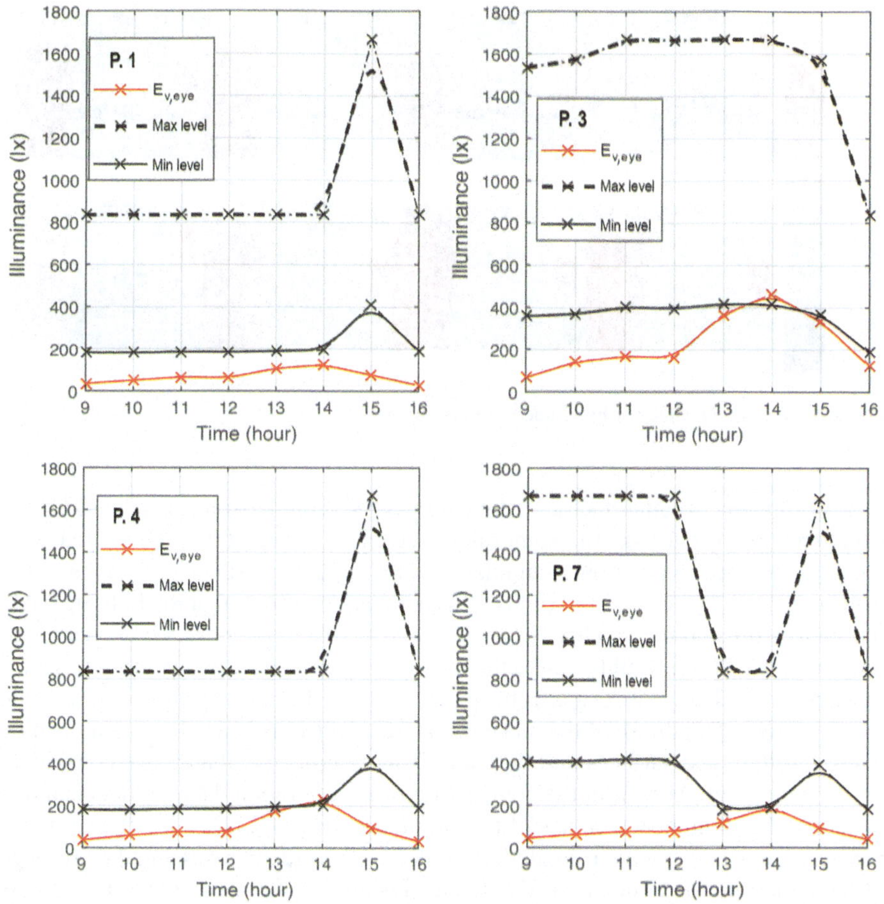


**Fig. 4.27** Measured values of eye illuminance in the kitchen/living room in winter

sky in the morning. While the second factor concerns the styles of use of the internal space. In fact, in the afternoon, in clear sky conditions, the occupants of the flat usually place a diffusing white curtain on the main window, to avoid disturbing glare due to natural light. From the measurements carried out, the presence of the curtain proves to be decisive in the modification of natural light, similar to what had been observed in CAD simulations. Also the luminance measurements carried out in the real space, at intervals of one hour from 9:00 to 16:00, have shown that the French window has a curtain that covers the glass area in order to reduce the luminance provided when observing the sky directly. Figure 4.28 shows that at 14:00 the luminance of the sky visible in the small window on the left is higher than 2,500 cd/m$^2$, while the luminance of the French window on the right, covered by the diffusing white curtain, is on average between 300 and 850 cd/m$^2$.

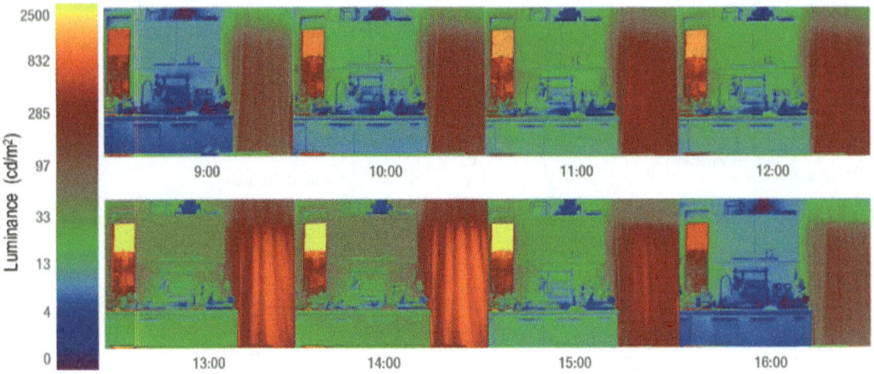


**Fig. 4.28** Measured values of luminance from position 3

Regarding the weather factor, in the winter days a comparison can be made of the situation with overcast sky in the morning, between the data of the lighting reaching the eyes, referred to position 3, simulated in the lighting CAD (Fig. 4.21) and that measured on site (Fig. 4.27 top right). This comparison shows that the lighting CAD has underestimated the levels of illuminance available to the observer's eyes in the real situation. This is mainly due to the fact that, in the simulation, in order to give a more general value to the results, reflectance values of the internal surfaces have been assumed lower than those of the real environment, which is instead characterised by white furniture and surfaces and therefore is more reflective than those considered in the simulation. Also from the comparison of the afternoon data, with a clear sky situation, the illuminance values of the simulation (Fig. 4.24) are lower than those measured on site, but with a smaller difference, still correctly assessing the presence of the curtain in the simulation. Finally, analysing the real measured values of the CCT, during the entire winter day in the interior, it can be observed that the values have been in the range of 4,229–5,643 K, which is however lower than the average values of overcast sky and even more with respect to the minimum values with a clear sky. Therefore, the CAD simulation has, as a starting point, overestimated the CCT really present in the interior.

Beyond these differences, mainly due to the reflection factors of the internal surfaces and partly also to the differences between the real SPD of the overcast and clear skies with respect to the SPD of the daylight illuminators D65 and D75, the measurements have confirmed that, in the winter period, natural light is not able to correctly stimulate the circadian system throughout the day, even in the presence of different weather conditions, similar to what has been determined by the simulation using a lighting CAD.

### 4.3.9 Measurement of the Circadian Light in Spring

The same measurements of natural light have been repeated four months later, during the *spring season*, on March 22, 2017, in the same case study used in winter. On that day, the sky has been mostly overcast, with slight openings in the afternoon from 15:00 to 17:00, when it went from intermediate to clear, then becoming overcast again. The measurement positions inside the kitchen/living room have been the same used in the previous winter period (p. 1, 3, 4 and 7 of Fig. 4.17 and Table 4.7), with the same instrumentation and measurement methodology. However, since the duration of the day has been greater, the measurements have been conducted for a longer period, from 8:00 to 18:00, at 1-h time intervals. Table 4.9 shows the CCT values measured in the kitchen/living room in the four locations considered on March 22, 2017. The daily averages for each individual position and the averages at each hour on all positions are also reported.

The measurements carried out for the CCT in the four observation positions are in a range from 3,877 to 6,413 K, with an average value of 4,876 K, slightly lower than the average value of the winter measures. The greater amplitude of the measured CCT values, compared to winter, derives from the contribution of indirect lighting, due to inter-reflections of light of internal surfaces in overcast sky conditions. In the case of the clear sky situation, at 16:00, the presence of direct sunlight brought a warmer light contribution, resulting in a lower CCT. In addition, the average values of the CCT measured in the positions that are not facing the window (p. 1 and 4) or in the innermost area of the room (p. 7) are lower than those measured in the position facing the window (p. 3).

**Table 4.9** The CCT values measured in the kitchen/living in spring

| Hour | Sky | P. 1 (K) | P. 3 (K) | P. 4 (K) | P. 7 (K) | Average (K) |
|---|---|---|---|---|---|---|
| 08:00 | Overcast | 4,984 | 5,668 | 5,146 | n.a. | 5,266 |
| 09:00 | Overcast | 4,816 | 5,417 | 4,929 | n.a. | 5,054 |
| 10:00 | Overcast | 4,723 | 5,264 | 4,792 | 4,856 | 4,909 |
| 11:00 | Overcast | 4,694 | 5,264 | 4,767 | 4,993 | 4,929 |
| 12:00 | Overcast | 4,654 | 5,212 | 4,724 | 4,683 | 4,818 |
| 13:00 | Overcast | 4,639 | 5,209 | 4,698 | 4,625 | 4,793 |
| 14:00 | Overcast | 4,514 | 5,075 | 4,561 | 4,657 | 4,702 |
| 15:00 | Intermediate | 4,490 | 5,175 | 4,607 | 4,583 | 4,714 |
| 16:00 | Clear | 3,941 | 4,704 | 3,877 (min) | 4,287 | 4,202 |
| 17:00 | Overcast | 4,693 | 5,112 | 4,736 | 4,797 | 4,834 |
| 18:00 | Overcast | 5,622 | 6,413 (max) | 5,890 | n.a. | 5,975 |
| Average | n.a. | 4,706 | 5,319 | 4,793 | 4,685 | 4,876 |

The measurement of the chromatic coordinates x, y CIE 1931 has confirmed the same results verified in the winter measurements. For the position furthest from the windows (p. 7) there is a greater difference in the colour of the light compared to the locus that describes the chromaticity of the daylight. This is due to the greater influence of the surfaces of the environment that modify the SPD of natural light. Also in this case the position that directly looks toward the windows (p. 3) is the one that has the closer chromaticity to that of natural light.

For the evaluation of the NIF effects of light on the elderly, the same methodology of the measurements performed in winter has been used. Figure 4.29 shows the measured values of eye illuminance in spring, in the four selected positions, compared with the minimum and maximum levels of circadian stimulation for an elderly observer, which are in turn expressed according to the CCT measured at each hour. In this case there has been a curtain in the afternoon, on the French window, to limit the discomfort glare of the natural light. In fact, on the spring day of the measurements, at 16:00 with clear sky, the luminance measured through the smaller window without a curtain has been over 8,000 cd/m$^2$. With the weather conditions on the day of the measurements, from Fig. 4.29 it is possible to observe that during the whole day the circadian stimulation has never been obtained when the subject has been in the observation position furthest from the windows (p. 7). In the other three observation positions (p. 1, 3 and 4) no circadian stimulation has been obtained during the whole morning and a limited action on the circadian system has been observed only between 15:00 and 18:00. This effect in the afternoon is due to the exposure of the windows oriented to the west, but it is inadequate for the correct functioning of the human biological clock. There is a short peak of high circadian stimulation at 16:00, from positions 3 and 4, which are those oriented towards the French window, since in that hour the sky has been clear, with a consequent increase in illuminance levels.

For the series of measures referring to the spring day, a comparison has been made of the situation with overcast sky for almost the whole day, at 15:00 with intermediate sky and at 16:00 with clear sky, between the data of the illuminance reaching the eyes referred to position 3, simulated in the lighting CAD and those actually measured (Fig. 4.29 top right). This comparison shows that the lighting CAD in this case has slightly overestimated the illuminance levels actually available to the eyes of the observer in overcast sky conditions. While in the afternoon it has underestimated the amount of light actually measured in intermediate and clear sky conditions. This could be due to the fact that the greater amount of light that enters from the windows in spring generates a greater amount of inter-reflections between the surfaces of the environment that in the simulation are darker than the real ones, thus leading to a smaller contribution of light coming from inter-reflections to the simulated observer.

In spring the real measured values of the CCT, during the entire day in the kitchen/living room have remained in the 3,877–6,413 K range, which is however lower than the average values of overcast sky and even more with respect to the minimum values with a clear sky. Therefore, the CAD simulation, also in this case, has overestimated the CCT really present in the interior, with the consequence of overestimating the real circadian stimulation of light.

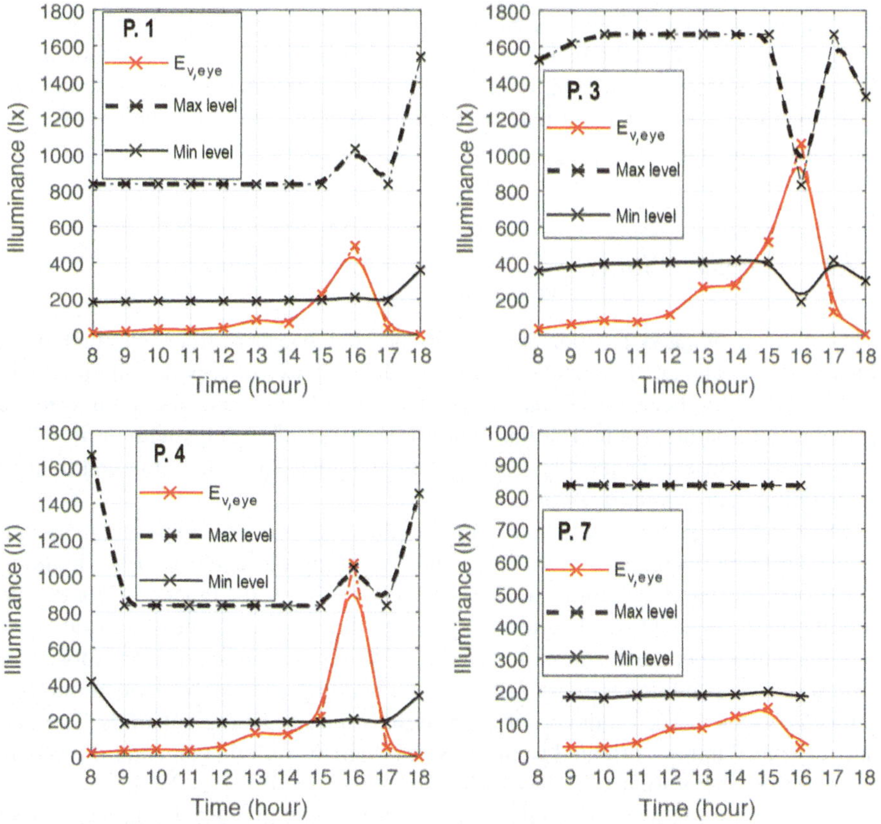


**Fig. 4.29** Measured values of eye illuminance in the kitchen/living room in spring

The fact that the CAD simulation on the one hand underestimates the amount of light in many situations, due to the choice of reflection factors set for the surfaces, but on the other hand overestimates the CCT of the light actually present in the environment, leads to two errors with an opposite sign, which tend to partially offset themselves, in the evaluation of the circadian stimulation through the CAD simulation.

## 4.4 Evaluation of the Light in Educational Spaces

In 2017, Laboratorio Luce of the Politecnico di Milano has been involved in a project research, aimed at the study of an existing educational building, with the purpose of assessing whether the natural and artificial light present in the school could be adequate to favour the training activities and the learning process. As introduced in

Chap. 2, the amount of light and SPD can have a significant impact on the cognitive performance of human beings. Therefore, an investigation has been conducted, designed to assess the existing situation of natural light and lighting installations in the building, to understand if lighting in the school could be beneficial for the activities of the pupils.

### *4.4.1 The Case Study*

The building is located in Milan at the ~45th parallel, has three floors and includes a nursery school and an elementary school. For the purposes of experimentation, only the lower floor has been considered, as it is the one with the greatest limitations in the availability of natural light. The building is composed of 4 wings with an irregular plan that have different exposures to the cardinal points. The plan of the first floor of the building is drawn in Fig. 4.30.

Most of the classrooms, with a rectangular plan, have a double exposure to natural light with large openings. A series of windows facing the exterior of the building and, on the side parallel to these, a series of windows facing the inner corridor that, in turn, has in its opposite side windows exposed to the outside, on the other side of the building with respect to the classrooms (Fig. 4.30). The classrooms have exposure to natural light depending on their location in the building and can be divided into two main categories, based on the orientation of the windows facing outwards. The first category has windows oriented to SE or SW, in which sunlight can enter directly in the morning or afternoon, causing overheating and glare problems. While the second category of classrooms is oriented to NW and NNW and in these the direct sunlight can enter only in the late afternoon on the longest days of the year. Furthermore, in most classrooms natural light can also come from the opposite side of the building through the windows facing the corridors. Users have available and use Venetian blinds to control natural light only on the windows of the classrooms facing the outside of the building. We can see in Fig. 4.31 a classroom on the first floor, with exposure to SW, with the Venetian blinds pulled down to control natural light.

### *4.4.2 The Chosen Classrooms*

For the light measurements, two classrooms have been chosen: the first, unlike the others, with a double exposure on two sides at 90° with NE and NW orientation (no. 1 in Fig. 4.30) and the second one with SW orientation (no. 2 in Fig. 4.30).

In the classrooms there is a classic artificial lighting system based on linear fluorescent lamps, with CCT of 4,000 K, inserted in $4 \times 18$ luminaires embedded in the false ceiling. This system does not adjust the flux or the CCT, nor a partial switching. The artificial light can only be switched on and off. For this reason, the illuminances

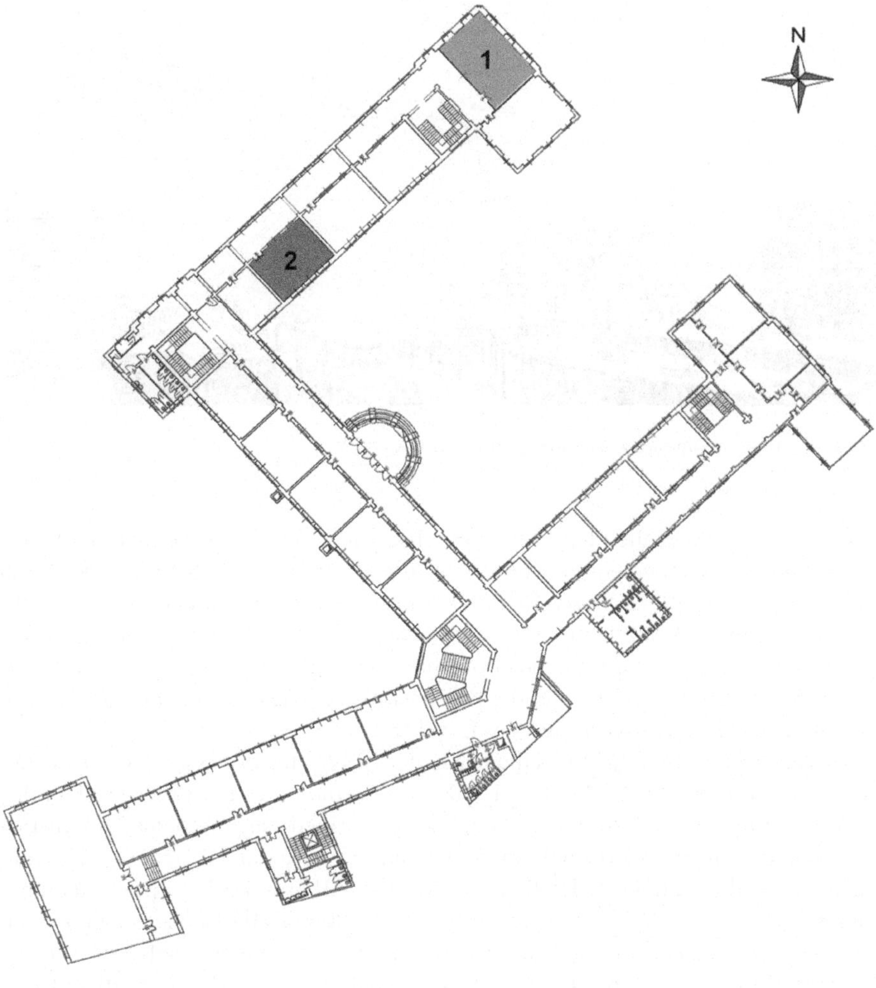


**Fig. 4.30** The plan of the first floor of the school

on the work plane have been measured in the presence of natural light only and with artificial light as well.

Inside the two selected classrooms the measurements have been taken in autumn, on October 11, 2017, from 13:00 to 13:30 during the pupils' lunch break, with overcast sky conditions. During the measurements the Venetian blinds have been raised to allow full entry of natural light. In order to evaluate the contribution of natural light, also in terms of NIF effects, the measurements of horizontal illuminance $E_h$ have been implemented on the work surface at a height of 70 cm and also at the eye level $E_{v,eye}$ in two positions for each classroom, with the pupils' gaze turned towards the teacher's desk and the teacher.

**Fig. 4.31** A classroom on the first floor with exposure to SW

Figure 4.32 shows the plan of the first classroom chosen for the measurements, with the positions of two possible observers turned towards the teacher's desk (A and B) and the three positions chosen for the measurement of the illuminances on the work surface (I, II and III). While Fig. 4.33 shows the plan of the second classroom chosen for the measurements, with the positions of two possible observers turned towards the teacher's desk (A and B) and the two positions chosen for the measurement of the illuminances on the work surface (I, and II).

Measurements also concerned the irradiance SPD to the eyes $E_{e,\lambda}$, the chromaticity and the CCT. A Konica Minolta CL-500 Spectra-radiometer has been used for this purpose. Luminance measurements have been performed using a Konica Minolta LS-100 luminance meter and a Canon Eos 550D camera with a standard lens (18–55 mm), together with the Luminance HDR (Kaneider et al. 2017) and HdrScope (Kumaragurubaran and Inanici 2013) software for the calibration of HDR images and the calculation of luminance images. In addition, the ColorCalculator software (Osram Sylvania 2017) has been used for colorimetric and circadian effects calculations.

### *4.4.3 The Results of the Measurement*

In Table 4.10 are illustrated the results of the measurements of mean illuminance values $E_h$ on the work surface, in the two chosen classrooms, in different positions with respect to the distance from the external windows of the building.

From the measurements carried out on the work plane it has been possible to find a great difference in lighting, due to natural light only, between the pupils' desks placed near the windows and those on the opposite side of the room. In classroom 1 there are 614 lx near the window and only 158 lx in the innermost position of the room. In classroom 2 the difference is more contained, 634 lx in the position towards

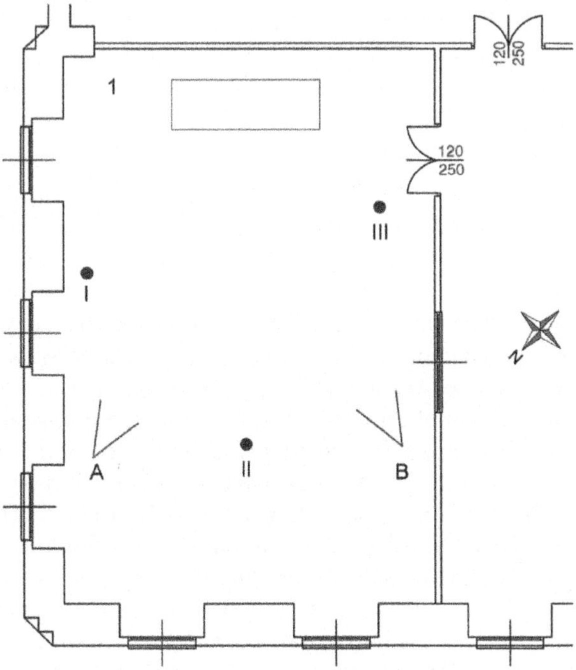


**Fig. 4.32** The plan of the first classroom chosen for the measurements

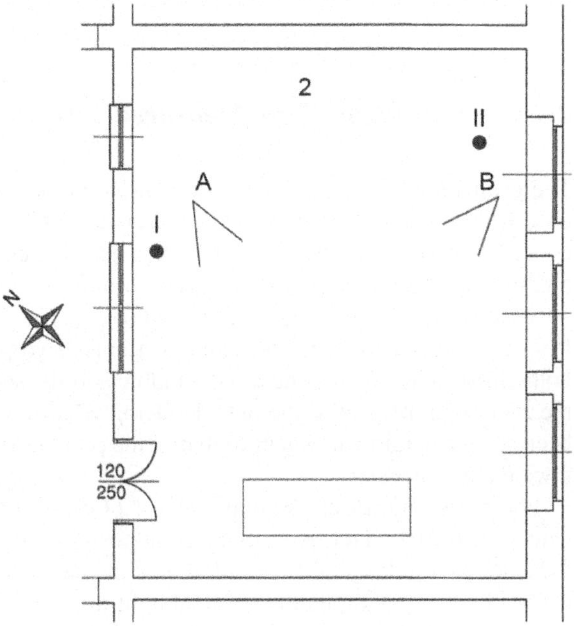


**Fig. 4.33** The plan of the second classroom chosen for the measurements

**Table 4.10** Results of the measurements of $E_h$ in the two classrooms

| Classroom | 1 (hour 13:00) | | | 2 (hour 13:30) | |
|---|---|---|---|---|---|
| Position | I window | II centre | III inner | I inner | II window |
| Daylight (lx) | 614 | 527 | 158 | 325 | 6,230 |
| Daylight and artificial light (lx) | 867 | 804 | 343 | 623 | 6,525 |
| Estimated artificial light (lx) | 253 | 277 | 185 | 298 | 300 |

the window and 325 lx in the one towards the inside, since the inner side of the room is in turn equipped with windows towards the corridor that receive natural light from the opposite side of the building. Measurements carried out in both classrooms with the addition of artificial light show that the illuminance on school desks complies with the European standard on the lighting of indoor workplaces EN 12464-1:2011.

Since it has been not possible to darken the rooms or access them at night, the artificial lighting contribution has been determined as the difference between the two measures. Furthermore, given that the installed luminaires are equipped with anti-glare blades, a ratio between vertical and horizontal illuminances $E_{v,eye}/E_h \approx 0.5$ has been estimated. That is to say that the contribution of artificial light to the eyes is about half of the horizontal one, measured in similar positions of the classroom on the work plane and that its SPD is that of the illuminant CIE F11 for fluorescent lamps, with CCT of 4,000 K, provided by the Philips TL84 Ultralume 40 lamp.

### *4.4.4 Discussion of the Measured Circadian Stimulus*

The evaluation of the circadian stimulation has considered the model of a young observer, with a lens that does not excessively filter the short wavelengths, aged under 19 (Barker and Brainard 1991; Rea and Figueiro 2016). For this purpose, the measurement has concerned the vertical illuminances to the eyes $E_{v,eye}$ due to the only natural light that enters from the windows. The CCT has been also determined from the SPD measure in the same positions. Two positions have been chosen in both classrooms, the first near the window and the second in the innermost area of the room with respect to the outside facing windows. The position of the eyes has been set at a height equivalent to that of the pupils sitting at their desks and looking towards the teacher.

For the two chosen classrooms, Table 4.11 illustrates the illuminance and the CCT values measured on the eyes, in different positions with respect to the distance from the external windows of the building and the calculated values of the CS factor for natural light and natural plus artificial lighting.

The analysis of the measured values has confirmed that in the positions furthest from the windows (1B and 2A) the CCT of overcast sky natural light is reduced, compared to its value in the exteriors, due to inter-reflections of the light in the

**Table 4.11** The circadian stimulus in the two classrooms

| Classroom | 1 (hour 13:00) | | 2 (hour 13:30) | |
|---|---|---|---|---|
| Position | A window | B inner | A inner | B window |
| Illuminance $E_{v,eye}$ (lx) | 221 | 176 | 463 | 1,080 |
| CCT (K) | 5,212 | 4,574 | 4,180 | 4,329 |
| CS (natural light) | 0.24 | 0.17 | 0.34 | 0.49 |
| Estimated total CS (natural + artificial light) | 0.27 | 0.20 | 0.35 | 0.50 |

internal surfaces. Moreover, in an overcast day, the time interval between 13:00 and 13:30, when measurements have been taken, is close to the maximum natural light contribution available outside, but the CS circadian stimulation factor is close to a full effect only in position 2B. In the other position of classroom 2 it is slightly higher than the minimum circadian stimulation level estimated at 0.29 (Fig. 4.19). In classroom 1, however, even if there is a double exposure of the NW and NE windows, the minimum level of light necessary to stimulate the circadian system is never reached.

By adding artificial light, the situation does not improve. Indeed, considering the lighting system present in the classrooms, with 4,000 K linear fluorescent lamps, the contribution of circadian light added in classroom 2 is irrelevant, given the predominant amount of natural light; while in classroom 1 the contribution is greater, but still not sufficient to reach the minimum level of circadian stimulation. It should be noted that the contribution of artificial light to the CS factor is also overestimated, given that a CCT of 4,000 K has been used in the calculations, coming out of the lamps as factory data, while that in the eyes is generally lower due to inter-reflections on internal surfaces.

## References

Andersen, M., Mardaljevic, J., & Lockley, S. (2012). A framework for predicting the non-visual effects of daylight—Part I: Photobiology-based model. *Lighting Research & Technology, 44*(1), 37–53. https://doi.org/10.1177/1477153511435961.

Artigas, J. M., et al. (2012). Spectral transmission of the human crystalline lens in adult and elderly persons: Color and total transmission of visible light. *Investigative Ophthalmology & Visual Science, 53*(7), 4076–4084. https://doi.org/10.1167/iovs.12-9471.

Barker, F., & Brainard, G. (1991). *The direct spectral transmittance of the excised human lens as a function of age (FDA 785345 0090 RA)*. US Food and Drug Administration. Available at: https://www.chemie.uni-wuerzburg.de/fileadmin/08020000/user_upload/makula/transmittance.pdf. Accessed May 15, 2018.

Barnard, K., & Funt, B. (2002). Camera characterization for colour research. *Color Research & Application, 27*(3), 152–163. https://doi.org/10.1002/col.10050.

Begemann, S. H. A., van den Beld, G. J., & Tenner, A. D. (1997). Daylight, artificial light and people in an office environment, overview of visual and biological responses. *International Journal of Industrial Ergonomics, 20*(3), 231–239. https://doi.org/10.1016/S0169-8141(96)00053-4.

Bellia, L., Pedace, A., & Fragliasso, F. (2017). Indoor lighting quality: Effects of different wall colours. *Lighting Research & Technology, 49*(1), 33–48. https://doi.org/10.1177/1477153515594654.

Borisuit, A., et al. (2015). Effects of realistic office daylighting and electric lighting conditions on visual comfort, alertness and mood. *Lighting Research & Technology, 47*(2), 192–209. https://doi.org/10.1177/1477153514531518.

Chain, C., Dumortier, D., & Fontoynont, M. (1999). A comprehensive model of luminance, correlated colour temperature and spectral distribution of skylight: Comparison with experimental data. *Solar Energy, 65*(5), 285–295. https://doi.org/10.1016/S0038-092X(98)00145-5.

CIBSE. (2014). LG10/14 Lighting Guide 10: Daylighting—A Guide for Designers. CIBSE.

CIE 015:2004 Colorimetry (3rd edn).

CIE 190:2010 Calculation and Presentation of United Glare Rating Tables for Indoor Lighting Luminaires.

Debevec, P. E., & Malik, J. (1997). Recovering high dynamic range radiance maps from photographs. In: *Proceedings of the 24th Annual Conference on Computer Graphics and Interactive Techniques* (pp. 369–378). New York, NY, USA: ACM Press/Addison-Wesley Publishing Co. (SIGGRAPH '97). https://doi.org/10.1145/258734.258884.

EN 12464-1:2011 Light and lighting—Lighting of work places—Part 1: Indoor work places.

Figueiro, M. G., Bierman, A., & Rea, M. S. (2008). Retinal mechanisms determine the subadditive response to polychromatic light by the human circadian system. *Neuroscience Letters, 438*(2), 242–245. https://doi.org/10.1016/j.neulet.2008.04.055.

Gall, D., & Bieske, K. (2004). Definition and measurement of circadian radiometric quantities. In *Proceedings of the CIE Symposium '04 on Light and Health* (pp. 129–32).

Hernández-Andrés, J., et al. (2001). Color and spectral analysis of daylight in southern Europe. *JOSA A, 18*(6), 1325–1335. https://doi.org/10.1364/JOSAA.18.001325.

Heschong, L., Wright, R. L., & Okura, S. (2002). Daylighting impacts on human performance in school. *Journal of the Illuminating Engineering Society, 31*(2), 101–114. https://doi.org/10.1080/00994480.2002.10748396.

Hosek, L., & Wilkie, A. (2012). An analytic model for full spectral sky-dome radiance. *ACM Transactions on Graphics, 31*(4), 95:1–95:9. https://doi.org/10.1145/2185520.2185591.

Hosek, L., & Wilkie, A. (2013). Adding a solar-radiance function to the Hošek-Wilkie skylight model. *IEEE Computer Graphics and Applications, 33*(3), 44–52. https://doi.org/10.1109/MCG.2013.18.

ISO 15469:2004(E)/CIE S 011/E:2003 Spatial distribution of daylight—CIE Standard General Sky.

Kaneider, D., et al. (2017). *Luminance HDR*. Available at: https://sourceforge.net/projects/qtpfsgui/. Accessed July 8, 2018.

Kobav, M. B., & Bizjak, G. (2005). Development of a substitutive light source for indoor daylight calculations. *Building and Environment, 40*(12), 1611–1618. https://doi.org/10.1016/j.buildenv.2004.12.013.

Kumaragurubaran, V., & Inanici, M. (2013). *hdrscope*. Available at: http://courses.washington.edu/hdrscope/. Accessed July 8, 2018.

Mardaljevic, J., et al. (2014). A framework for predicting the non-visual effects of daylight—Part II: The simulation model. *Lighting Research & Technology, 46*(4), 388–406. https://doi.org/10.1177/1477153513491873.

Mccurdy, T., & Graham, S. E. (2003). Using human activity data in exposure models: Analysis of discriminating factors. *Journal of Exposure Science & Environmental Epidemiology, 13*(4), 294–317. https://doi.org/10.1038/sj.jea.7500281.

Mclntyre, I. M., et al. (1989). Human melatonin suppression by light is intensity dependent. *Journal of Pineal Research, 6*(2), 149–156. https://doi.org/10.1111/j.1600-079X.1989.tb00412.x.

Ng, E. Y.-Y., et al. (2001). Advanced lighting simulation in architectural design in the tropics. *Automation in Construction. (CAADRIA), 10*(3), 365–379. https://doi.org/10.1016/S0926-5805(00)00053-4.

NRC. (1981). *Indoor Pollutants*. Washington, DC: The National Academies Press. https://doi.org/10.17226/1711.

Osram Sylvania. (2017). *LED ColorCalculator*. Available at: https://www.osram.us/cb/tools-and-resources/applications/led-colorcalculator/index.jsp. Accessed September 10, 2017.

Rea, M. S., Figueiro, M. G., & Bullough, J. D. (2002). Circadian photobiology: an emerging framework for lighting practice and research. *Lighting Research & Technology, 34*(3), 177–187. https://doi.org/10.1191/1365782802lt057oa.

Rea, M. S., et al. (2005). A model of phototransduction by the human circadian system. *Brain Research Reviews, 50*(2), 213–228. https://doi.org/10.1016/j.brainresrev.2005.07.002.

Rea, M. S., et al. (2010), Circadian light. *Journal of Circadian Rhythms, 8*. https://doi.org/10.1186/1740-3391-8-2.

Rea, M. S., et al. (2012). Modelling the spectral sensitivity of the human circadian system. *Lighting Research & Technology, 44*(4), 386–396. https://doi.org/10.1177/1477153511430474.

Rea, M. S., & Figueiro, M. G. (2013). A working threshold for acute nocturnal melatonin suppression from white light sources used in architectural applications. *Journal of Carcinogenesis & Mutagenesis, 4*(3), 1–6. https://doi.org/10.4172/2157-2518.1000150.

Rea, M. S., & Figueiro, M. G. (2016). Light as a circadian stimulus for architectural lighting. *Lighting Research & Technology, 50*(4), 497–510. https://doi.org/10.1177/1477153516682368.

Satel-Light. (2018). *The European database of daylight and solar radiation*. Available at: http://www.satel-light.com/. Accessed July 5, 2018.

Shao, L., Elmualim, A. A., & Yohannes, I. (1998). Mirror lightpipes: Daylighting performance in real buildings. *International Journal of Lighting Research and Technology, 30*(1), 37–44. https://doi.org/10.1177/096032719803000106.

Solli, M., et al. (2005). Digital camera characterization for colour measurements. In *2005 Beijing International Conference on Imaging: Technology and Applications for the 21st Century* (pp. 278–279). Science Press.

Spitzglas, M. (1984). Defining daylighting from windows in terms of candlepower distribution curves. In *IEEE/IAS 1984 Annual Meeting*. Chicago: IEEE Computer Society. Available at: http://gaia.lbl.gov/btech/papers/18087.pdf. Accessed April 3, 2018.

Wandachowicz, K. (2006). Calculation of circadian illuminance distribution with Radiance. In *5th International Radiance Scientific Workshop*. Leicester, UK: Radsite. Available at: https://radiance-online.org/community/workshops/2006-leicester/Presentations/Wandachowicz-2_RW2006.pdf. Accessed May 5, 2018.

# Chapter 5
# LEDs and New Technologies for Circadian Lighting

## 5.1 Products for Circadian Lighting

As shown by many years of physiology research, the topic of circadian lighting is important for our health. From the standpoint of the design approach, we should avoid extreme simplifications. *To say that a light source or a lighting product produces a circadian light or is human-centric is inherently false*. This because circadian lighting must always be assessed at the level of the eyes, in the typical positions of human beings in a specific internal environment. Moreover, it depends on natural light, artificial light and also on the way in which all the elements of the project, including walls and furnishings, influence the light present in the environment. *Circadian lighting therefore depends on the whole lighting project, in relation to human beings, the spaces they are in, the activities they perform and the time factor.*

Technologies can anyway provide useful tools to design a circadian lighting system. In this sense, LEDs provide fundamental characteristics to create lighting products that previous light sources did not make possible, or made only partially possible, with scarce efficiency and high costs. In particular, the characteristics of LED light sources that are useful for circadian lighting purposes can be summarised with these key elements:

- Easy to adjust luminous flux.
- Wide availability of different light CCTs.
- Availability of many light colours.
- Good design possibilities to control light intensity distribution.
- Electronic control of the power supply.
- High energy efficiency.
- Long operating life.
- Decreasing costs over the years and, in perspective, in line with those of the old light sources.

Alongside LEDs, there are also other technological elements that contribute decisively to the possibility of creating products that can be used for circadian lighting



M. Rossi, *Circadian Lighting Design in the LED Era*, Research for Development,
https://doi.org/10.1007/978-3-030-11087-1_5

design. In particular, *light sensors* that make it possible to receive information about the light actually present in the illuminated environment. Therefore, not only the contributions of natural light, but also the changes induced, on the quantity and SPD of the light, by the reflection factors of all the elements and objects present in the interior. *Presence sensors* too are obviously useful for energy saving purposes. However, compared to classical presence detectors, an innovative trend is underway aimed at a more accurate analysis of people's presence and activities, which is no longer based solely on the classic presence-detecting sensors. The most advanced experiments even get to use actigraphs able to monitor the activity and some physiological parameters of the users. Lastly, new systems are also being tested for the *detection of presence and position* of people in interior spaces. This is thanks to wireless systems, smartphones and the intelligence of the new electronic *light management systems* (LMS), in a perspective of integration into the smart home and the smart office of the future.

In order for these new products and systems to facilitate the development and commercial success of a new way of designing lighting, there are two key characteristics that should be taken into account in the integration with the increasingly pervasive building automation (Brodrick 2015). The first is compatibility, i.e. the possibility of multiple devices or systems acting in cooperation on the same environment, interacting positively rather than interfering with each other. The second characteristic is interchangeability, both to facilitate effective maintenance and to reassure the customer that the lighting system installed today will not become obsolete tomorrow and is upgradable over time.

## 5.2 LED Light Sources

### *5.2.1 The Revolution of LEDs*

As explained in the first chapter, from the end of the 1800s to the beginning of 2000, the field of electric lighting lamps has evolved according to the scientific and engineering development typical of the electromechanical sector, and this has been true until the advent of LED technology.

A functional feature of the LED is that of emitting light in a limited wavelength range, i.e. it emits coloured light. The first LED with a visible red light was made by Nick Holonyak Jr in 1962. Later LEDs with emissions of other colours were created, but there was no LED that was able to emit bright blue light. This was made in 1994 by scientists Shuji Nakamura, Hiroshi Amano and Isamu Akasaki who received the Nobel Prize in physics in 2014, thanks to this discovery (Overbye 2014).

Although even before the invention of the blue LED it was possible to create white light by putting together LEDs of various colours, that technique had numerous drawbacks. The main limit of a white LED light created by the addition of various colours is the restricted real colour rendering of the obtainable light, much lower than with the other light sources. In addition, LEDs of different colours are also made of different materials and require different powering systems according to the dissipated heat, thus posing problems linked to the complexity and costs of power systems. A white light of acceptable quality, without increasing the complexity of the light source, was achieved by using the blue LED coated with yellow phosphorescent material capable of transforming the emitted light spectrum, thus obtaining a broader emission spectrum that is perceived as white light by the human eye.

At the beginning of the third millennium, there were white LEDs on the market, which could be used for lighting, but had a limited luminous flux, a lower efficiency than fluorescent lamps and very high costs. For these reasons, most lighting manufacturers did not consider them a viable alternative to the more common traditional light sources on the market for interior lighting, and fluorescent and discharge lamps for the exterior. However, around 2010, the first white LEDs with an energy efficiency superior to fluorescent lamps and metal halide lamps were proposed on the market. In the following years, new LED sources with increasing efficiency, higher light output, decreasing costs and longer life than traditional light sources were proposed on the market; so much so that in 2014 CREE announced that it had made the first white LED with an efficiency of over 300 lm/W, an unthinkable result only two years before (CREE 2014), and even more extraordinary when compared to the efficiency of 10–15 lm/W typical of classic incandescent lamps. The speed of this innovation took place according to the typical times of the development of consumer electronics, which have been introducing new products, with improved performances, every six months in the last ten years. In a few years, the producers of light sources have changed their commercial idea regarding LEDs from declarations stating that "... LEDs are a technology that is only to supplement other light sources in low-power lighting" to the declarations issued from 2013 onwards stating that "... LEDs will progressively replace almost all other light sources". Indeed, the entry into the market of LED lamps has meant that within a few years, from 2010 to 2016, the common use lamps that could be purchased even in supermarkets have been gradually replaced, i.e. incandescent lamps with tungsten filament and this category also includes halogen lamps.

There were consequences of the disappearance of incandescent lamps. In a first phase, there was an increase in the sale of compact fluorescent lamps that were already on the market. However, these, although much more energy-efficient than incandescent lamps, have never been much appreciated by the consumer market due to a series of technological limitations such as size, shape, non-immediate ignition, premature degradation possible in case of frequent switching on/off and limited colour rendering. Furthermore, these lamps cause disposal problems at the end of their life cycle, as they contain mercury vapour. In a second phase, from 2010 onwards, there was a progressive introduction of LED lamps on the market. These initially had limited luminous fluxes and a high purchase cost, but progressively their luminous performances have improved and there has been a simultaneous decrease in prices, so much so that today they have almost completely replaced the compact fluorescent lamps. Indeed, LED lamps are even better than compact fluorescent ones in terms of durability, energy efficiency, robustness and reliability. Figure 5.1 shows, from

**Fig. 5.1** Eight years of evolution of LED lamps

left to right, eight years of evolution of LED lamps used as a retrofit of the classic incandescent lamp.

These first two phases of the introduction of LEDs in the lighting market have been characterised by a "retrofit" approach, in which the new light sources have been created to try to replace in shape and function the classic domestic lamps that are fitted in the holder of luminaires, often with final results of poor design and aesthetic value. Today we are in a phase where the process of retrofitting LED lamps is quite extreme, whereby internal filaments have been made with semiconductors that try to reproduce both the distribution of the luminous intensities and the aesthetic appearance of the old incandescent lamps.

At the same time, however, a process of innovation has begun in the design of the lighting products for street lighting, for workplaces and also for home environments. Actually, since 2010, new luminaires have been developed in which the lamp holder has been eliminated and a design of the LED light sources integrated into the lighting product has been developed. This design approach leads to the loss of objective separation between lamp and luminaire. With the advent of LEDs, the historical differentiation between lamp manufacturers and luminaire manufacturers has begun to fade and this process was also highlighted by the birth in 2012 of the industrial association LightingEurope, born from the merger of CELMA (European Federation of National Associations of Luminaires and Components Manufacturers) and ELC (European Federation of Lamp Manufacturers). This new line of product development has led to the marketing of new luminaires, in which the long-lasting LED source is no longer replaceable as it was in the past. Figure 5.2 shows a street lighting fixture with integrated LED light sources designed by Lab. Luce—Politecnico di Milano in 2011.

In these cases, the end of life of the product involves the dismantling of the entire luminaire. However, if in the past the lamp life ranged from 1,000 to 10,000 h, today a LED luminaire can have a duration that varies from 25,000 to 50,000 h, and most of the time it is limited by the life of its internal power supply rather than by the LEDs. This means that a LED luminaire running eight hours a day can last from 8 to 17 years and, if it is used for a few hours a day, it is expected to work for up to fifty years.

Fig. 5.2 A street lighting fixture with integrated LED light sources

### 5.2.2 Fundamentals on LED Technology

The LED is an optoelectronic component made by coupling two semiconductors. In the two parts of the semiconductors, indicated with P and N, impurities are added, made up of other chemical elements, to increase (P) or decrease (N) the number of electrons in the bonds of the crystal mesh. The absence of electrons in the atomic bonds generates holes. When an electron current flows from the N-zone, which is rich in electrons, to the P-zone, which is rich in holes, the electrons fill the holes, decreasing their potential energy and emitting photons. The wavelength of the emitted light depends on the difference in potential between zone P and zone N, i.e. the materials used (Table 5.1). Furthermore, this emission occurs in a narrow band of the electromagnetic spectrum. This means that LEDs emit only coloured light, which has a high saturation level. In addition to producing light, the electrical current that passes through the coupling also produces heat and this must be dissipated by the semiconductor chip that contains the PN coupling.

The light is emitted by the LED only when this is directly polarised, i.e. when the electric current of positive charges flows from P to N. This means that this light source must be powered by direct current (DC). The current must also be stabilised to check its operating regime and avoid an uncontrolled increase in the current, which could damage the LED. In some cases, it is powered by alternating current (AC). In this situation, the LED remains on only for less than half of the time, when the current flows in the direct half-wave and could generate flicker, a phenomenon explained earlier in Chap. 3 addressing flickering issues.

**Table 5.1** Colour of the light produced by LEDs according to the materials used

| Radiation emission | Material |
|---|---|
| Near to far ultraviolet | Aluminium gallium nitrate (AlGaN) |
| Ultraviolet | Diamond (C) |
| Blue | Sapphire ($Al_2O_3$) as substrate |
| Blue | Silicon carbide (SiC) |
| Near ultraviolet, blue, bluish green | Indium gallium nitrate (InGaN) |
| Green | Aluminium gallium phosphide (AlGaP) |
| Green, emerald green | Gallium nitrate (GaN) |
| Green, yellow, red | Gallium phosphide (GaP) |
| Yellow, orange, bright orange red | Aluminium gallium indium phosphide (AlGaInP) |
| Yellow, orange and red, orange, red | Gallium arsenide phosphide (GaAsP) |
| Red and infrared | Aluminium gallium arsenide (AlGaAs) |

Since LEDs only emit coloured light, various technologies are used to obtain white light. The simplest and most common technology is to use a blue LED in combination with a wavelength-converting material. In this case, the blue light generated by the LED excites a phosphorescent material that converts it, emitting it at longer wavelengths, with a peak typically in the yellow zone of the spectrum. In this technology, the chip is covered with yellow or orange phosphors, able to modify the spectrum of the emitted light, which is thus perceived as white by humans. A rather widespread problem in the early versions of these white LEDs, but which today has been reduced thanks to the improvement of production technologies, is the phenomenon of the so-called yellow-ring (Yang et al. 2013), due to a different distribution in the space of the blue light compared to the yellow light.

The choice of the blue LED and the phosphorescent material determine the CCT, the colour rendering and the efficiency of the light source. The most widespread and cost convenient solutions for the manufacture of white LEDs are based on the use of near-UV or blue LEDs (InGaN), coated with yellow phosphors (Ce:YAG). The phosphor conversion method makes it possible to achieve high levels of colour rendering, but this is achieved at the expense of efficiency. Figure 5.3 shows the measurement of the relative SPD of one of the first LEDs with white light emission, placed on the market at the beginning of the third millennium. This emits a white light with a CCT of about 10,000 K and is made by a blue LED, with a peak wavelength of 460 nm, covered with yellow phosphors. The hump visible in the SPD at 560 nm is due to the light converted by the yellow phosphors. From the perceptive standpoint, this source emits a very cool white light, tending towards light blue.

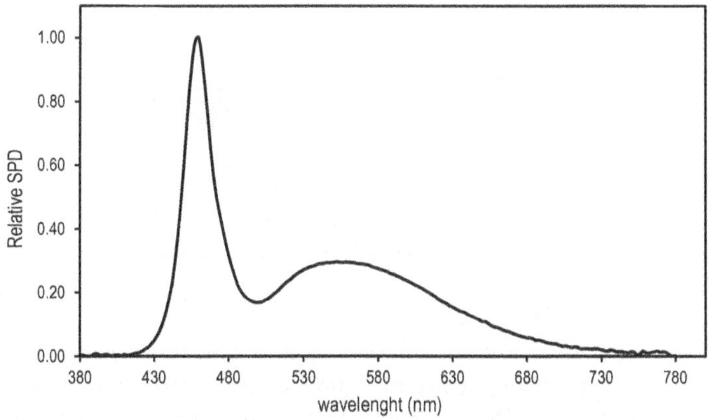


**Fig. 5.3** Spectrum emitted by a blue LED covered with yellow phosphors

## 5.3 LEDs for Lighting

For reasons due to the construction limits of semiconductors, the PN coupling is always very limited in size and therefore a single LED is not able to emit the same amount of light emitted by other electric light sources. Moreover, the electric current cannot increase beyond a certain limit because otherwise the LED is no longer able to dissipate the heat and burns. Even the most powerful single LED is not able, by itself, to emit a luminous flux comparable to that of the previous electric light sources. Research in this area has therefore developed composite light sources, belonging to the LED family, which are larger and more powerful than a single LED.

Alongside power LEDs for lighting (Fig. 5.4a), LED arrays and multichip LEDs have been created, as well as other even more advanced systems whose nomenclature is sometimes confused. The correct nomenclature has been defined and updated over time in the ANSI/IES RP-16-17 (2018) standard. These light sources generate both a stronger light flux compared to LEDs made with a single PN coupling and a qualitatively superior light, in terms of SPD and colour rendering. They are indeed light sources that contain many LEDs, even if the way in which they are aggregated to obtain the composite source of light is different. This type of product has given a decisive boost to the success of LEDs in the lighting sector. It has also provided greater possibilities for the design of lighting products, making it possible to obtain functions and results that have been impossible with previous light sources. Multichip LEDs, as well as power LEDs, are electronic products that need a supporting PCB, therefore the design of a lighting product using this technology also requires skills in the field of electronics.

The *multichip LEDs* are characterised by a series of high-power LEDs, made with separate chips and inserted on the same PCB in the same package. The same light source can feature LEDs with different colours and different CCTs. They are generally small in size and make it possible to obtain high luminous fluxes. Due to

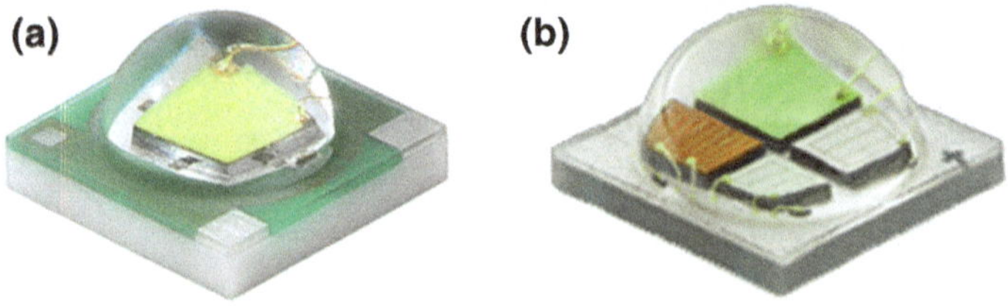


**Fig. 5.4** Comparison between a power LED (**a**) and a multichip LED (**b**)

their small size, these light sources can concentrate a large flux in a minimum housing space and for this reason they are used in the design of projectors, in combination with refracting optics or reflectors. Another interesting feature is that since they are able to contain different types of LEDs, they can emit light with a richer SPD in the various wavelengths, but also modify the emitted SPD by appropriately controlling the LEDs driver.

LEDs made of different materials also require different power supply characteristics. This means that in the manufacturing of lighting products, which envisage the inclusion of LEDs with different chromatic light emissions, separated drivers for the different types of LEDs must be provided. So, if on the one hand having different coloured LEDs in the same product makes it possible to mix these colours and then produce light with many different colour hues, on the other hand this leads to greater electronic complexity of the system and increased costs.

Multichip LEDs are also an alternative method to using blue LEDs and yellow phosphors to obtain white light and various colours. This is achieved by combining multiple LED chips of different colours: typically red, green and blue (RGB). Indeed, the use of the multichip solution achieves good results in terms of colour mixing. The advantage of this solution lies in the possibility of creating a range of colours within the gamut area defined by the chromatic coordinates of the individual RGB LEDs, with the consequent possibility of creating also a variety of whites, ranging from cool white to warm white. On the other hand, this method has the disadvantage of generating a white light with poor colour rendering (Viénot et al. 2005). To overcome this problem, some manufacturers have created multichips of the RGBW type (red, green, blue and white, Fig. 5.4b). A further disadvantage is the high production cost, since the electronic supply system is substantially more complex. This is the reason why, in the case of applications requiring white light only, this method is generally disregarded.

The *LED arrays* are made with a chip on board technology (COB, Fig. 5.5), have a high number of power LEDs, equal to each other and are characterised by a single phosphor layer, so the light emission area is larger than in the multichip systems. The LED arrays must be fixed directly in contact with the heat sink, using screws or suitable adhesives. This is a design and production simplification of the source in order to improve its performance, both in indoor and outdoor lighting fixtures. LED

**Fig. 5.5** COB LED array

arrays emit luminous fluxes comparable to traditional sources, but with efficiencies that are much higher today.

Subsequently, the need to obtain even better performing light sources and more adaptable to different types of applications has led to the creation of *LED modules*, sometimes also called light engines. These make it possible to create complete and standardised lighting engines, yet simpler to use from a design standpoint for those who do not have expertise in electronics. These sources integrate lighting functions with high light fluxes and quality, along with electronic, power, management and control functions, in a compact package of a few centimetres in size. The idea behind these products is providing a light source that can be easily integrated into projects, which in addition to the LEDs already include the PCB, a coupling system, connectors and secondary optics. Today, linear formats with lengths of over one metre are also available. These modular solutions facilitate the task of lighting designers and lighting product designers, who do not need to intervene on the source, since its characteristics are already predetermined by the manufacturer. With these sources, the lighting product design can be a work of formal styling, which can assume again the shape of traditional luminaires, or invent completely new formal solutions. At the international level, the Zhaga® manufacturer consortium was also established with the aim of defining a standard for the size and electrical and thermal interfaces of LED modules and LED arrays (Zhaga 2010).

In some LED modules, remote phosphor technology has been introduced (Narendran et al. 2005). In a remote phosphor source there are generally many blue LEDs on the PCB, but sometimes also of other colours, fitted in a mixing chamber. This is made of diffusing materials with high light reflectance and a transparent surface coated internally with yellow phosphors is applied at its output. Indeed, if the phosphors are positioned at a sufficient distance from the LED chip, the probability of a ray of light being reabsorbed by the LED is greatly reduced. In addition, the remote phosphor configuration reduces the operating temperature of the phosphors by increasing the life, efficiency and reliability of the white light LEDs.

An early example of LED module introduced on the market has been the Fortimo Philips, also available in a remote phosphor version, which has evolved over the years with increasingly high-performance models, with dynamic and better-quality light (Philips 2018). Osram has introduced on the market the Prevaled system that, in the latest versions, has been adapted to different application areas such as retail and food lighting (Osram 2018). The same has been done by CREE (2018), which has introduced LED modules whose product family has evolved over time and that today are characterised by high colour renderings and efficiency. Xicato (2013) has made LED modules, based on the remote phosphor technology, which have been the first to achieve a $R_a > 95$ colour rendering with commercial LED sources.

A further evolution, useful to obtain white light with high colour rendering and high efficiency, are the *hybrid LEDs*. In these sources, white light is obtained thanks to the combination of two types of LEDs: one red or amber and the other with a white hue, which however has an SPD with chromaticity shifted to the green area. These white-green LEDs are produced employing traditional phosphor conversion technology, using a blue emitter coated with a green phosphor. The advantage is that the latter has a very low light loss conversion factor, given that green and blue are closer in the spectrum than blue and yellow. This LED is therefore able to produce, in combination with the blue emission of the chip, a very efficient light source. The addition of the red or amber component, necessary to obtain a white light with a neutral-warm hue and a high colour rendering value, is generated by an LED that directly emits coloured light without the need of converting light through phosphors. Since two types of distinct sources are used, whose emissions must be mixed together, the product can have an optical component, such as a mixing chamber, in order to guarantee the uniformity of the chromatic properties even in near field conditions. Alternative optical solutions use reflectors, which ensure a fair colour uniformity, provided that the emission solid angle is not too narrow. Examples of hybrid sources have been developed by CREE (2015), with the TrueWhite® technology and by Osram (2013), with the Brilliant Mix technology. In a measurement of the light emitted with the latter solution, it has been possible to evaluate the effect on the SPD of two different mixes made to obtain white light, with a CCT of 3,000 and 4,000 K. In Fig. 5.6 it is possible to observe the variations of the emitted relative SPD. In the case of the 3,000 K, there is a greater contribution of the amber LED, while in the case of the 4,000 K there is an increase in the contribution of the white-green LED.

Recently, Toshiba has introduced a new LED technology called TRI-R (Toshiba 2017), with the aim of trying to reproduce the SPD of sunlight, without the ultraviolet and infrared components. According to the characteristics presented by Toshiba, these LEDs would emit an SPD in the range between 380 and 780 nm, able to reproduce that of sunlight for 95%. This technology uses an LED that emits purple light, covered with three different types of phosphors: blue, green and red (Toshiba 2018). This light source can obtain a CCT in the 2,000–6,500 K range and has a colour rendering $R_a = 97$. Seoul Semiconductor (2018) has also launched a new family of medium-power LEDs, of the Sunlike product line, based on the TRI-R technology with a luminous flux of 23 lm, with various CCT available from 2,700 to 6,500 K and with a colour rendering $R_a = 97$.

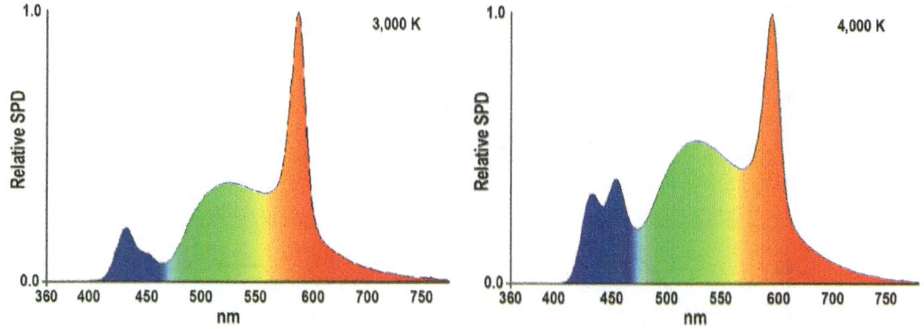


**Fig. 5.6** The SPD of the Osram Brilliant Mix, set at 3,000 and 4,000 K

## *5.3.1 Tunable White LEDs*

As widely explained in this book, the human circadian system is synchronised in nature by the variation of natural light in quantity, direction and CCT, from sunrise to sunset. Starting from the observation that the CCT of natural light changes during the hours of the day, LED arrays, LED modules, lamps and luminaires have been available for some years with the characteristic of being able to vary the CCT through a manual or programmed control system. In this section of the book, we are particularly interested in considering the light sources, i.e. the LED arrays and the LED modules, because they are the basic elements with which the luminaires are built.

The use of COB LED arrays represents a valid possibility for the creation of tunable white systems for lighting applications. Among various products available on the market, it is interesting, for example, to analyse the technology of the Tiger Zenigata product by Sharp (2014). This source consists of a COB in which two types of different phosphors, with alternating stripes, are placed on the matrix of the blue LEDs, with the aim of producing a variable emission, from warm to cool light, within the same and compact light source. In the matrix, the LEDs covered by the two types of phosphors are connected to separate electric circuits and therefore can be controlled independently on two channels. As shown in Fig. 5.7, by appropriately adjusting the percentage of the currents in the two channels, a white light source can be obtained, with a colour temperature ranging between 2,800 and 5,500 K.

For example, by mixing the two channels to obtain a CCT of 3,000 and 4,000 K, the main values of colour rendering and efficiency have been measured, obtaining a very high colour rendering: $R_a > 96$ (Table 5.2). Furthermore, the rendering indices of the two colours that are usually problematic for LEDs, red ($R_9$) and pink ($R_{13}$), provide good results. Figure 5.8 shows the relative SPD of this source when it is adjusted to provide a CCT of 3,700 K. From the shape of the spectrum, it is possible to observe the emission peak due to the blue LED and the two emission humps, one in the yellow-green zone and one in the orange zone, due to the two different types of phosphors applied on the LED matrix.

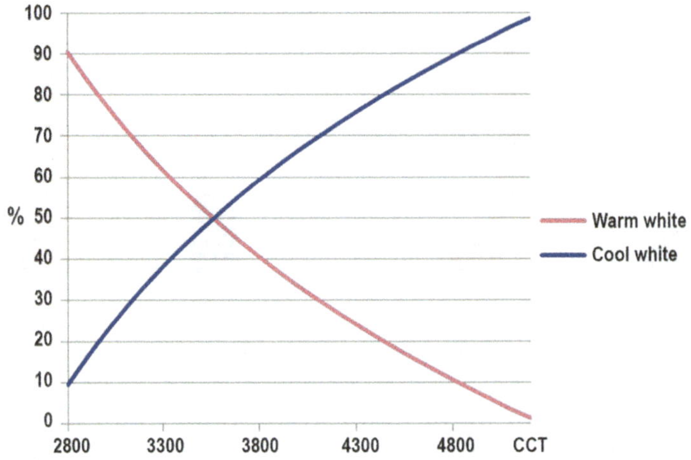


**Fig. 5.7** Percentage of warm white and cool white light needed to obtain the desired CCT

**Table 5.2** LER values and colour rendering measured for the Sharp Tiger Zenigata LED

| LED | Target CCT (K) | Measured CCT (K) | Warm (%) | Cool (%) | $R_a$ | $R_9$ | $R_{13}$ | LER (lm/W) |
|---|---|---|---|---|---|---|---|---|
| Tiger Zenigata | 3,000 | 2,929 | 83.43 | 16.57 | 97 | 85 | 100 | 269 |
| Tiger Zenigata | 4,000 | 3,946 | 38.37 | 61.63 | 96 | 86 | 98 | 279 |

**Fig. 5.8** SPD of the Sharp Tiger Zenigata LED array at 3,700 K

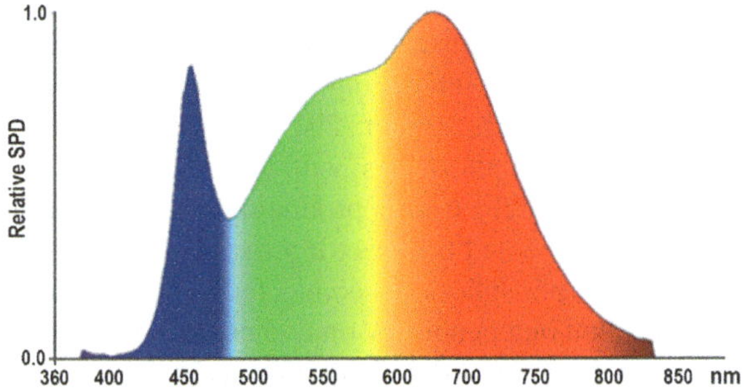


The operating principle of the tunable white LEDs is therefore based on the controlled mixing of at least two sources, which emit warm white light and cool white light, in order to obtain variable CCT (Fig. 5.9) values according to a manual or automatic control. In addition to LED arrays, LED modules have also been available on the market for a long time, with very advanced management functions, which make it possible to obtain the tunable white.

Figure 5.9 shows the measured SPDs, normalised at 560 nm, of the Tridonic Tunable White system. This product family has been one of the first to be launched

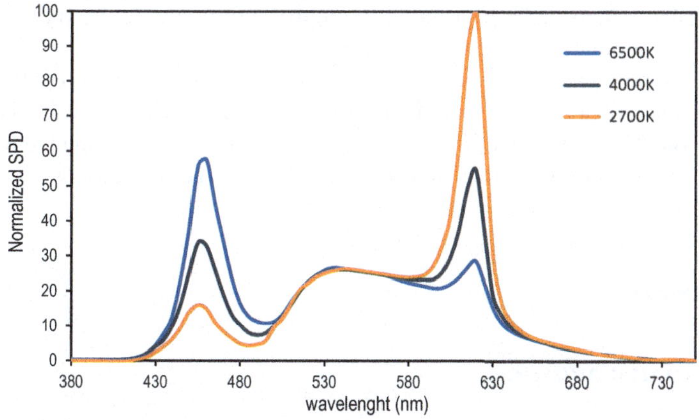


**Fig. 5.9** SPDs of the Tridonic tunable white system

on the market (Tridonic 2013) and today has a colour rendering that can be greater than 80 or 90, depending on the model. The control electronics in the LED module system makes it possible to adjust the CCT while keeping the flux constant, which in turn can be adjusted, from 3 to 100%, independently of the CCT. This family of products offers the classic compact LED modules, with a diameter of 5 cm, linear solutions up to almost 30 cm long, and square area sources with a side of 27 cm or round ones with diameters up to 31 cm. There are also modular components that can be assembled to build longer lengths and wider areas. The variation range of the CCT depends on the models and may range between 2,700 and 6,500 K. An interesting aspect of this lighting system is that it can also be controlled via wireless technology with the digital addressable lighting interface standard (DALI 2017) used for a long time in building automation.

Another interesting LED module system is the LED Engine LuxiTune by Osram (2017), which provides a CCT and a flux that are independently adjustable. This system makes it possible to emulate the behaviour of a halogen incandescent lamp, with a CCT variation between 3,000 and 1,600 K, when it is dimmed, or a general setting of the CCT in the range 1,800–6,500 K. The system is composed by an LED module and optical accessories that make it possible to obtain light beams with different apertures, from 21° to 45°. For a CCT of 3,000 K this source has a colour rendering $R_a > 90$, while the average is 85. The module can be integrated into an LMS thanks to a series of electronic control and communication accessories. This feature is one of the key elements for the integration of LED lighting products in building automation and human centric lighting of the future. Indeed, in addition to the classic connectivity with the DALI and digital multiplex standards DMX 512a with RDM (USITT 2013), it also offers the possibility of communication and control through wireless networks. Thanks to configurable accessories, the system can communicate over low-power wireless networks, such as Zigbee (2006) and Bluetooth (2010) low energy (BLE), which make it possible to control and integrate the lighting system into building automation projects.

Thanks to these characteristics, artificial lighting can be integrated with light and presence sensors, but also sensors identifying people in interior spaces and detecting their position. This network of light sources and sensors can be managed by a smart LMS to promote energy savings and also an indoor circadian lighting.

## 5.4 Sensors

### 5.4.1 Passive Sensors

*Light sensors* are devices based on electronic components such as photoresistors, photodiodes and phototransistors, able to detect the presence of ambient light. They have been used for many years both for adjusting the screen brightness of smartphones and tablets and for setting the levels of artificial lighting. Thanks to these sensors positioned on the ceiling, but also on tables (Caicedo et al. 2017), it is possible to detect the presence of natural light in an environment and adjust artificial light to the minimum necessary to reach a planned illumination level. This makes it possible to achieve an energy saving target. The information received from these sensors, in addition to the control of electricity consumption, can also be useful for determining the presence of people in the home environment (Makonin and Popowich 2011).

The family of the *chromaticity sensors* includes both RGB light sensors and sensors able to measure the XYZ tri-stimulus values, which are used for the accurate measurement of the chromaticity of light sources. The XYZ sensors are able to provide, through a simple calculation, the values of the Yxy CIE 1931 colour coordinates of an incident light flux, within the standard colour space (CIE 015:2004). Colour sensors can include filters to block unwanted infrared emissions (IR), thus providing extremely accurate colour measurements in the visible spectrum only. Their high sensitivity combined with a wide dynamic range makes them suitable for continuous measurement of the ambient light CCT in display management systems. They are also suitable for managing automatic white balance in cameras (Valencia et al. 2013). Some sensors feature an additional IR channel to facilitate the identification of light sources with IR content, such as natural light. This option is useful for distinguishing the component of natural light compared to that of artificial light from LED sources, since these do not emit IR. This type of sensor is scarcely used in lighting control systems. Nowadays, its use is growing due to the possibility for an LMS to manage also the CCT of light, in a human centric lighting perspective. In this kind of scenario, the possibility of controlling the illumination level in the environment, but also the CCT, becomes an important factor for the success of a control strategy.

*Passive infrared sensors* (PIR) respond to temperature variations within their field of view and are defined as passive because they do not emit any form of energy, but only detect variations in infrared emissions in the environment. The PIR is more sensitive to moving objects that emit thermal energy with a wavelength of approximately 8–14 μm, typical of the human body (Moghavvemi and Seng 2004). The Fresnel lens applied to the sensor has a profile designed in such a way that the visual field of the sensor is divided into discrete wedge-shaped areas. The sensor responds when it detects a change in temperature within two or more discrete zones, so it is less likely that this type of sensor detects small movements when a subject moves away from the sensor. Indicatively, the PIRs are sensitive to the movement of a hand up to ~4.5 m, of an arm and of the torso up to ~6 m and of the body up to ~12 m. The main limitation of PIRs concerns their need for a completely free view. Indeed, if the movement takes place behind furniture elements, the lighting can switch off even if the room is busy (Benya et al. 2003). These sensors therefore detect people's movements but cannot directly report their presence. If a person is sitting motionless for enough time, will become invisible to the sensor. Some models of PIR sensors also incorporate a light sensor. Figure 5.10 shows the principle of operation of the Steniel presence detector LuxMaster BLS with three PIR sensors and an integrated ambient light sensor. In this way, the light is switched on only when there is movement at the same time as a level of light below a predetermined threshold. This is to avoid activating the lighting system in the presence of natural light or other sources of illumination.

The *sound sensors* are passive sensors that do not emit any kind of signal. These use microphones or other transducers to determine if an environment is busy or not, measuring the noise present. They are used in industries and storage warehouses. They react to all noises, including those coming from spaces adjacent to those controlled and consequently they tend to turn on the light even in unoccupied environments. This technology is almost never applied alone but is used in conjunction with PIR sensors.

Another type of passive sensors are the *video cameras*, used in conjunction with image processing software. This method makes it possible to reveal the presence of individuals in a space. The images detected by a video camera are a very reliable way not only to determine if an environment is occupied or not, but also to detect the number of occupants and possibly their identities. The images acquired by the camera can be processed in real time by a software that makes it possible to quantify and recognise the people present, thanks to face recognition, thus making it possible to activate different lighting scenarios according to the detected conditions (Chun and Lee 2013). However, in public places or workplaces, but also at home, this type of device generates problems of acceptance in users who may feel their privacy to be violated. The result is that these systems are rarely used in building management systems, and especially for lighting control.

An interesting alternative to video cameras, which does not pose privacy problems, are the *thermopile array* and *microbolometer array* thermal sensors of new generation that represent a very promising technology for the future (Kimata 2017), also usable in addition to PIRs. These sensors make it possible to locate people inside an interior space by measuring the temperature in different areas, whose dimensions vary depending on the angle of view and the matrix resolution of the sensor used. The detection of people within the scene framed by the sensor takes place by subtracting

**Fig. 5.10** Principle of operation of the LuxMaster BLS sensor. *Courtesy* Steinel Vertrieb GmbH

the background thermal image of the environment. Sensors can also be equipped with optical systems to adapt the sensor's visual width to the environment in which it is used, home, office or other interior space (Gonzalez et al. 2013; Kimata 2013). This type of sensors can also detect the presence of people who are motionless (Beltran et al. 2013) as well as the user's position or movements within a room, the first step for recognising the activities carried out (Basu and Rowe 2015). The new generation of these sensors makes it possible to recognise objects and object-occupant interactions through the analysis of the respective heat models, managing to detect a fair number of different activities (Kallur 2014; Hayashida et al. 2017).

## 5.4.2 *Active Sensors*

*Ultrasonic sensors* are another category of devices used for presence detection. They are active devices that emit ultrasound frequencies in the 25–40 kHz range and detect the waves reflected by the environment and any occupants. The waves reflected by moving objects have a frequency variation, compared to the emitted one, known as Doppler effect, due to the relative movement between source and sensor. Unlike PIR sensors, these sensors do not need a free view, since ultrasonic waves can be reflected by the surfaces present in the environment reaching, theoretically, all points of the controlled space. However, this greater visibility makes these sensors more sensitive to false detection in unoccupied spaces due to movements in areas adjacent to the one to be monitored. Some examples of false detection are the activation of the sensor due to air turbulence produced by an air conditioning system, or the movement of paper sheets coming out of a printer (Floyd et al. 1995). PIR sensors have wedge-shaped sensitivity zones and activation occurs when an object moves from one zone to another, resulting in a discrete-type activation mechanism. Ultrasonic sensors, on the other hand, have a detection mechanism that can be defined as continuous. Since the intensity of the reflected wave decreases with the square of the distance, in ultrasonic sensors the sensitivity decreases as the distance between the detector and the moving object increases. These sensors, in the presence of particular operating conditions, which we could define critical, tend to turn on the light in unoccupied spaces, contrary to what happens with the PIR sensors.

*Microwave sensors* are active sensors, conceptually similar to ultrasonic sensors. They emit a radio signal at the frequency of 10 GHz and measure the frequency variation of the signal reflected by moving objects or people. They are used in large public spaces, where movements are large, such as in hallways, sports halls and in school lobbies. The long range of these devices makes them suitable for outdoor applications, such as for monitoring the passage of vehicles, or for automatic door opening systems, in which the door opens only when it detects movement near the entrance to the building. Microwave sensors have the ability to see through non-metallic materials, such as plastic, fiberglass, and even not very thick brick walls. However, this feature makes them very sensitive and can generate undue activations of the lighting system, due to the detection of movements in spaces adjacent to the controlled one.

### *5.4.3 Wearable Sensors*

The use of *smartwatches*, *actigraphs* and other wearable devices with growing features is an increasing trend in recent years (Marr 2016). These devices can inform users regarding a series of data concerning the organisation of their daily life, also with reference to certain physiological parameters. The reason why there is a need, especially in more advanced countries, to monitor oneself, derives from the fact that the self-improvement is one of the most deeply rooted motivation factors in human nature. In a futuristic scenario, a wearable sensor could record the activities and the light to which the wearer is subject during the day and transfer this information to the LMS to obtain the most appropriate lighting for a correct circadian stimulation. For now, some of these products have been used to analyse users' activities in medical research (Henriksen et al. 2018). Although not exactly part of the wearable sensors

**Fig. 5.11** The Daysimeter in its jacket version. *Courtesy* Lighting Research Center—Rensselaer Polytechnic Institute

category, in the research carried out by Heise and Skubic (2010) these scientists have experimented with a non-invasive system, placed under the mattress, which makes it possible to monitor the different phases of sleep. During the night this sensor is able to monitor breathing, heartbeat, time spent in bed and the level of agitation of people when they are lying down.

Conversely, a research conducted by Figueiro et al. (2014) has used a device to measure the daily exposure to light of people during their daily activities.

The interesting aspect of this research is that the activity and the amount of light received by the patients over the 24 h have been evaluated using a measurement instrument known as Daysimeter (Figueiro et al. 2013). This is a small wearable instrument that continuously records the light, its chromaticity and the levels of activity of the person wearing it (Fig. 5.11). Various versions of the Daysimeter, wearable on clothes or on the head, have been developed at the Lighting Research Center of the Rensselaer Polytechnic Institute, Troy, NY, USA.

A different product has been developed by Philips, the Actiwatch Spectrum Plus, which can be worn on the wrist. The Actiwatch, in the PRO version, can also generate audible and vibrating alarms to remind the subject wearing it to enter subjective scores based on a programmed system or on a manual basis (Philips Respironics 2018). In another research Thorne et al. (2009), however, have shown that these wrist products risk to make incorrect recordings of exposure to light because, particularly in cold seasons, they may be covered by clothing. Later, other actigraphs have been developed that could also record exposure to light. Price et al. (2017) have compared the performance of seven wearable products, such as badges, which therefore do not present the problem of risking being covered with clothes like in the case of wristwatches.

After many years of empirical experiments in the field of circadian lighting, where in the absence of precise and shared reference standards there is a great diversity of methods for measuring light on human subjects (Hubalek et al. 2006), the issue of correctly evaluating, in the experiments, the quantity of light, its spectral content, its timing and the history of its daily exposure, has become of primary importance because there is now a widespread criticism of the so-called in vitro experiments or experiments conducted on animal samples (Figueiro 2013).

If we want to verify the effects of artificial light on the life and well-being of human beings, it is essential that these effects are evaluated on the field, in the usual spaces of everyday life, just as it was for research on the relationship between light and Alzheimer's disease presented in the second chapter, which has been conducted in the places where patients live. In the past, however, much research has been done by placing human subjects in closed and controlled rooms, in front of luminous screens, with various monitoring systems installed and used for that purpose. If this experimental methodology is certainly convenient and controllable by physiologists, it is also true that the results obtained are generally too limited to be able to infer guidelines for designing the lighting of everyday living spaces. Indeed, in closed experiments, the subject is removed from the context of his or her real lifestyle, because even the colours of the surrounding surfaces must be considered, since they influence the spectrum of light that reaches the eyes, and because the timings and degrees of freedom, which come into play when we are in everyday life spaces, are much larger than the controlled rooms of in vitro experiments.

## 5.5 Indoor Positioning Systems

The indoor positioning systems (IPS) make it possible to estimate the position of users and objects in interior spaces. The global positioning system (GPS) used outside is not usable for this purpose, since the satellite signal is not available inside buildings. To create an IPS system, a structure consisting of receivers and transmitters is needed, which can be organised according to two possible approaches.

The first method is based on a client-based principle in which a smart mobile device (SMD), like a smartphone or a smartwatch, is equipped with a special app and accompanies the user. This device is able to continuously determine its position within an enclosed space thanks to the reception of an external signal (Wi-Fi, BLE, Li-Fi) and to the integrated sensors: accelerometer, gyroscope and magnetic sensor. Through a wireless network, the SMD then transmits its position to the server that can manage it within the LMS.

In the second method, called server-based, there is no processing within the SMD client, which merely transmits wireless presence signals. These signals are received by a network of sensors connected to the server, which can thus determine the position of the client.

The main advantage of these IPS systems is based on the fact that smart mobile devices are now widely distributed among the population. However, the use of smartphones as a system for detecting the position of a person, has its main limit in the fact that the position that is detected is that of the smartphone itself and therefore if the user does not take it with him or her, the system does not work. In the future, for the purposes of the LMS, in a server-based configuration, wearable devices such as smartwatches or actigraphs with the possibility of connection in a wireless network may be more suitable.

### 5.5.1 Position Detection via Wi-Fi

Wi-Fi-based IPS have been used in many research projects, as today there are plenty of Wi-Fi hotspots installed (Dortz et al. 2012; Park and Rhee 2017). Positioning also works when users have Wi-Fi enabled on their SMD but have not logged into the network through an authentication procedure. However, there are some disadvantages in detecting indoor position with Wi-Fi. Accuracy is 5–15 m, less than BLE, and moreover, iOS devices are excluded from client-based positioning via Wi-Fi. The signal detection range for a Wi-Fi-based system is less than 150 m, but it is about 5 times higher than that obtained through a BLE signal. The accuracy of the position depends on many factors, such as the number of hotspots available in a specific building area, the reflections affecting the signal, the materials of floors, ceilings, walls and also the shielding effect of the user's body. Position accuracy can be slightly improved through information provided by smartphone sensors.

The operating principle, called fingerprinting, is based on the analysis of the signals that the hotspots transmit continuously to communicate their presence. An app running on the user's SMD compares the intensity of these signals with a database associated with the interior environment to determine the position. Position identification occurs faster when the SMD is connected to the Wi-Fi network, otherwise there may be delays of up to five minutes. In order to use this system, in addition to the app on the user's device, the database used by the fingerprinting must be determined and calibrated by the system installer.

### 5.5.2 Position Detection via BLE

BLE was introduced in the Bluetooth v4.0 standard in 2010 to facilitate short-range communication between devices, with reduced power consumption. The use of this method for positioning was possible with the introduction of Beacons on the market (Hern 2014; Kalbandhe and Patil 2016). Beacons are small and compact radio transmitters sending a BLE signal, covering an area up to about 70 m around their position, powered by a battery that can last a few years. The emitted signal can be detected by the users' smart mobile devices in a client-based architecture, or by a specific hardware in a server-based approach.

An SMD must be equipped with an app capable of interpreting the signal strength of at least three Beacons through which, with the mathematical triangulation method, it can determine its own position. Using BLE Beacons, it is possible to obtain a precision of about 1–2 m in determining the user's SMD position. Also in this case, in order for the IPS to work, the position of the Beacons in the 3D space of the interior environment must be known to the server, following an initial IPS installation and calibration procedure.

### 5.5.3 Position Detection via Li-Fi

The introduction of LEDs for lighting has made it possible to develop a new communication system, light fidelity (Li-Fi), for short-range data transmission through artificial lighting (Dimitrov and Haas 2015). It is currently being developed worldwide by many manufacturers. With this system, data is transmitted, using a protocol similar to Wi-Fi, via light at a frequency that is not visible to the human eye.

In addition to data transmission, this system also lends itself very well to IPS. Indeed, every light source is assigned to a precise position in the lighting project of an interior space. The light produced by a source identifies a delimited area, in which the illumination obtained from the light source is greater than the others. A first solution of this type has been proposed by Signify (2017) with the Indoor Positioning system for retail lighting. Thanks to a specific app, a smartphone can, for example, provide information on products in a supermarket, identified with a precision of 30 cm. For user positioning, the system uses both Li-Fi and BLE (Fig. 5.12).

For Li-Fi communication systems, an international standard is being defined (Pretz 2018). From the standpoint of functioning as IPS, it is essential to underline that this system works only with the light on.

### 5.5.4 Open Source Solutions

The IPS topic is also covered by some open source software development projects. These are based on the detection of Wi-Fi or BLE signals. The use of such software can therefore avoid the use of sensors but requires a computer to be run. This can be a server in the building where the IPS is needed, or also a single-board computer like the Raspberry Pi (2012), whose cost is very low. These open source solutions require engineering and computer skills to be installed and configured. The configuration typically includes the three-dimensional identification of the Raspberry Pis and the fingerprinting of the considered environment.

**Fig. 5.12** The IPS based on Li-Fi BLE for retail lighting. *Courtesy* Signify/Philips Lighting

The *Room Assistant* (Rothe 2018) software makes it possible to estimate the position of BLE devices such as SMDs and Beacons with respect to the system on which the software is running, which obviously must be equipped with a Bluetooth RF transceiver. An interesting aspect of this IPS is that it integrates into the larger open source Home Assistant project (Home Assistant 2018). This software is proposed as an alternative to commercial home management devices, making the topic of local control and privacy a priority in managing the system.

The project *Find3* (2018) also offers a software library that performs an IPS function and interfaces with open source home automation systems, such as Home Assistant and openHAB (2018). Its position locating method is based on both BLE and Wi-Fi signals generated by the user's SMD. The software *Find-lf* (Schollz 2018) uses Find3 libraries to perform a server-based scan of the users' SMDs and then identify their position in different rooms. The advantage of the server-based signal scanning mode is that no specific app is required on the SMD, nor any authentication on the Wi-Fi network, but only the Wi-Fi activation on the device. The client-based scanning method is instead used by the Datanoise (2018) software in which there is an app running on the user's SMD to be located and uses both the Wi-Fi and the BLE signals to determine the user's position.

An alternative, based only on the Wi-Fi signal, is the software *whereami* (van Kooten 2018). This system requires the presence of multiple Wi-Fi detectors, such as the Raspberry Pi, in the environment considered and based on these it determines the position of the user through his or her SMD, provided that this has the Wi-Fi function enabled.

## 5.6 Smart Lighting

A currently emerging idea is smart lighting, i.e. the creation of lighting systems with the ability to control, communicate and interconnect data, able to provide new ways of interacting with the luminous performances in new luminaires, equipped in turn to offer additional services. This idea of modularity, integration and customisation of the lighting system also provides a remarkable ability to adapt the lighting system, which must have its own intelligence, according to the lighting needs of people in interior spaces. This design scenario is facilitated by progress in various technological sectors that are now part of the lighting world including, in particular, the ability to manage and exchange information and the connection of the digital and analogue worlds through the Internet of Things (IoT). Additional elements that facilitate this innovation are the miniaturisation and diffusion of components that can be integrated into the LED module systems, such as sensors and wireless communication, which have already been discussed.

In this scenario, lighting from a commodity can become a service integrating itself in the development of the ICT sector. Luminaires can become hubs of data from other sources and must therefore be able to adapt luminous performance based on the processing of multiple information. Compared to the past, the transformation of lighting systems into more intelligent and network-connected objects also entails a simplification of the control interfaces, which are becoming simpler and more enabling. Sensors are also less expensive and the possibilities for innovation in light control are increasingly widespread.

Interior lighting can be designed in a flexible and customisable way, by means of autonomous or automatic systems that can adapt the lighting to increase productivity during the activity hours, but also change its direction and the SPD to improve the well-being, safety and health of users. Lighting will be able to support the performance of those who feel tired, to calm down in moments of particular stress and to adapt better to the daily rhythms. Already today, the relationship between human beings and lighting is looked at through new interaction paradigms. New applications and possible uses are also being developed thanks to the availability of smartphones, tablets and other SMDs as new active light control interfaces.

### *5.6.1 Control Methods*

In the control of lighting with the new active interfaces, the user is directly involved in the interaction through the element that allows lighting control. This means that the information transformed into light derives from the manipulation or action on the object used for the control. The active control is directly linked to the customisation of the surroundings, guaranteeing the principle of self-determination in the way the user relates to the illuminated surroundings. In the case of a passive relationship, the user is not directly required to interact with the product. In this case, the feedback is designed to send digital information about the status of the system, so that the user can decide whether to act to restore or change the lighting status.

Today, thanks to LED-based systems and wireless communications, manual control of the quantity, CCT and light direction makes it possible to adjust the lighting to suit the condition desired by the user in that particular moment. This is a factor that improves the light quality perception, as well as the mood and environmental satisfaction of people (Newsham et al. 2004; Boyce et al. 2006; Veitch et al. 2008). However, a completely manual control, in more complex situations, could lead to chaotic lighting with results that may be not completely positive. This happens where many individuals modify the lighting and this may be also influenced by factors independent of people, such as the variation of natural light.

In the case of a completely automatic control, the lighting system makes autonomous decisions, based on information obtained from the context and does not allow the individual to interact with the illuminated atmosphere. Lighting changes according to the parameters set at the design stage and is adjusted according to the information it receives from various types of sensors. An automated lighting system is useful in situations where several luminaires must be controlled at the same time and they have different lighting characteristics. Furthermore, an automated system could select a particular lighting based on the experience gained from data over time, or on pre-sets that can improve the lighting atmosphere, setting the light with new potential features that users often ignore. However, a fully automated system can cause unwanted light situations, misinterpreting the signals received from the sensors. This eventuality could produce a sense of diminished control by the users and a poor acceptance of the environment, resulting in a feeling of discomfort.

Today, we look for a combination between manual or automatic systems for light control. We try to find a balance between a lighting that preserves some autonomous and automated behaviours, pre-set during the design phase, and the possibility, if necessary, of a manual intervention by the user, providing a level of customisation and interaction. This hybrid system makes the individual able to interact by determining the behaviour of lighting in specific places and times and with the desired characteristics. At the same time, it must make it possible to choose different predetermined scenarios, or completely delegate to the lighting system an autonomous behaviour, although interruptible and controllable at any time. Some studies have also shown that lighting customisation provides the advantage of saving energy (Moore et al.

2003; Galasiu et al. 2007). However, the real extent of this saving is a controversial topic (Veitch et al. 2010; Williams et al. 2012).

This is what today is meant as intelligent LMS, i.e. a system capable of learning directly from user behaviour, in order to create possible lighting scenarios that are not previously pre-determined, but which are nevertheless able to better adapt to the environment, to activities and to users.

### *5.6.2 Control Strategies*

A control strategy is a description of how lighting can be modified to suit the needs of the stakeholders, such as users, owners and managers. In the simplest strategy, users can turn the light on and off and choose intermediate levels of brightness through dimmers. Additional control possibilities are offered by panels that make it possible to select the lighting scene that is most suited to the task at hand. For example, in a meeting room the scenes could be: projection, concentration and discussion.

Another control strategy is the creation of a timed list of activations and level adjustments, defined for the entire day. With this method, consumption is lower because, during the periods in which the building is assumed to be empty, lights are off, or only surveillance lights are on. The LMS restores the lighting when the building is in use, just before the first user enters the premises. This strategy lends itself very well to some types of buildings: commercial and institutional ones and all the times when it is possible to define a precise presence schedule.

Many modern buildings integrate occupancy monitoring systems to control their technical facilities. Current systems usually use a single device, placed in a well-defined position in the environment and without any interaction with other management systems able to control space occupancy. This produces results that are subject to a high degree of uncertainty. To compensate for this uncertainty, the time for switching off the light is usually delayed and sensors are set with a high sensitivity. However, these choices lead to a reduction in the energy savings achievable with a more accurate occupancy measurement. Despite the delay in the switching off phase, light is sometimes turned off in areas that are still occupied, since there is only one sensor and the user could occupy a position outside of its sensitivity range. Moreover, in some cases the area actually covered by the sensor is smaller than that declared by the manufacturer. A better performance can only be achieved by considering a larger sensor network and an accurate processing of the information received from the entire network of detectors, which also makes it possible to determine the exact position of the users.

The adaptation of lighting to space occupancy is the strategy used in buildings where such presence is temporary or sporadic. The most used technology today is based on passive infrared sensors (PIR), due to their cost-effectiveness and ease of use. This technology leads to an estimated energy saving of ~24% (Williams et al. 2012). In early systems based on this method, the illuminance level was returned to the initial value when a user was detected. In more recent ones, the control strategy is of the auto-off/manual-on type and guarantees a better degree of user satisfaction (ANSI/ASHRAE/IES Standard 90.1-2016). Indeed, it has been observed that the excessive use of LMS without any manual option can lead to wasting energy due to people's behaviour (Pigg et al. 1996).

The flux of luminaires into the environment can be adjusted, according to the amount of natural light, to maintain a pre-set illuminance threshold. In this case, artificial light acts as an addition to natural light and the achievable savings have been estimated at around 28% (Williams et al. 2012). The LMS is based on a light sensor that measures the level of illuminance within the environment and the light emission of the light sources is modified to reach the pre-set threshold. The sensor is usually placed on the ceiling of the environment to be monitored. This is a weakness, since the sensor does not measure the level of illuminance on the visual task, but rather a magnitude proportional to the luminance in its opening angle. The early systems only used one sensor per room (Pigg et al. 1996). Later, LMS with multiple sensors were proposed, capable of controlling many luminaires (Meugheuvel et al. 2014; Borile et al. 2017). Space sensor networks have also been used to provide greater control and stability for systems consisting of a series of luminaires with built-in sensors (Caicedo et al. 2017). The information flow from the sensors on the work stations can be used to estimate, when training the control algorithm, the way in which natural light is mapped within the environment. Natural light decreases with the distance from windows and the illuminance level also depends on possible obstructions (Borile et al. 2017). Another approach is to use mobile sensors or sensors integrated in the workspace (Newsham et al. 2008; Yeh et al. 2010). In this case, the problems are the temporary darkening of the sensor caused by the user or by objects on the work surface and the power supply of the sensors, provided by batteries.

Personal lighting control must be considered during the design and installation of an LMS. The simple solution of guaranteeing only one lighting value inside the room, without taking into account the specific needs of individual users or their needs for specific tasks, often leads to uncomfortable solutions, disaffection of users and energy waste. The LMS should always offer users the chance to control the lighting around them, depending on the activity carried out, because this possibility ensures greater user satisfaction (Veitch et al. 2008; Tan et al. 2018).

In order to meet energy saving and comfort requirements, the control strategies seen above can be used in combination: timing, light and presence sensors. In this way, energy savings could reach ~38% (Williams et al. 2012). Estimating energy savings is quite complex, because they depend on factors such as the configuration of the sensors and the space occupancy dynamics (Richman et al. 1996; Neida et al. 2001). Natural light sensors offer savings that depend on building configuration, urban context, orography, weather conditions and the sensor network used.

### 5.6.3 Main LMS Available on the Market

Today the market already offers wireless LMS for service, hospitality and domestic environments.

One of the long-established systems in the field of building automation is the *Konnex* (1999), although it is not strictly just an LMS. Briefly called KNX, it is an open standard (ISO/IEC 14543-2-1:2006; EN 50090-1:2011) aimed at managing comfort and energy savings in all types of buildings: from homes to offices, from shopping centres to industrial spaces. The KNX system can manage energy, heating, ventilation and air conditioning, lighting, blinds and shutters, security systems, audio-video and home appliances. The most used means of communication for the KNX is the wired twisted-pairs, but using specific gateways there are also versions based on powerline, Ethernet and even wireless, with the Z-Wave (2001) network protocol. The KNX can manage over 57,000 connected devices. In addition to the classic wall controls, today there are dozens of smartphone apps, free or commercial, that can play the role of interfaces, even remote, for the management and control of all the connected devices. KNX has also opened up the connection to the Internet of Things. It has made available free software tools, the KNX Web service exporter and the KNX Web service gateway, which lower the barrier to access the IoT.

The *Casambi* (2018) system is based on wireless communication and control via BLE. This platform makes it possible to control and configure the LMS from a common smartphone through a friendly graphical interface. For the communication between the devices of the lighting system, Casambi uses a wireless mesh network based on a proprietary protocol that can be integrated into the IoT via a gateway (Fig. 5.13). This protocol carries the information packets containing the commands and the status of the luminaires between the different devices of the system. The intelligence of the LMS is replicated in every node of the mesh. This means that if one or more nodes fail, the system continues to operate. Furthermore, the system provides for the possibility of updating the firmware present in the nodes via the network itself, thus containing installation and maintenance costs. In addition to a smartphone, the lighting can be controlled with accessories from the Casambi network, such as the Xpress wall switch, which is battery operated and requires no electrical connection (Fig. 5.14). The system also makes it possible to turn an old wall switch into an intelligent control, by adding a BLE module.

The possibilities offered by the configuration and control interfaces are many (Fig. 5.15) and may include:

- Adjusting the luminous flux of the luminaires.
- Grouping the luminaires.
- Configuring, saving and recalling lighting scenes.
- Configuring, saving and recalling dynamic lighting settings.
- Adjusting the CCT of luminaires equipped with tunable white.
- Adjusting the colour for luminaires with RGB sources.
- Setting a timer to recall scenes or dynamic lighting based on date and time.
- Considering the time of sunrise and sunset in programming.

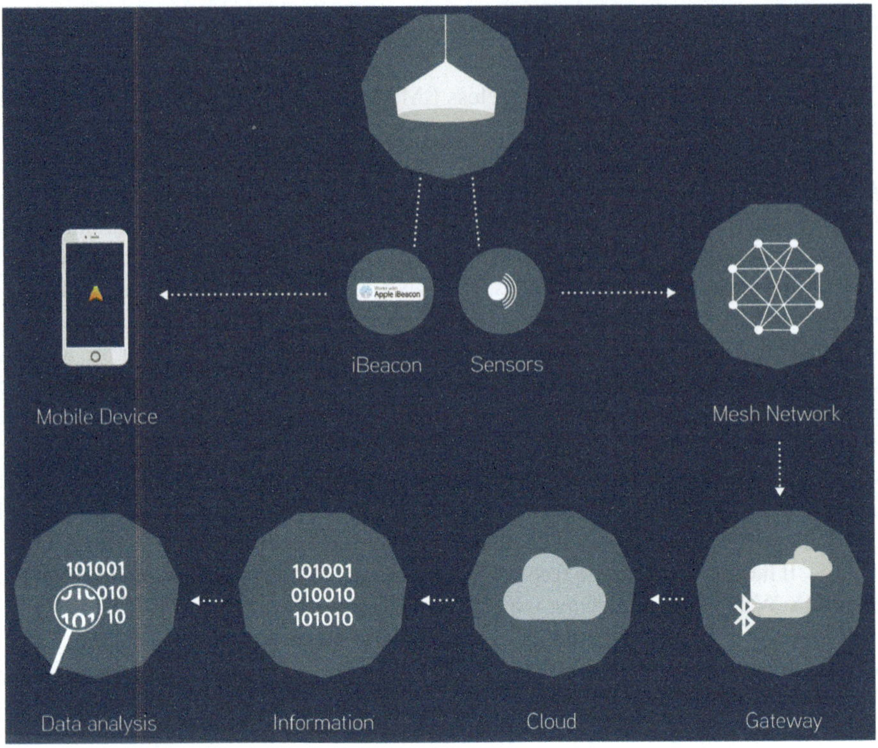


**Fig. 5.13** The Casambi BLE mesh and IoT. *Courtesy* Casambi Technologies Oy

- Using light and motion sensors.
- Using a photo of the spaces to add lighting control directly on the photo. This helps in the configuration phase of the system when it is necessary to arrange the controls, the luminaires and the sensors and define the links between them.
- Integrating custom elements from other manufacturers to control different parts of the interior space, such as blinds, curtains, doors and other building automation systems.
- Obtaining at any time all information on the status of the system, being able to monitor it and configure it remotely.

In order to offer all these features, the Casambi system provides a variety of driver accessories controlled directly by BLE, or interfaced with the DALI system, or the 0–10 V analogue control. Since its establishment in 2011, the Casambi system has also been included in the luminaires of many manufacturers. Finally, third-party devices have also been developed, such as sensors, drivers, actuators and other lighting control accessories, all compatible with the Casambi system and app. The Casambi system has been used in many projects: residential, retail, hospitality, museums and offices; such as the project of the Ratos Headquarter in Stockholm by

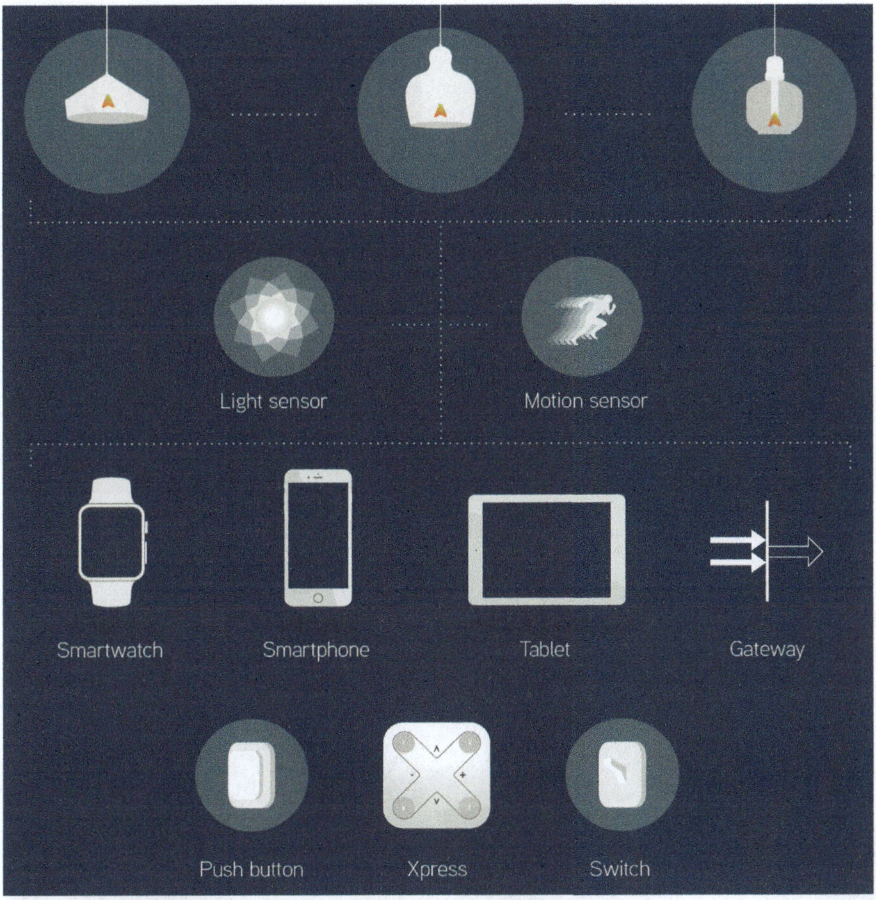


**Fig. 5.14** The Casambi system. *Courtesy* Casambi Technologies Oy

**Fig. 5.15** The Casambi App. *Courtesy* Casambi Technologies Oy

Fig. 5.16 The Ratos headquarter in Stockholm. *Courtesy* Annell Ljus + Form AB

Fig. 5.17 The Ratos headquarter lighting controlled by Casambi App. *Courtesy* Annell Ljus + Form AB

Annell and Rickard Eremar, with lighting design by Jonas Lindahl (Figs. 5.16 and 5.17).

Another LMS based on a BLE mesh is the Blue Light Link (BLL), realized by TCI (2017), to be inserted into the luminaires in order to integrate them into the IoT. The system allows luminaires, sensors and other devices to be managed in more than 500 units. This system has been recently integrated into Artemide luminaires with a custom app that allows various types of adjustments on individual lamps or

**Fig. 5.18** The Artemide App. *Courtesy* Artemide SpA

groups of lamps, in manual or automatic mode. The Artemide APP, shown on the left of Fig. 5.18, allows the management of lighting scenarios with individual or group control of luminaires, which are automatically recognised with their characteristics.

The light parameters that can be managed by this system are the direct and indirect light components, the tunable white for the variation of the CCT and also the colour (Fig. 5.19). In addition, through the sensors, the PRO version of the app, shown on the right of Fig. 5.18, can detect the presence and analyse data on the activities of people in the interior to determine the most appropriate features of lighting. Many products of Artemide today integrate this new possibility of control (Figs. 5.20 and 5.21).

A further wireless control and monitoring system is E*yeNut* (Harvard 2018), which offers users complete control of their interior lighting, with the main purpose of achieving a reduction in electricity consumption. The system is equipped with a web-based interface, i.e. accessed via a browser. The operating mode of the control system is based on the possibility, through a single gateway, to manage up to 500 devices through a reliable and secure network, based on the Zigbee wireless communication system. Luminaires can be defined as groups, also creating logical control sets. The devices can be shown on a plan of the interior spaces in which the system is installed. This facilitates its configuration and control. This system too makes it possible to plan and control a lighting program based on time and date. A function for the management and verification of emergency lighting is also available. From the management point of view, there is no limit to the number of gateways that can be defined and therefore to the number of devices that can be controlled and managed. The system also makes

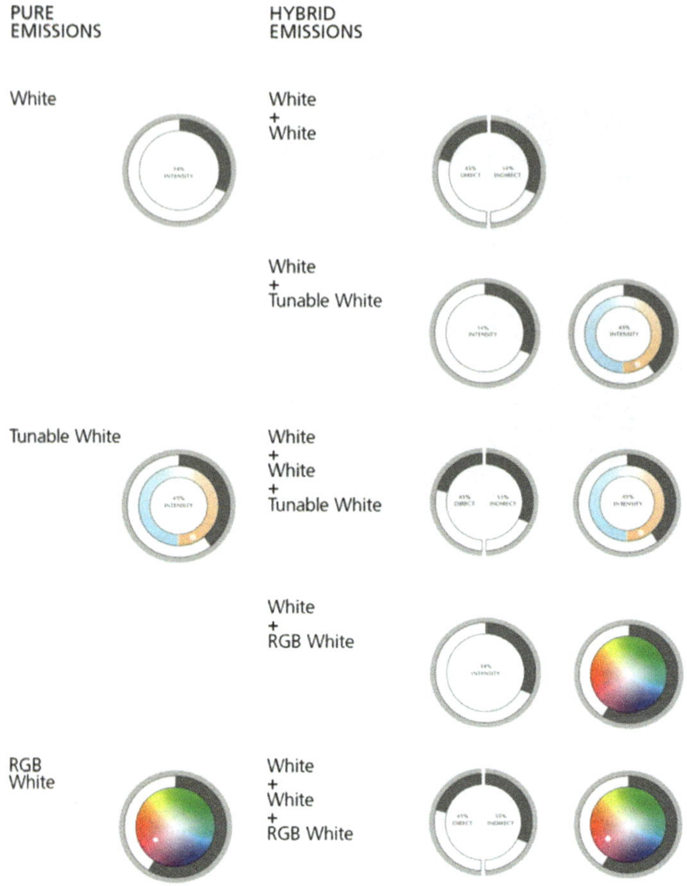


**Fig. 5.19** The light emission control with the Artemide App. *Courtesy* Artemide SpA

it possible to manage the access of multiple users. The gateway can be placed in the user's site or installed and managed in the cloud by the Eyenut manufacturer. The interface between the wireless system and the LED driver can be of the analogue 0–10 V or digital DALI type. The system also allows managing presence and daylight sensors and to transform a traditional switch or a capacitive sensor to be installed on the wall in a wireless control.

The *Silvair* (2016) LMS is also dedicated to the management of lighting systems in professional applications, such as commercial spaces and offices. This system is based on wireless BLE and uses a double protocol: the first to access the system via central devices such as computers, smartphones or tablets, the second to implement the BLE mesh for the communication and control between all the devices in the lighting system, which are considered the peripheral nodes of the mesh. The commissioning of the system includes two distinct phases:

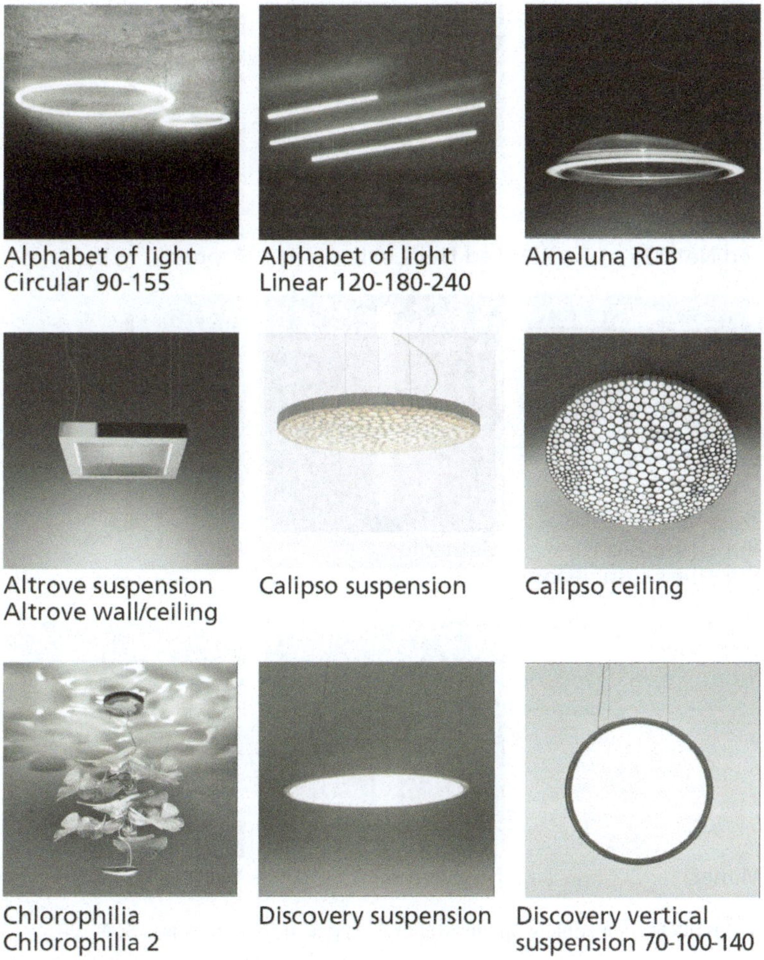


**Fig. 5.20** Products compatible with the Artemide App. *Courtesy* Artemide SpA

- The project phase, which can be carried out remotely, uses a PC and the system is accessed via a browser. In this phase the building's plans are imported, the areas to be illuminated are delimited and the lighting scenarios are defined.
- The implementation phase, which is carried out on site, uses a smartphone with an app dedicated to commissioning. In this phase, the lighting system devices are added to the network, the lighting zones and scenes are implemented, and the configurations are optimised if necessary.

With its monitoring function, the system can provide valuable information not only on the lighting system controlled by the mesh, but also on the processes that take place in specific spaces. In this way, facility managers have access to the plant's diagnostic tools, but also to data on space occupancy and energy consumption. The

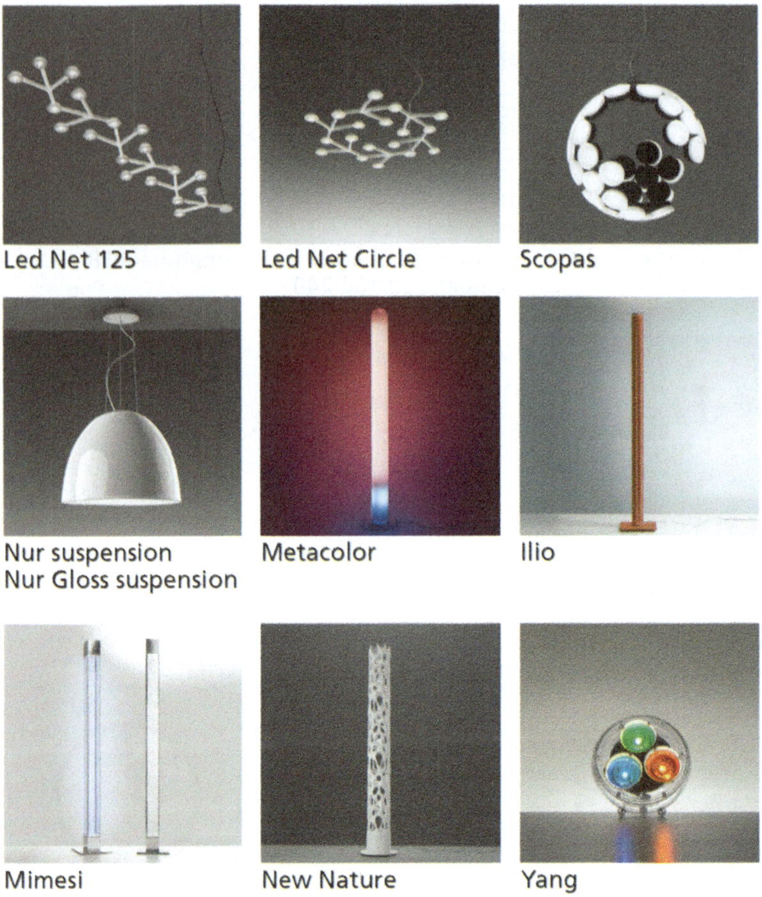


**Fig. 5.21** Products compatible with the Artemide App. *Courtesy* Artemide SpA

system obviously makes it possible to integrate light and presence sensors, and also to use Beacon or other BLE sources to activate lighting. Any BLE equipped device can be added to the mesh. Fans, shutters and any other device enabled in the network can be controlled by the user. For instance, on/off controls, colour, temperature and fan speed of a fan coil. Every change of status made by a user on a device is immediately made available on the network and each central device of the mesh is updated with the new status.

The *e-Sense Tune* (Fagerhult 2015) system has been developed with the aim of providing a completely independent, personal and wireless control for single offices, individual work spaces, conference rooms and recreational spaces. The goal is to be able to adapt the light based on the needs and preferences of the user, even in different spaces. Lighting control and customisation are obtained through smartphones with a BLE connection. The system is also supplied with sensors for measuring the illuminance level, so that the emission of the luminaires can be adjusted according to the natural light present in the environment. The idea behind this system is that, once configured, it is able to automatically recognise a user (carrying a smartphone) entering a room and adapt the lighting according to his/her profile. Another interesting aspect of this system is the fact that it is supplied with a series of pre-set scenes, including the ones for energy saving and for the simulation of natural light. This last mode is obtained through a predefined dynamic variation of both the CCT and the amount of light during the day. The system automatically changes the scene synchronising with the smartphone clock and the user can choose between two different scenes or even change the CCT of the emitted light through the smartphone app at any time.

The idea of having a lighting system based on a smartphone or another SMD is spreading rapidly in the lighting industry. The SMD can be seen both as a tool to identify the presence of the user, who can be provided with an interface for accessing the lighting control, and as a tool to configure and control the lighting system and other building automation devices. The integrated management of the building systems, in addition to lighting, can include heating/cooling control, natural light control, system monitoring, access control, surveillance, and the transmission of information based on the location within the building and the management of spaces. LMS are becoming increasingly integrated into other building automation systems, such as, among many others, the solutions proposed by Crestron (2017) and Eelectron (2018).

## 5.7 Regulatory Aspects on the Energy Saving of Lighting Products

The introduction of LED technologies in the lighting sector has posed the fundamental question of how to include them in a social context in which the issue of energy sustainability of products is currently very important. This applies to both LED lamps used for retrofitting purposes, and new luminaires, based on LED arrays, LED modules or other solutions integrated into the lighting product.

The European Parliament and Council Directive 2005/32/EC of July 6, 2005 established a framework for defining eco-design requirements in the design of Energy-using Products (EuP). This directive defines a regulatory framework for the determination of the technical requirements with which the manufacturer of energy-using devices must comply: from the design phase to the production, operation and disposal phases, i.e. the entire life cycle of the product. These types of products also include lamps and luminaires, as they consume electricity. With its energy saving objectives, this directive has also contributed to the affirmation and diffusion of LED lighting products in the European continent and indirectly also worldwide, stimulating European companies in the sector.

This directive has been updated by Directive 2009/125/EC of October 21, 2009, which extended the scope of intervention of the previous directive to products that do not consume energy directly but are connected to energy consumption, such as water taps, doors and windows and thermal insulation materials. However, this update did not affect the technical aspects of lighting products.

The technical aspects of the directive, with reference to luminaires, were initially defined by three main implementing regulations issued by the European Commission, followed by others later. The first is dedicated to non-directional household lamps, the second focuses on lighting in the service sector and on street lighting and the third deals with issues related to directional lamps and LEDs.

Regulation 244/2009/EC established the requirements for the design and introduction on the markets of European countries of non-directional lamps, used at home, but also for all lamps of this type when they are used outside the home or are integrated into other products. This regulation was amended shortly after its publication (859/2009/EC), due to an editorial error, which was noticed after its publication in the Official Journal of the European Union, which would have prevented the sale of halogen lamps, whose final ban, except for some particular products, took place nine years later in September 2018.

Regarding the requirements for environmentally friendly design and the introduction of general lighting products on European markets, used in office lighting and street lighting, these have been defined by regulation 245/2009/EC. After the publication of this regulation in the Official Journal of the European Union, it was realised that there were several technical errors in the published text and therefore the regulation was amended by another one issued about a year later. The changes introduced by this supplementary and corrective regulation concerned some technical definitions, the requirements for eco-design and the exemptions (347/2010/EU). At the date of its publication, prior to the advent of the LED era, Regulation 245/2009/EC mainly concerned fluorescent lamps without integrated power supply, high intensity discharge lamps, the power supplies of these lamps and the luminaires that contain such lamps. The regulation also applies when these lighting products are actually integrated into other products. Among its objectives, this regulation aims at achieving the decrease in the amount of energy used as well as the decrease in the mercury content of the lamps. The first objective, together with the contribution of other sectors, aims at reducing energy consumption in the European Union by 20% by 2020. The second objective concerns mercury pollution in the environment. On this point, there has been a commitment from some producers, for the elimination of obsolete mercury vapour discharge lamps from the market and for the reduction of mercury in fluorescent lamps. This objective has been intrinsically achieved later thanks to the widespread use of LEDs, which do not use mercury in their materials.

In line with the development of the market, the European Commission has promulgated Regulation 1194/2012/EU, which deals with the issues related to directional lamps and LEDs, the new light sources, which in that period, after a decade of evolution, began to be introduced on the market as real alternatives to many of the previous light sources. This regulation has been effective since September 2013 and concerns the aspects of environmentally sustainable design for LED lamps and LED modules in general, for all the technologies concerning directional lamps (spotlights and projectors), for control devices such as switches, dimmers and sensors, for incandescent

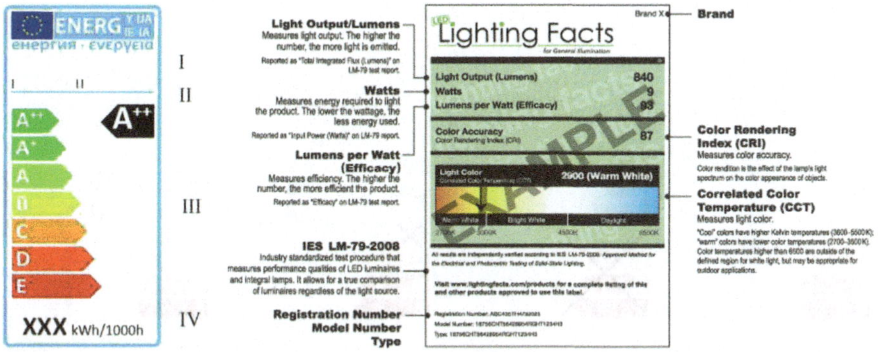


**Fig. 5.22** Energy labels in Europe and North American market

lamps and LEDs, and for the design of luminaires containing incandescent, compact fluorescent and LED sources.

This regulation was defined a few months after the publication of Regulation 874/2012/EU, which introduced the mandatory Energy Label for lamps and also for luminaires including information on compatibility with the energy classes of the lamps. The purpose of this directive was also to make the final consumer aware of the real energy efficiency of the lamps he/she buys. The label must report these 4 basic pieces of information:

I. The name or trademark of the supplier.
II. The supplier's lamp model identification code.
III. The energy efficiency class identified by the arrowhead.
IV. The weighted annual energy consumption in kWh for 1,000 h.

Figure 5.22 shows on the left the Energy Label for Lamps introduced by Regulation 874/2012/EU and on the right the "Lighting Facts" Energy Label in the North American market.

Since 2008, the voluntary marking for LED lamps known as "Lighting Facts" has been introduced in the North American market. This is aimed at promoting the accuracy and consistency of the actual performances presented on the packaging of LED lighting products, which therefore does not only concern the energy aspects (D.O.E. 2008). This label also includes other typical lighting engineering data such as luminous flux, efficiency, electrical power consumed, the CCT and the colour rendering index (CRI). Indeed, today this data is provided, albeit in different formats, also on almost all lamps sold in the European market.

Regulation 874/2012/EU introduced the labelling principle also for luminaires intended for sale, with reference to the definition of the energy classes of lamps compatible with the luminaire. In this case, the label has various different formats that cover different product categories (Fig. 5.23):

- Luminaires with lamp holders for lamps of all energy classes.

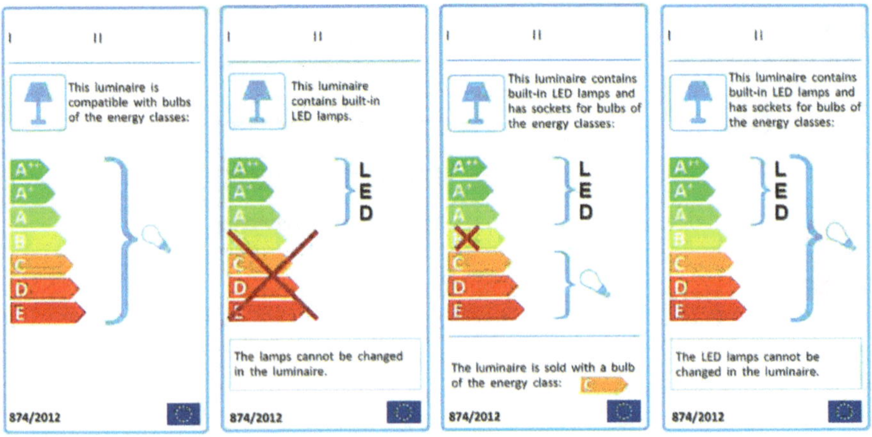


**Fig. 5.23** Four energy label examples for different types of luminaires

- Luminaires containing LED modules integrated into the product and therefore not replaceable.
- Luminaires containing both non-replaceable LED modules and lamp holders, provided with lamps.
- Luminaires containing both non-replaceable LED modules and lamp holders, not provided with lamps.

In addition, the Energy Label of the luminaire must provide these five basic pieces of information:

I. The name or trademark of the product supplier.
II. The supplier's luminaire model identification code.
III. A phrase informing about the compatibility between the luminaire and the lamps, and/or the presence of LEDs integrated in the product.
IV. The range of energy efficiency classes of lamps that can be inserted into the luminaire or LED modules integrated into the luminaire.
V. A phrase specifying if the luminaire is sold with a lamp and its energy level, or if the luminaire is sold without a lamp, or if the luminaire is sold with integrated LED sources that cannot be replaced.

Following the innovation trend brought about by the introduction of LEDs on the market, Regulation 1428/2015/EU was then published, which since February 2016 has integrated and updated the previous regulations implemented by the European Union. This regulation makes two amendments to Regulation 245/2009/EC, which in fact make it possible to continue the production and marketing of high pressure sodium discharge lamps and high power metal halide lamps, used in large areas such as sports fields, airports, seaports and stations, because to date there is no valid alternative, both from a technical and economic standpoint, achievable with LED sources. Furthermore, Regulation 1194/2012/EU was also amended and integrated. Actually, since September 2016, the introduction of luminaires has been allowed only if fully compatible with A+ energy rating lamps; in practice, this almost exclusively means LED light sources, with the exception of the high power ones already mentioned above, thereby accelerating the transition to LED light sources. A limitation of this regulation is that compatibility with A+ lamps is mentioned regarding the electrical and mechanical aspects, but not with regard to photometric characteristics. Furthermore, since September 2016, the placing on the market of low-efficiency directional lamps, such as halogen spotlights operating directly with mains voltage, has been also banned.

## 5.8 Smart Lighting in the Smart Home

From the advent of the new LED technologies to today, in the course of a few years, the approach of manufacturers and designers to the creation of new lighting products has profoundly changed. The initial retrofitting approach, with the new light sources inserted in traditional lamps and luminaires has been gradually accompanied by new, high-quality and innovative products, both at the formal level and in terms of applications. LED arrays, LED modules, tunable white LEDs embedded in light engines with interchangeable optical systems have provided designers with a new possibility for design experimentation. From a formal aesthetic standpoint, the union between LED sources and increasingly performing materials makes possible a design experimentation for the generation of products with unexplored potential. In the opinion of De Lucchi (2011) today there are new and fascinating degrees of freedom in the design of lighting products, which makes it possible to better consider the context in which they will be inserted. There is the possibility of designing lighting by defining characteristics that are more attuned to the well-being of the individual, also thanks to the new interaction features. The innovative potential of these new technologies, which embrace the world of electronics and information technology, is undergoing continuous experimentation and development, and is expressed not only in the aesthetic research, but also in dynamism and interactivity.

As illustrated in the previous paragraphs, innovation in lighting design is not only limited to the use of light sources that consume less energy, but most importantly provides the possibility of making lighting intelligent, through forms and systems that are more suited to the specific space and to the end user, integrating itself into the smart home.

The smart home concept is normally referred to a home that includes audio-video entertainment systems, lighting control, temperature management, security systems and all home devices that can be connected in a IoT network and communicate with each other. Thanks to the Internet and smartphones, these devices can also be controlled remotely. Prior to the widespread use of the Internet and smartphones, the terms home automation or domotics were used to indicate the automatic control of household devices and systems. Although today the two terms are used almost equally in many contexts, there is still a subtle difference (Young and Young 2018). A home equipped with domotics can operate independently, it is able to adapt the systems to users' preferences. In this sense, domotics can be seen as one of the possible functions of the smart home, able to operate autonomously. The smart home offers the additional possibility of using interfaces to access home systems, ranging from the smartphone to voice control, as well as the ability to easily integrate new devices and new control functions. While home automation involves the external intervention of professionals for the design and installation of systems, the smart home could, at least in some aspects, fall within the do-it-yourself (DIY) method. The basic idea of the smart home is to make possible the electronic control of temperature, lighting, appliances, audio-visual equipment, energy and all the other devices in the home that can be connected and managed on the network, including surveillance systems (Harper 2003). The goal is to provide an environment that is comfortable, safe, energy-efficient and, most importantly, easy to configure (Leitner 2015).

One of the first DIY smart lighting systems for the home introduced on the market has been the Philips Hue (2012). It has been initially proposed with a retrofit approach, using the smart bulb Hue. This makes it possible to control the emitted flux, the CCT and the colour of the light, through a smartphone app or a remote control. Now the system has evolved and offers a complete set of luminaires, LED strips and control devices. It is based on Wi-Fi and requires the use of a bridge in order to connect all the devices to the network. The manufacturer has opened the application programming interface (API) of the system, allowing anyone to develop new apps in addition to the official one. Today, dozens of apps are available to manage this system (Ludlow 2018), including some that modify artificial lighting, with circadian purposes, based on the normal cycle of natural light (Zunn Labs Inc. 2016).

## 5.8.1 Smart Bulbs

LIFX (2015) has created smart bulbs featuring a master light that connects to the Wi-Fi network and provides access to other light sources via a low-power wireless mesh network. In addition to light bulbs, LED strips are now available. The system is controlled with the app developed by the manufacturer, but also with other apps offering many different functions.

Figure 5.24 shows the Megaman (2018) RGBW Colour Changing LED Classic Bulb. This 9 W smart lamp has a luminous flux of 806 lm and can connect to a smartphone via the Ingenium® ZB gateway. Producers like Ilumi (2015) and many others (Colon 2018) have also proposed smart bulbs, which connect directly to a smartphone via BLE, thus avoiding the need for a further device, such as a bridge, to obtain a network connected lighting.

At first, smart bulbs have been proposed for the purpose of energy saving, even if the real energy efficiency of smart systems has been immediately controversial (Kofod 2016). Indeed, a series of tests carried out by SSL Annex (2016) on light bulbs for domestic use, with an assumed use of 2 h per day, has found that the average consumption of these products in the hours when they are on stand-by is

**Fig. 5.24** The RGBW colour changing bulb. *Courtesy* Neonlite Distribution Ltd.—Megaman

greater than the consumption necessary to produce artificial light. Considering this data, in addition to the consumption of any accessories necessary for their operation on the network, the real efficacy of these products for the typical daily use, at the date of the tests, was between 12 and 53 lm/W. This value is much lower than the efficacies available today for white light LED bulbs. The added value of these products should not therefore be sought in efficacy but rather in the additional performances, useful both for the purposes of circadian lighting and for the possibility of integrating them more easily in domestic contexts thanks to the retrofit approach.

### *5.8.2 Smart Homes*

In the smart home, lighting becomes one of the elements capable of integrating with other objects and systems in the home. With this in mind, manufacturers of building automation systems have started to introduce software tools that provide a configuration and access interface to all connected systems, in addition to lighting. An example is the Pyng app developed by Crestron (2014) for its products, which has emphasised the DIY concept. Also Lutron (Clauser 2014), a manufacturer of lighting control systems, has proposed a system that uses a bridge to connect to the Wi-Fi network and thus allows the connection and control, in addition to lighting, of other systems in the building. A similar system has also been proposed by Vimar (2011) with the By-me system and its app. Many other manufacturers in the electrical sector offer similar products.

Following the DIY principles, in recent years giants of the new economy such as Amazon, Apple, Google and Samsung have also entered the smart home arena. Their systems are based on the IoT, i.e. they can configure, manage and control devices that are connected to Wi-Fi and the Internet, either directly or through a bridge.

The Alexa virtual assistant is a software initially developed by Amazon for its Echo product and then integrated in other products and also accessible through the Reverb app (Clauser 2018). Alexa was born as a system based on a voice recognition interface to manage music, update the agenda, set alarms and provide information on weather, traffic and news. The system is also being extended to the control of many smart home devices, including lighting. The idea is to allow companies and professional developers to create new functions for Alexa, through software elements called Skills, which can be implemented with a special development library (Amazon 2015). Subsequently, the ability to add Skills has been also made available to end users in a simplified form, using the Skills Blueprints (Amazon 2018). The number of devices available on the Amazon store compatible with Alexa is in fact constantly increasing.

The Home Kit system proposed by Apple (2014) provides a series of accessories for the control of smart home devices, including lighting. Configuration and management are carried out via iOS devices. This system also provides the ability to give voice commands via the Siri virtual assistant. Apple has also launched, through the MFi program, the possibility for other manufacturers to develop other home accessories compatible with Home Kit under license.

Google Assistant is a virtual assistant able to control the Google Home system through voice recognition (Google 2016). Born for the management of music and access to Google's online services, it is also being extended to the control of a variety of smart home devices, including lighting. To this end, Google has launched a program of agreements with smart home device manufacturers.

With the acquisition of the SmartThings company, initially born from a crowd funding project, Samsung has also entered the smart home scene (Null 2018). The SmartThing hub does not have Wi-Fi, but must be connected to the home router, and is able to connect wirelessly with all ZigBee and Z-Wave compatible devices, including lighting. The system is configurable and manageable through the SmartThings app, which is able to recognise compatible devices. Samsung has initiated collaboration agreements with smart home device manufacturers to make their devices compatible with this system.

Today, the manufacturers of devices for the home, such as light bulbs, thermostats, home appliances and many others, are increasingly interested in being compatible with multiple smart home management systems, in line with the DIY philosophy.

All lighting, control and management systems illustrated up to now can be potentially useful for planning, installing and configuring a circadian lighting system. However, they are not sufficient on their own, because they have to be integrated into a project that, in addition to lighting, considers interior spaces and most importantly the human being.

From this point of view, a research still underway conducted jointly by the California Lighting Technology Center (CLTC of the University of California Davis) and American Honda Motor Co. has designed a circadian lighting system in a smart home (CLTC 2014). The Honda Smart Home (HSH) is an open project, everyone can download the house project to build their zero emission and circadian-friendly home (Fig. 5.25). This project was completed in 2014 (Fig. 5.26) (Honda 2016).

The HSH is designed to achieve a circadian lighting aimed at minimising the alteration of people's circadian rhythm. This is to better support cognitive skills during the day, improve people's safety and ensure the quality of sleep at night. The

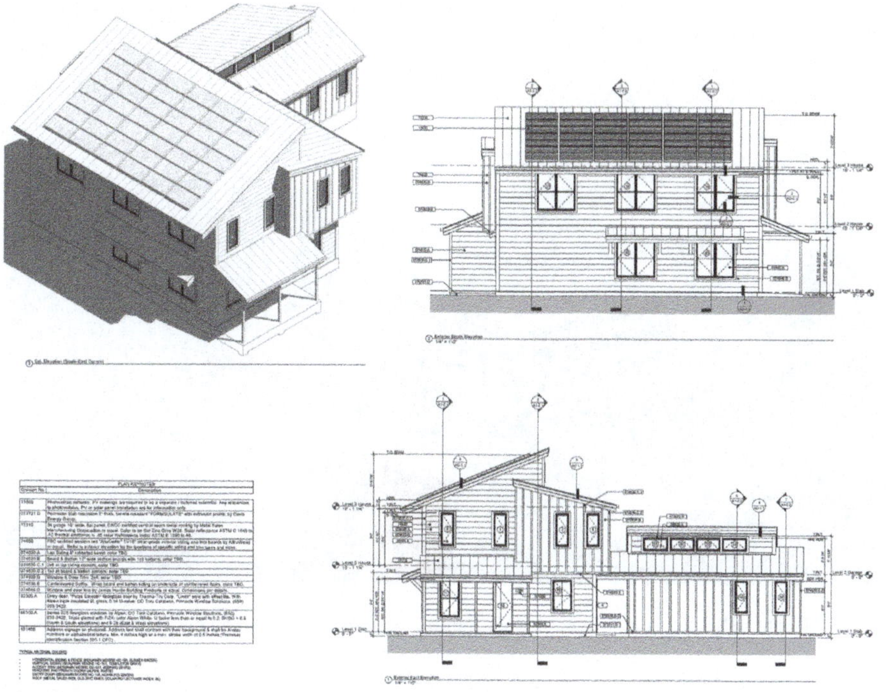

**Fig. 5.25** The Honda smart home open project. *Courtesy* American Honda Motor Co.

**Fig. 5.26** The photograph of the Honda smart home. *Courtesy* American Honda Motor Co.

**Fig. 5.27** The HSH interior with cool CCT lighting. *Courtesy* American Honda Motor Co.

HSH has been equipped with a high-efficiency LED lighting system. A new logic for controlling the CCT of light in interior spaces has been implemented: artificial lighting imitates the changes in natural light during the day. The occupants of the house can select light scenes that can support the correct timing of the circadian rhythm and at the same time guarantee sufficient vision for night movements.

The house is energy-independent and is equipped with photovoltaic panels on the roof for the production of electricity. To minimise energy consumption, the LMS adapts the lighting levels according to the occupancy of the various rooms. For the circadian purposes, they have tried to use natural light as much as possible. This goal has been achieved with a careful design of windows and skylights in order to have plenty of natural light inside but without creating glare. In addition, artificial LED lighting have been used, with a CCT in the 5,000–6,500 K range, to supplement daylighting only when needed (Fig. 5.27). After sunset, the LED lighting only intervenes with a low CCT in the 2,200–2,700 K range and with lower lighting levels (Fig. 5.28).

At night, amber LEDs guarantee the visibility needed for orientation and movement in the hallways, without affecting the human circadian system. In addition to the LMS, the house features smart systems managing heating and cooling and controlling energy and water, with a zero impact perspective. The HSH is therefore an

**Fig. 5.28** The HSH interior with warm CCT lighting. *Courtesy* American Honda Motor Co.

example of design and construction of a house that adopts and uses many of the elements illustrated in this chapter for the purposes of circadian lighting, integrating them into the broader scope of the smart home.

## References

244/2009/EC Commission Regulation (EC) No. 244/2009, of March 18, 2009, implementing Directive 2005/32/EC of the European Parliament and of the Council with regard to ecodesign requirements for non-directional household lamps. Available at: https://eur-lex.europa.eu/legal-content/EN/ALL/?uri=CELEX%3A32009R0244 (Retrieved: July 15, 2018).

245/2009/EC Commission Regulation (EC) No. 245/2009, of March 18, 2009, implementing Directive 2005/32/EC of the European Parliament and of the Council with regard to ecodesign requirements for fluorescent lamps without integrated ballast, for high intensity discharge lamps, and for ballasts and luminaires able to operate such lamps, and repealing Directive 2000/55/EC of the European Parliament and of the Council. Available at: https://eur-lex.europa.eu/legal-content/EN/ALL/?uri=CELEX%3A32009R0245 (Retrieved: July 15, 2018).

347/2010/EU Commission Regulation (EU) No. 347/2010, of April 21, 2010, amending Commission Regulation (EC) No. 245/2009 as regards the ecodesign requirements for fluorescent lamps without integrated ballast, for high intensity discharge lamps, and for ballasts and luminaires able to operate such lamps. Available at: https://eur-lex.europa.eu/legal-content/EN/ALL/?uri=CELEX%3A32010R0347 (Retrieved: July 16, 2018).

859/2009/EC Commission Regulation (EC) No. 859/2009 of September 18, 2009 amending Regulation (EC) No. 244/2009 as regards the ecodesign requirements on ultraviolet radiation of non-directional household lamps. Available at: http://data.europa.eu/eli/reg/2009/859/oj/eng (Retrieved: July 16, 2018).

874/2012/EU Commission Delegated Regulation (EU) No. 874/2012 of July 12, 2012 supplementing Directive 2010/30/EU of the European Parliament and of the Council with regard to energy labelling of electrical lamps and luminaires Text with EEA relevance. Available at: http://data.europa.eu/eli/reg_del/2012/874/oj/eng (Retrieved: October 12, 2018).

1194/2012/EU Commission Regulation (EU) No. 1194/2012, of December 12, 2012, implementing Directive 2009/125/EC of the European Parliament and of the Council with regard to ecodesign requirements for directional lamps, light emitting diode lamps and related equipment. Available at: https://eur-lex.europa.eu/legal-content/EN/ALL/?uri=CELEX%3A32012R1194 (Retrieved: July 16, 2018).

2005/32/EC Directive of the European Parliament and of the Council of July 6, 2005 establishing a framework for the setting of ecodesign requirements for energy-using products and amending Council Directive 92/42/EEC and Directives 96/57/EC and 2000/55/EC of the European Parliament and of the Council. Available at: http://data.europa.eu/eli/dir/2005/32/oj/eng (Retrieved: July 15, 2018).

2009/125/EC Directive of the European Parliament and of the Council of October 21, 2009 establishing a framework for the setting of ecodesign requirements for energy-related products (Text with EEA relevance) (2009) 285. Available at: http://data.europa.eu/eli/dir/2009/125/oj/eng (Retrieved: July 15, 2018).

2015/1428/EU Commission Regulation (EU) 2015/1428 of August 25, 2015 amending Commission Regulation (EC) No. 244/2009 with regard to ecodesign requirements for non-directional household lamps and Commission Regulation (EC) No. 245/2009 with regard to ecodesign requirements for fluorescent lamps without integrated ballast, for high intensity discharge lamps, and for ballasts and luminaires able to operate such lamps and repealing Directive 2000/55/EC of the European Parliament and of the Council and Commission Regulation (EU) No. 1194/2012 with regard to ecodesign requirements for directional lamps, light emitting diode lamps and related equipment (Text with EEA relevance). Available at: http://data.europa.eu/eli/reg/2015/1428/oj/eng (Retrieved: July 29, 2018).

Amazon. (2015). *Alexa skills kit—Build for voice with amazon*. Available at: https://developer.amazon.com/alexa-skills-kit (Retrieved: October 25, 2018).

Amazon. (2018). *Alexa blueprints*. Available at: https://blueprints.amazon.com/ (Retrieved: October 25, 2018).

ANSI/ASHRAE/IES Standard 90.1-2016. *Energy standard for buildings except low-rise residential buildings* (I-P ed.).

ANSI/IES RP-16-17. *Nomenclature and definitions for illuminating engineering*. Available at: https://www.ies.org/standards/ansi-ies-rp-16/ (Retrieved: October 17, 2018).

Apple. (2014). *HomeKit*. Available at: https://www.apple.com/shop/accessories/all-accessories/homekit (Retrieved: October 25, 2018).

Basu, C., & Rowe, A. (2015). *Tracking motion and proxemics using thermal-sensor array*. CoRR, abs/1511.08166. Available at: http://arxiv.org/abs/1511.08166.

Beltran, A., Erickson, V. L., & Cerpa, A. E. (2013). ThermoSense: Occupancy thermal based sensing for HVAC control. In *Proceedings of the 5th ACM Workshop on Embedded Systems for Energy-Efficient Buildings* (pp. 11:1–11:8). New York, NY, USA: ACM (BuildSys '13). https://doi.org/10.1145/2528282.2528301.

Benya, J., et al. (2003) *Advanced lighting guidelines*. White Salmon, Washington: New Buildings Institute. Available at: https://www.lightingassociates.org/i/u/2127806/f/tech_sheets/Advanced_Lighting_Guidelines_2003.pdf (Retrieved: July 16, 2018).

Bluetooth. (2010). *New bluetooth standard built for the universal remote control, bluetooth technology website*. Available at: https://www.bluetooth.com/news/pressreleases/2008/07/22/new-bluetoothstandard-built-for-the-universal-remote-control (Retrieved: October 18, 2018).

Borile, S., et al. (2017). A data-driven daylight estimation approach to lighting control. *IEEE Access, 5*, 21461–21471. https://doi.org/10.1109/access.2017.2679807.

Boyce, P. R., et al. (2006). Occupant use of switching and dimming controls in offices. *Lighting Research & Technology., 38*(4), 358–376. https://doi.org/10.1177/1477153506070994.

Brodrick, J. (2015). Unlocking the full potential of indoor systems: A source that is controllable for light output, CCT and chromaticity is the promise of SSL. *LD + A Magazine*, June 1.

Caicedo, D., Li, S., & Pandharipande, A. (2017). Smart lighting control with workspace and ceiling sensors. *Lighting Research and Technology, 49*(4), 446–460. https://doi.org/10.1177/1477153516629531.

Casambi. (2018). *Lighting control for the modern world. The most robust, cost effective and future proof wireless lighting control solution*. Available at: https://casambi.com/ (Retrieved: October 21, 2018).

Chun, S. Y., & Lee, C. -S. (2013). Applications of human motion tracking: Smart lighting control. In *Proceedings of the 2013 IEEE Conference on Computer Vision and Pattern Recognition Workshops* (pp. 387–392). Washington, DC, USA: IEEE Computer Society (CVPRW '13). https://doi.org/10.1109/cvprw.2013.65.

CIE 015:2004. *Colourimetry* (3rd ed.).

Clauser, G. (2014). Lutron's new smart bridge pro makes Lutron lighting easier to integrate. *Electronic House*, April 8.

Clauser, G. (2018). *What is Alexa? What is the amazon echo, and should you get one?* Wirecutter: Reviews for the Real World, September 21.

CLTC. (2014). *Adaptive circadian lighting demonstration: Honda smart home US, UCDavis*. Available at: https://cltc.ucdavis.edu/project/adaptive-circadian-lighting-demonstration-honda-smart-home-us (Retrieved: October 25, 2018).

Colon, A. (2018). *The best smart light bulbs of 2018*. PCMAG, October 5. Available at: https://www.pcmag.com/article2/0,2817,2483488,00.asp (Retrieved: October 25, 2018).

CREE. (2014). *Cree first to break 300 lumens-per-watt barrier*. Available at: http://www.cree.com/news-media/news/article/cree-first-to-break-300-lumens-per-watt-barrier (Retrieved: July 21, 2018).

CREE. (2015). *Cree advances high-quality lighting performance with New TrueWhite LEDs*. Available at: https://www.cree.com/news-media/news/article/cree-advances-high-quality-lighting-performance-with-new-truewhite-leds (Retrieved: October 17, 2018).

CREE. (2018). *LMH2 + LED module for indoor lighting fixtures*. Available at: https://www.cree.com/led-components/products/led-modules-non-integrated/lmh2-led-module-ca5406 (Retrieved: October 16, 2018).

Crestron. (2014). *Pyng™ transforms home automation*. Available at: https://www.crestron.com/en-US/News/Press-Releases/2014/Crestron-Pyng-153;-Transforms-Home-Automation (Retrieved: October 25, 2018).

Crestron. (2017). *Zūm™*. Available at: https://www2.crestron.com/lead-gen (Retrieved: October 22, 2018).

DALI. (2017). *Digital illumination interface alliance*. Available at: https://www.digitalilluminationinterface.org/ (Retrieved: October 18, 2018).

Datanoise. (2018). *Esp-find3-client: 3rd party Indoor Location using ESP8266/ESP32 and Find3* (https://github.com/schollz/find3). Available at: https://github.com/DatanoiseTV/esp-find3-client (Retrieved: October 14, 2018).

De Lucchi, M. (2011). *La Venaria Reale—Michele De Lucchi—iGuzzini@Triennale 2011*. Available at: https://www.youtube.com/watch?v=ZEwrdy0IsR8 (Retrieved: October 23, 2018).

Dimitrov, S., & Haas, H. (2015). *Principles of LED Light Communications: Towards Networked Li-Fi* (1st ed.). Cambridge, United Kingdom: Cambridge University Press.

D.O.E. (2008). LED Lighting Facts. Available at: https://www.lightingfacts.com/ (Accessed: 19 July 2018).

Dortz, N. L., Gain, F., & Zetterberg, P. (2012) WiFi fingerprint indoor positioning system using probability distribution comparison. In *2012 IEEE International Conference on Acoustics, Speech and Signal Processing (ICASSP)* (pp. 2301–2304). https://doi.org/10.1109/icassp.2012.6288374.

Eelectron. (2018). *Otomo—Optimized office automation, Eelectron S.p.A.* Available at: http://otomo.eelectron.com/?lang=en (Retrieved: October 22, 2018).
EN 50090-1:2011. *Home and building electronic systems (HBES)—Part 1: Standardization structure.*
Fagerhult. (2015). *e-Sense tune, Fagerhult.* Available at: https://www.fagerhult.com/Lighting-control/e-sense/e-sense-tune/ (Retrieved: October 22, 2018).
Figueiro, M. G. (2013). Why field measurements of circadian light exposure are important. *Lighting Research & Technology, 45*(1), 6. https://doi.org/10.1177/1477153512473709.
Figueiro, M. G., et al. (2013). Comparisons of three practical field devices used to measure personal light exposures and activity levels. *Lighting research & technology., 45*(4), 421–434. https://doi.org/10.1177/1477153512450453.
Figueiro, M. G., et al. (2014). Tailored lighting intervention improves measures of sleep, depression, and agitation in persons with Alzheimer's disease and related dementia living in long-term care facilities. *Clinical Interventions in Aging, 9,* 1527–1537. https://doi.org/10.2147/cia.s68557.
Find3. (2018). *The framework for internal navigation and discovery.* Available at: https://www.internalpositioning.com/ (Retrieved: October 14, 2018).
Floyd, D., et al. (1995). *Energy efficiency technology demonstration project for Florida educational facilities: Occupancy sensors.* FSEC-CR-867-95.
Galasiu, A. D., et al. (2007). Energy saving lighting control systems for open-plan offices: A field study. *LEUKOS, 4*(1), 7–29. https://doi.org/10.1582/leukos.2007.04.01.001.
Gonzalez, L. I. L., Troost, M., & Amft, O. (2013) Using a thermopile matrix sensor to recognize energy-related activities in offices. *Procedia Computer Science, 19*, 678–685. [In *The 4th International Conference on Ambient Systems, Networks and Technologies (ANT 2013), the 3rd International Conference on Sustainable Energy Information Technology (SEIT-2013)*]. https://doi.org/10.1016/j.procs.2013.06.090.
Google. (2016). *Explore what you can do with Google Home.* Available at: https://support.google.com/googlehome/answer/7130274?hl=en (Retrieved: October 25, 2018).
Harper, R. (2003). *Inside the Smart Home.* Softcover reprint of the original 1st ed. 2003 ed. London, New York: Springer.
Harvard. (2018). *EyeNut: Wireless management solution for interior lighting, Harvard technology.* Available at: https://www.harvardtechnology.com/products/eyenut/ (Retrieved: October 22, 2018).
Hayashida, A., Moshnyaga, V., & Hashimoto, K. (2017) The use of thermal ir array sensor for indoor fall detection. In *2017 IEEE International Conference on Systems, Man, and Cybernetics (SMC)* (pp. 594–599). https://doi.org/10.1109/smc.2017.8122671.
Heise, D., & Skubic, M. (2010). Monitoring pulse and respiration with a non-invasive hydraulic bed sensor. In *2010 Annual International Conference of the IEEE Engineering in Medicine and Biology* (pp. 2119–2123). https://doi.org/10.1109/iembs.2010.5627219.
Henriksen, A., et al. (2018). Using fitness trackers and smartwatches to measure physical activity in research: Analysis of consumer wrist-worn wearables. *Journal of Medical Internet Research, 20*(3), e110. https://doi.org/10.2196/jmir.9157.
Hern, A. (2014). What is Apple's iBeacon? *The Guardian*, January 13.
Home Assistant. (2018). *Awaken your home, Home Assistant.* Available at: https://www.home-assistant.io/ (Retrieved: October 14, 2018).
Honda. (2016). *Full year data release: HSH continues to meet and beat expectations, Honda smart home US.* Available at: http://www.hondasmarthome.com (Retrieved: October 25, 2018).
Hubalek, S., Zoschg, D., & Schierz, C. (2006). Ambulant recording of light for vision and non-visual biological effects. *Lighting Research & Technology, 38,* 314–324. https://doi.org/10.1177/1477153506070687.
Ilumi. (2015). *LED smart light bulbs with bluetooth mesh, ilumi.* Available at: https://ilumi.co/ (Retrieved: October 25, 2018).
ISO/IEC 14543-2-1:2006. *Information technology—Home electronic systems (HES) architecture—-Part 2-1: Introduction and device modularity.*

Kalbandhe, A. A., & Patil, S. C. (2016). Indoor positioning system using bluetooth low energy. In *2016 International Conference on Computing, Analytics and Security Trends (CAST)* (pp. 451–455). https://doi.org/10.1109/cast.2016.7915011.

Kallur, D. C. (2014). *Human localization and activity recognition using distributed motion sensors*. Available at: https://shareok.org/handle/11244/14924 (Retrieved: October 21, 2018).

Kimata, M. (2013). Trends in small-format infrared array sensors. In *2013 IEEE Sensors* (pp. 1–4). https://doi.org/10.1109/icsens.2013.6688495.

Kimata, M. (2017). Uncooled infrared focal plane arrays. *IEEJ Transactions on Electrical and Electronic Engineering, 13*(1), 4–12. https://doi.org/10.1002/tee.22563.

Kofod, C. (2016). Is smart lighting energy smart. In *Proceedings of the 8th International Conference on Energy Efficiency in Domestic Appliances and Lighting EEDAL '15 re-edition, JRC Science Hub* (pp. 1215–1221). https://doi.org/10.2790/012477.

Konnex. (1999). *KNX association*. Available at: https://www.knx.org/knx-en/for-professionals/index.php (Retrieved: October 25, 2018).

Leitner, G. (2015). *The future home is wise, not smart: A Human-centric perspective on next generation domestic technologies* (1st ed. 2015 ed). Cham u.a: Springer.

LIFX. (2015). *Lighting reimagined, LIFX Europe*. Available at: https://eu.lifx.com/ (Retrieved: October 25, 2018).

Ludlow, D. (2018). *The best Philips Hue apps to shake up your smart lighting, The Ambient*. Available at: https://www.the-ambient.com/guides/best-philips-hue-apps-284 (Retrieved: October 25, 2018).

Makonin, S., & Popowich, F. (2011). An intelligent agent for determining home occupancy using power monitors and light sensors. In B. Abdulrazak, et al. (Eds.), *Toward Useful Services for Elderly and People with Disabilities (Lecture Notes in Computer Science)* (pp. 236–240). Berlin Heidelberg: Springer.

Marr, B. (2016). 15 noteworthy facts about wearables in 2016. *Forbes*, March 18.

Megaman. (2018). *MEGAMAN® RGBW colour changing LED classic bulb, Megaman*. Available at: http://www.megaman.cc/news/top-news/20181027/megaman-rgbw-colour-changing-led-classic-bulb-new-addition-to-ingenium-zb-smart-lighting-solution (Retrieved: November 4, 2018).

Meugheuvel, N., et al. (2014). Distributed lighting control with daylight and occupancy adaptation. *Energy and Buildings, 75*, 321–329. https://doi.org/10.1016/j.enbuild.2014.02.016.

Moghavvemi, M., & Seng, L. C. (2004). Pyroelectric infrared sensor for intruder detection. In *2004 IEEE Region 10 Conference TENCON 2004* (Vol. 4, pp. 656–659). https://doi.org/10.1109/tencon.2004.1415018.

Moore, T., Carter, D., & Slater, A. (2003). Long-term patterns of use of occupant controlled office lighting. *Lighting Research & Technology, 35*(1), 43–57. https://doi.org/10.1191/1477153503li061oa.

Narendran, N., et al. (2005). Extracting phosphor-scattered photons to improve white LED efficiency. *Physica Status Solidi (A), 202*(6), R60–R62. https://doi.org/10.1002/pssa.200510015.

Neida, B. V., Manicria, D., & Tweed, A. (2001). An analysis of the energy and cost savings potential of occupancy sensors for commercial lighting systems. *Journal of the Illuminating Engineering Society, 30*(2), 111–125. https://doi.org/10.1080/00994480.2001.10748357.

Newsham, G., et al. (2004). Effect of dimming control on office worker satisfaction and performance. In *Proceedings of the IESNA Annual Conference*. Tampa, New York: Illuminating Engineering Society of North America.

Newsham, G. R., et al. (2008). Individual control of electric lighting in a Daylit space. *Lighting Research & Technology, 40*(1), 25–41. https://doi.org/10.1177/1477153507081560.

Null, C. (2018). Samsung SmartThings hub review: One of the best in the business. *TechHive from IDG*, April 18.

Openhab. (2018). *Empowering the smart home*. Available at: https://www.openhab.org/ (Retrieved: October 14, 2018).

Osram. (2013). *The Osram Brilliant Mix concept—All under control*. Available at: https://www.osram-group.com/en/media/press-releases/pr-2013/24-01-2013 (Retrieved: October 17, 2018).

Osram. (2017). *LED Engine LuxiTune. Tunable white light engine for HCL*. Available at: https://www.osram.us/ledengin/products/luxitune/index.jsp (Retrieved: October 18, 2018).

Osram. (2018). *Special light colours for fashion and food: PrevaLED core G7 LED modules*. Available at: https://www.osram.com/ds/news/special_light_colours_for_fashion_and_food_prevaled_core_g7_led_modules/index.jsp (Retrieved: October 15, 2018).

Overbye, D. (2014). American and 2 Japanese physicists share nobel for work on LED lights. *The New York Times*, October 8, p. A7.

Park, C., & Rhee, S. H. (2017). Indoor positioning using Wi-Fi fingerprint with signal clustering. In *2017 International Conference on Information and Communication Technology Convergence (ICTC)* (pp. 820–822). https://doi.org/10.1109/ictc.2017.8190791.

Philips Hue. (2012). *The definition of smart lighting*. Available at: https://www.google.com/search?q=philips+hue&ie=utf-8&oe=utf-8&client=firefox-b (Retrieved: October 25, 2018).

Philips. (2018). *Fortimo LED system, Philips lighting*. Available at: http://www.lighting.philips.co.uk/oem-emea/products/house-of-brands/fortimo-led-system (Retrieved: October 15, 2018).

Philips Respironics. (2018). *Actiwatch spectrum PRO*. Available at: http://www.actigraphy.com/solutions/actiwatch/actiwatch-pro.html (Retrieved: August 4, 2018).

Pigg, S., Reed, J., & Works, T. (1996). Behavioral aspects of lighting and occupancy sensors in private offices: A case study of a university office building. In *ACEEE Summer Study on Energy Efficiency in Buildings*. Pacific Grove, CA: ACEEE.

Pretz, K. (2018). *IEEE task force works on standard for light communications*. IEEE. Available at: http://theinstitute.ieee.org/resources/standards/ieee-task-force-works-on-standard-for-light-communications (Retrieved: October 21, 2018).

Price, L. L. A., Lyachev, A., & Khazova, M. (2017). Optical performance characterization of light-logging actigraphy dosimeters. *Journal of the Optical Society of America. A, Optics, Image Science, and Vision, 34*(4), 545–557. https://doi.org/10.1364/JOSAA.34.000545.

Raspberry Pi. (2012). *Teach, learn, and make with Raspberry Pi*. Available at: https://www.raspberrypi.org (Retrieved: October 22, 2018).

Richman, E. E., Dittmer, A. L., & Keller, J. M. (1996). Field analysis of occupancy sensor operation: Parameters affecting lighting energy savings. *Journal of the Illuminating Engineering Society, 25*(1), 83–92. https://doi.org/10.1080/00994480.1996.10748136.

Rothe, H. (2018). *A companion client for Home Assistant to handle sensors in multiple rooms: mKeRix/room-assistant*. Available at: https://github.com/mKeRix/room-assistant (Retrieved: October 14, 2018).

Schollz, Z. (2018). *Find-lf: Track the location of every Wi-Fi device (:iphone:) in your house using Raspberry Pis and FIND*. Available at: https://github.com/schollz/find-lf (Retrieved: October 14, 2018).

Seoul Semiconductor. (2018). *SunLike*. Available at: http://www.seoulsemicon.com/en/technology/Sunlike/ (Retrieved: October 17, 2018).

Sharp. (2014). Sharp extends MegaZenigata LEDs with 15–50 W device. *LEDs Magazine*. Available at: https://www.ledsmagazine.com/ugc/2014/02/18/sharp-extends-megazenigata-leds-with-15-50w-device.html (Retrieved: October 17, 2018).

Signify. (2017). *Philips Lighting installs first supermarket with indoor positioning in Germany, Philips*. Available at: https://www.signify.com/global/about/news/press-release-archive/2017/20170306-philips-lighting-installs-first-supermarket-with-indoor-positioning-in-germany (Retrieved: October 21, 2018).

Silvair. (2016). *Interoperable wireless solution for lighting control*. Available at: https://silvair.com/en/ (Retrieved: October 22, 2018).

SSL Annex. (2016). *Stand-by of "smart lamps" can be larger than their energy use for lighting*, September 6. Available at: https://ssl.iea-4e.org/news/stand-by-of-smart-lamps (Retrieved: October 25, 2018).

Tan, F., et al. (2018). Sensor-driven, human-in-the-loop lighting control. *Lighting Research & Technology, 50*(5), 660–680. https://doi.org/10.1177/1477153517693887.

TCI. (2017). *BLL—blue light link*. BlueLightLink. Available at: https://www.bluelightlink.com/ (Retrieved: October 17, 2018).

Thorne, H. C., et al. (2009). Daily and seasonal variation in the spectral composition of light exposure in humans. *Chronobiology International, 26*(5), 854–866. https://doi.org/10.1080/07420520903044315.

Toshiba. (2017). *TRI-R technology*. Available at: https://www.toshiba-tmat.co.jp/en/tri-r/technology/ (Retrieved: October 17, 2018).

Toshiba. (2018). *Sunlight spectrum reproducible white LED (TRI-R)*. Available at: https://www.toshiba-tmat.co.jp/en/case_tri_r/index.htm (Retrieved: October 17, 2018).

Tridonic. (2013). *Tunable white*. Available at: https://www.tridonic.com/com/en/products/tunable-white.asp (Retrieved: October 18, 2018).

USITT. (2013). *DMX512 FAQ*. Available at: http://archive.is/5rp0n (Retrieved: October 18, 2018).

Valencia, J. B., Giraldo, F. L., & Bonilla, J. V. (2013) Calibration method for correlated colour temperature (CCT) measurement using RGB colour sensors. In: *Symposium of Signals, Images and Artificial Vision—2013: STSIVA—2013* (pp. 1–6). https://doi.org/10.1109/stsiva.2013.6644921.

van Kooten, P. (2018). *Uses WiFi signals, signal_strength, and machine learning to predict where you are: Whereami*. Available at: https://github.com/kootenpv/whereami (Retrieved: October 14, 2018).

Veitch, J. A., et al. (2008). Lighting appraisal, well-being and performance in open-plan offices: A linked mechanisms approach. *Lighting Research & Technology, 40*(2), 133–151. https://doi.org/10.1177/1477153507086279.

Veitch, J. A., et al. (2010) *Office Occupants? Evaluations of an Individually-Controllable Lighting System*. National Research Council Canada. https://doi.org/10.4224/20374060.

Viénot, F., et al. (2005). Discriminating colours under LED illumination. In *Proceedings of 10th Congress of the International Colour Association* (pp. 33–36). Granada (ES): AIC.

Vimar. (2011). *By-me home automation on the iPhone*. Available at: https://www.vimar.com/en/int/by-me-home-automation-on-the-iphone-575132.html (Retrieved: October 25, 2018).

Williams, A., et al. (2012). Lighting controls in commercial buildings. *LEUKOS, 8*(3), 161–180. https://doi.org/10.1582/leukos.2012.08.03.001.

Xicato. (2013). *Xicato broadens artist series® of LED modules*. Xicato. Available at: http://xicato.com/xicato-broadens-artist-series%C2%AE-led-modules (Retrieved: October 17, 2018).

Yang, C.-C., Tsai, H.-Y., & Huang, K.-C. (2013). Yellow-ring measurement of white LED in various lighting environments. *Optical Review, 20*(2), 232–235. https://doi.org/10.1007/s10043-013-0042-1.

Yeh, L., et al. (2010). Autonomous light control by wireless sensor and actuator networks. *IEEE Sensors Journal, 10*(6), 1029–1041. https://doi.org/10.1109/jsen.2010.2042442.

Young, M. S., & Young, C. (2018) *Smart Home: Digital Assistants, Home Automation, and the Internet of Things*. Independently published.

Zhaga. (2010). *The Zhaga consortium*. Available at: https://www.zhagastandard.org/ (Retrieved: October 17, 2018).

Zigbee. (2006). *Zigbee alliance*. Available at: https://www.zigbee.org/ (Retrieved: October 18, 2018).

Zunn Labs Inc. (2016). *Sunn—Circadian lighting for Philips Hue, app store*. Available at: https://itunes.apple.com/us/app/sunn-circadian-lighting-for-philips-hue/id935320028?mt=8 (Retrieved: October 25, 2018).

Z-Wave. (2001). *The internet of things is powered by Z-wave. Z-Wave alliance*. Available at: https://z-wavealliance.org/ (Retrieved: October 25, 2018).

# Chapter 6
# Case Study: An Applied Research for Circadian Lighting Design

## 6.1 Background of the Research

As part of an industrial research project, co-funded by the Lombardy Region and the Italian Ministry of Education and Research (MIUR), the Laboratorio Luce of the Politecnico di Milano has participated, along with industrial partners from the sector, in the research for the development of an artificial lighting system with circadian purposes for the elderly.

This research has been inspired by the fact that lighting technologies have evolved very rapidly in recent years. The advent of LEDs, as a valid alternative to traditional light sources, and the corresponding evolution of light management and control technologies, have opened up new scenarios in the lighting sector. In particular, the possibility, provided by the LED technology, to precisely adjust the light flux and to manage the spectral power distribution (SPD), has given impetus to the research activities that investigate the effects of artificial light on human beings. With the perspective of using new technologies to prevent or encourage the containment of specific disorders.

Another aspect that has been a stimulus for this research is the fact that in Europe, and not only, the phenomenon of population ageing (Eurostat 2012, 2015) will have an increasingly significant influence on the market in the not too distant future (Renesas 2013; DesMarais 2014; Huff 2017). The products and services of the future will have to be designed for this ever-increasing segment of the population. The Organization for Economic Co-operation and Development (OECD 2012) has predicted that the number of elderly people will increase significantly in the future, expecting from 2012 to 2030 a 37% increase in the age group between 65 and 79 and a 54% increase in the age group over 80. This progressive mutation of society brings social implications that will have a growing impact on the public policy of distributing resources between generations, on the state budget and on the state tax system. Already for years there have been tensions regarding pension and social security systems in the European continent. The implications also concern the need to rethink the workplace for an ageing workforce and at the same time the possible lack of manpower given that the



M. Rossi, *Circadian Lighting Design in the LED Era*, Research for Development,
https://doi.org/10.1007/978-3-030-11087-1_6

number of people of working age will decrease. Another impact concerns the effects on the healthcare system and the need for workers specialised in the geriatric field, as well as the inevitable increase in the demand for specific healthcare services and care for long-term residents.

The term active ageing, as defined by the World Health Organization (WHO 2002), indicates a process for improving the opportunities for psychophysical health, social inclusion and safety, with the aim of maintaining a good quality of life during the ageing phase. This definition includes the possibility of extending the work activity of the elderly, delaying their retirement, and also the participation in a series of social, economic, civil and cultural activities. It is expected that the older generation of the future will be more active, qualified, attentive to the quality of life, with economic resources, highly educated and with an increasingly developed interest in the use of Information and Communication Technology products (ICT) (Chéron 2011). However, as highlighted in the review of Peek et al. (2014), until now, the acceptance of new technologies, to promote the quality of life of the elderly in their home, has not been a given. This is influenced by various factors, such as costs, privacy implications, ease of use, influence of others, but also the possibility to continue to live and grow old at home.

The design for the ageing society will have to increase the awareness of the productive sectors and public bodies regarding the needs and potentialities of this consumer base. Both in order to trigger innovation processes useful for economic development and to contribute to the increase in the quality of life of society as a whole. This research has also aimed at taking into account the variety of needs of potential users, focusing on disadvantaged users, primarily the elderly, but also those who have lightly visual and cognitive impairments, by applying the principles of universal design. The term universal design, coined by Ronald L. Mace, invites us to think about a project in an inclusive way since inception, taking into consideration all individuals, including those with disabilities. Universal design, starting from the needs of disadvantaged users, designs better products for all users and not just for the disabled. Indeed, if the needs of users with disabilities are taken into account, general improvements can be obtained for the benefit of all users (Steinfeld and Maisel 2012). By applying these principles to lighting design, lighting systems should align with concepts such as equity, flexibility, simplicity, perceptibility, fault tolerance, reduced physical effort, and the ability to adapt to sizes and spaces sufficiently accessible (Rizzo 2007). The theoretical interest of the research has been to understand the specificities and needs of user groups such as the elderly and the cognitive disabled. But also to investigate the way in which light can improve their quality of life and translate the guidelines into lighting systems. Therefore, the aim has been to relate the quality of life to smart lighting, able to improve the living experience in interior spaces, both in functional and in perceptual and emotional terms, in an environment adapted to the needs of a disadvantaged user but well suited to everyone. The need to improve the living comfort of disadvantaged users could be met by this research by providing a correct lighting of interior spaces, achievable through a lighting system design supported by scientific findings, also from other sectors.

## 6.2 Analysis of the Target

Although with the increase in the average lifespan the old age limit has risen over time and the perception of this limit also depends on the age group that is asked when a person can be defined as old (Taylor et al. 2009), today, people over 65–70 who are no longer active in the labour market are considered as elderly (Laham 2015).

The research has also been inspired by the known fact that natural light may be scarce in interior spaces, as discussed in Chap. 4, identifying the elderly subjects as those who are more likely to spend all day indoors, in their own homes or in nursing homes (NH). Indeed, some studies have shown that even in the NH the elderly are not exposed to adequate lighting for a correct stimulation of the circadian system (Figueiro and Rea 2005; Noell-Waggoner 2006). One of the biggest problems of the elderly is the poor quality of sleep. In particular, 40–60% of the elderly over 65 suffer from sleep-related problems. This involves frequent nocturnal awakenings, difficulty in getting to sleep and repeated naps during the day. These sleep disorders are also contributing to a reduction in the general health of the body, with cardiovascular problems, irregularities in the functions of the endocrine system, and affect the decline of the immune system. The limited exposure to light in the elderly is due both to the normal physiological changes involved in the ageing of the visual system and to daily habits that cause them to stay in interior spaces, often poorly lit.

### *6.2.1 Ageing in Place Versus Assisted Living*

The ageing in place concept indicates the possibility of being able to live in one's own home, inserted in one's own community, in comfort and safety, with a good degree of independence, regardless of age, income and personal abilities. Many argue that ageing in place can promote the psychophysical well-being of the elderly; even more if assisted by a support aimed at re-thinking the design of the house to eliminate possibly dangerous architectural barriers, and assign new meanings to the home, also to facilitate the independence of the elderly (Tanner et al. 2008). In some situations, pressured by relatives or as a consequence of a traumatic episode, but also increasingly by a conscious and long-meditated choice, the elder accepts or decides independently to leave his or her home and move to an assisted living facility (ALF). These structures are evolving rapidly compared to the past, getting away from the hospitalisation concept and trying to develop favourable environments that are more similar to homes, in their furnishings and the general appearance of their spaces. The aim is to make less traumatic the displacement of the elderly, no longer completely independent, from their home to the assisted residence (Rowles and Bernard 2013). In these complexes there are single rooms and small apartments, inside buildings equipped with paramedic personnel able to assist the users. The difference between ALFs and NHs generally lies in the increasing level of healthcare that in the latter can be offered to the elderly 24 h a day. As reported by several researches, it is interesting to note that no significant differences have been found in life expectancy in ALFs compared to NHs (Pruchno and Rose 2000; Hedrick et al. 2009; Wysocki et al. 2015). In the USA, there are newly developed structures called Continuing Care Retirement Communities for the elderly (CCRCs) in the form of real villages, sometimes considered part of the ageing in place concept. In these communities, the elderly can take advantage of various levels of assistance, if necessary, directly in their residences: from the support for daily tasks to specialist healthcare, in the case of disabling diseases of a motor or neurological nature.

### 6.2.2 The Level of Independence

One method for assessing the level of independence of an elderly person is primarily based on his or her ability to carry out the main six Activities of Daily Living (ADL), i.e. eating, washing, taking care of personal hygiene, dressing, moving and being able to manage physiological needs. An elderly person who is not able to independently carry out at least three of these activities is not considered capable to live alone without assistance (Noelker and Browdie 2014). In addition to these activities, there are those that make it possible to assess the ability of the elderly to relate with society, called Instrumental Activities of Daily Living (IADL). These include the ability to make a telephone call, shop, cook, clean the house, wash clothes and linen, use transport, manage medications and handle money. These activities require the elderly to still have physical dexterity and good cognitive skills.

Which housing solution is best for the elderly, between their own home and other types of residence, must be assessed on a case by case basis depending on the person, his or her degree of independence, the family and the social context (Löfqvist et al. 2013; Kendig et al. 2017). In most cases, the idea of having to leave home, even more if after a long time, is a source of concern for the elderly, because this means leaving behind a comfortable and stable family environment and many memories (Robison and Moen 2000). In addition, there is also the feeling of losing control of one's life, because one's own home is seen, sometimes erroneously, as a source of security. However, if on the one hand family and social ties, the economic value of one's home and financial resources are positively associated with ageing in place, on the other hand, property taxes, management costs, changes in the household and the decline in the physical well-being of residents may be associated with factors in contrast to ageing in place (Gaymu et al. 2006; Sabia 2008).

### 6.2.3 Accidents of the Elderly

The elderly, due to the physiological decline of senility, are often the victim of domestic accidents that in many cases prevent the return to previous levels of mobility and independence. This can have a strong negative psychological impact, which contributes to a general decline of the individual, who tends to isolation and loneliness and is likely to suffer from depressive disorders, which often exacerbate senile diseases (Schaie and Willis 2015). Numerous medical studies on the elderly indicate that chronic diseases and disabilities mainly affect sensory sharpness, psychomotor speed, mobility, balance and the cognitive system (Bengtson and Settersten 2016). In the USA, the National Center for Injury Prevention and Control (NCIPC 2016) has established that falls are the leading cause of domestic accidents. In particular, the most frequent domestic accidents are falls, burns and poisonings. The elderly are twice likely victims of fatal accidents due to fire compared to the rest of the population. Poisonings are mostly caused by the inappropriate use of medications, often caused by visual and cognitive problems; the elderly are six times more subject to physical reactions due to the wrong use of drugs than other social groups.

The reduction of visual skills is one of the main causes of fall (Bell 2006). In fact, in the elderly, as in the young, vision plays a very important role in both movement and balance. Particularly, in older people there is a decrease in the speed of movement and a shortening of the steps, both because of age, and due to the characteristics of the space in which they move and live. The risk of falling is therefore higher due to a series of factors: first of all, there is a loss of efficiency of the visual system, in addition there is a decrease in the performance of the somatosensory cerebral cortex, and finally there is also a decline in the motor and vestibular systems. Sight is reduced in terms of amplitude of the visual field, image sharpness and loss of contrast due to the veling glare caused by the scattering of light in the aged lens. These problems determine not only a decrease in vision, but also difficulty in relating to the surrounding space, due to the decrease in the adaptation speed of the eye and the reduced contrast. As highlighted by Brundle et al. (2015), the factors that come into play in increasing the probability of falling are related to balance problems typical of ageing, to cognitive problems, and also to visual impairments linked to the ability to perceive and negotiate the interaction with the surrounding environment. At night, with scarce lighting, items like a door handle or a door frame may not be totally visible, causing disorientation and falls (Figueiro and Rea 2005). Older people often have difficulty moving around the house or experience discomfort and insecurity in doing so because they are aware of the risk related to falls and injuries. In the home environment the risk of falling is related to the type of floors, lighting, possible clutter and to all those elements that can divert the route. For instance, obstacles such as exposed or loose cables, carpets, broken or unstable handrails, furnishings and other objects placed on the ground. For an elderly person it can be particularly problematic to go up and down stairs, get on and off the bed, get up from a chair or a sofa, go out and get back into the house (Campbell et al. 1999; Haslam et al. 2001).

Falls can therefore be caused by factors linked to the home environment or by visual impairments. Indeed, many studies have shown the relationship between the visual impairment of the elderly and the risk of falls in a residential context (Tinetti 2003; Lord 2006). An epidemiological research conducted in the UK by Scuffham et al. (2002) has found that 89% of falls requiring hospital treatment involved people over 75 and that 21% of these were people with visual impairments. Falls can cause serious fractures, such as the femur, with disabling consequences that are often permanent, and sometimes even fatal. Lighting can help prevent home accidents and improve the quality of life of older people who suffer from disabling limitations, both for those living in private homes and those living in ALFs and NHs (Figueiro et al. 2012b).

### *6.2.4 Visual Impairment in the Elderly*

An epidemiological research carried out worldwide by Stevens et al. (2013), in the period 1990–2010, on moderate and severe vision impairment (MSVI) and blindness, has found that the prevalence of the age factor on these disorders is slightly decreasing. However, the increase in the average age of the population means that the number of subjects suffering from MSVI has increased, while the level of blindness has remained constant. Varma et al. (2016) have highlighted the importance of social actions to screen the visual deficits of the elderly in order to contain the percentage of individuals who in the period 2015–2050 will suffer from MSVI and blindness as a consequence of population ageing.

From the everyday life experience of the elderly, many studies have analysed the decline of visual functions in this ever-broader segment of the population (West et al. 1997; Evans et al. 2002; Addis et al. 2013) while others have highlighted the relationship between the decline of visual functions and the ability of the elderly to perform daily tasks (Voleti and Hubschman 2013). Moreover, Ong et al. (2018) have highlighted a relationship between limited physical activity and the onset of diseases of the visual system and a mutual influence between these two factors.

Analysing the visual deficits that occur with the normal ageing process, the first element emerging is the loss of near vision, or presbyopia. This, in turn, causes limitations in daily activities, such as low-light vision, the transition from a high light condition to dark conditions, reading texts printed with small characters and the ability to distinguish dark colours. Along with presbyopia there are other diseases, such as senile cataract, which causes an excessive yellowing of the lens, blurring vision and decreasing the amount of short wavelengths, useful for stimulating the circadian system, which can reach the retina (Turner and Mainster 2008).

Another pathology is senile miosis, which causes a decrease in the size of the pupil, due to a partial atrophy of the iris muscle, which occurs progressively after 60 years of age (Nugent 2013). This pathology causes a decrease in the amount of light reaching the retina in low luminance conditions; however, it does not seem to cause a decrease in spatial sensitivity to contrast levels, which seems rather due to the ageing of the neurological system responsible for vision in the cerebral cortex (Elliott 1987; Sloane et al. 1988).

In industrialised countries, age-related macular degeneration (AMD) is the first cause of low vision for people over 50 years of age. This pathology affects the central part of the retina and leads to a progressive loss of functionality of the photoreceptors. The main causes are genetic in nature, but other causes have also been identified such as cigarette smoking, diabetes, obesity, hypertension, a diet rich in saturated fats, and excessive exposure to sunlight throughout life and a very fair complexion. In the early stages, AMD is asymptomatic, but later it leads to blurred vision and complete loss of vision in the central part of the visual field. In the worst cases AMD can lead to blindness. In less advanced forms, the resulting low vision makes it difficult to read, drive cars, recognise faces and even perform a number of daily activities (NEI 2015).

Disorders similar to those of AMD are due to diabetic retinopathy, which particularly affects people with diabetes mellitus at a rate 25 times greater than other people. This disease of the retina is more likely to develop as the time from which one suffers from diabetes and for this reason it may occur already after 20 years of age, but with a higher incidence and a greater degenerative state of the retina in the elderly (Jenkins et al. 2015; Hendrick et al. 2015).

Glaucoma is also a very common visual disease. This causes a progressive damage to the optic nerve, starting from the part that is visible on the retina, which may impair vision and even lead to blindness. The effect is a decrease in peripheral vision with a progressive reduction of the visual field towards to the central region. Unfortunately, this disease, in addition to being irreversible, is often diagnosed only in an advanced stage, and is the second cause of acquired blindness in more developed countries (Gupta and Chen 2016; Mantravadi and Vadhar 2015).

The degenerations of the visual and neurological system in the elderly also involve variations in colour perception and loss of colour saturation. The study by Wijk et al. (2002) has shown that the degeneration of visual functions affects the ability to distinguish and name colours. However, visual functions are always associated with cognitive functions. For this reason, people over 90 years also show a decline in the ability to perceive colours. The study also provides practical guidance on how to design the colours of residences for the elderly in accordance to their limited colour perception. The use of strong colour contrasts, for instance, can be useful to facilitate spatial perception and depth, but also to simplify the recognition of objects.

The research carried out by Cheng et al. (2011) has verified the perception of colours by the elderly with LED lighting, with two different CCTs (2,800 and 6,000 K) and three different illuminance levels (30, 100 and 1,000 lx). The results obtained have led to deduce that older people have better colour discrimination with a greater amount of light, but also with a higher CCT, which compensates for the decreased capacity by the lens of transmitting short wavelengths. The elderly subjects involved in the test have said to prefer higher levels of illuminance and CCT also for reading books, but not in the evening.

All pathologies of the visual system involve deficits that add to the degenerative diseases of the neurological system of the elderly (Jackson and Owsley 2003). The perceptual consequences are a decrease in contrast sensitivity and an increase in sensitivity to glare, a decrease in visual acuity, a reduction in the amplitude of the visual field, an increase in the time necessary for adapting to luminance variations, and the reduction in the perception of depth (Boyce 2014), especially in patients with senile dementia (Mendez et al. 1996).

### 6.2.5 Senile Dementia

Senile dementia is a syndrome characterised by the presence of a severe progressive cognitive impairment in the elderly and is due to various degenerative and vascular diseases, which determine a progressive damage of the brain tissue. The main risk factor is age. Over 65, the probability of getting sick increases exponentially. Before reaching a real senile dementia stage, the term used is Mild Cognitive Impairment (MCI) in subjects with cognitive deficiencies that cannot be attributed to their age and level of education (Petersen et al. 1999). Although Alzheimer's may be one of the main causes of MCI and dementia, in the latter case the symptoms are more severe and affect more cognitive areas (Knopman and Petersen 2014). Older people with dementia suffer from loss of memory and impairment in other cognitive functions such as language, visual and spatial skills, executive functions and reasoning abilities. A cognitive decline, even minor, causes time and space disorientation and is a major disabling factor because it is related to daily mobility, making elderly individuals often unable to organise a daily routine and leading them to a stage of confusion and immobility. Despite pharmacological therapies, this type of illness increases over time, leading to a total lack of autonomy. Unfortunately, alongside the cognitive symptoms there are other serious disorders such as: mood swings, apathy, excitement, malnutrition, negative verbalisation, problems of interaction with others, depression or psychiatric disorders such as aggressiveness, suspicious attitude, the switching of the day-night rhythm, wandering, hallucinations and delusions.

The Dementia Alzheimer's type (DAT) is the most common form of senile dementia. This disease involves an incurable and progressively degenerative condition that damages the nervous system. From the standpoint of the relationship with the surrounding environment, it reduces cognitive abilities and increases spatial disorientation. DAT, however, also involves other serious deficits such as loss of memory, sense of confusion, language problems, loss of identity, emotional and personality changes and disturbances of the sleep-wake rhythm. In its most advanced stage, DAT also involves abnormal motor and verbal activities, hallucinations, incontinence, agitation and irritability, aggressiveness, wandering and exclusion from social life. In these patients, excessive lighting, associated with loud and sudden noises, can cause an over-stimulation that leads to confusion, agitation, and distraction. On the other hand, sensory deprivation leads to problems that hinder communication with the outside world and orientation. As a further complication, DAT also involves problems in the visual system that affect the perception of colours and contrasts, the stereoscopic vision and the visual backward masking (Cronin-Golomb et al. 1991).

Although there is no unanimous consensus in the medical field (Vilela et al. 2017; Bauer et al. 2015) some studies suggest that the redesign of the living environment of the patient may bring some benefits and mitigate the symptoms of dementia (Weert et al. 2005; Collier and Jakob 2017; Duchi et al. 2018). The most recent studies on the design of spaces for dementia patients, also called Snoezelen, focus on the characteristics of an environment that meets the needs of the patient, with the addition of a specific multi-sensory behaviour therapy (MSBT) that can help improve the patients' quality of life (Staal et al. 2007; Chan et al. 2010; Maseda et al. 2018). Multi Sensory Environments (MSE) provide stimulation for the 5 senses, sight, hearing, touch, taste and smell, and encourage movement, in order to promote a sense of well-being and comfort, relieving stress and suffering (Todder et al. 2016). MSEs also seek to improve communicative relationships and cognitive and mnemonic skills. The review by Livingston et al. (2014) has shown that the use of MSEs in the treatment of various types of dementia can improve psychological and behavioural symptoms and reduce patient agitation.

## 6.3 Lighting for the Elderly

The relationship between light and users is one of the fundamental topics for understanding the influence of artificial lighting on the quality of life of the elderly with respect to issues such as independence, well-being, comfort and health at home. As a consequence of the visual changes due to the normal ageing of the visual system, lighting is a fundamental design element for improving daily activities, preventing falls, limiting accidents, especially in the kitchen, and increasing the mobility and autonomy of the elderly person. As highlighted above, other elements to consider are diseases other than those affecting sight, such as senile dementia and seasonal depression (SAD), which often affect the elderly and have been already discussed in Chap. 2.

Based on these pathologies, we must ask ourselves how artificial lighting can help improve the quality of life of the elderly and the visually impaired. As we have seen, the various pathologies have different effects on the visual and the nervous systems, so there is no optimal solution for all the problems, and indeed the solutions to one issue could worsen another. It is indeed necessary to underline that every specific pathology, due to ageing or different forms of dementia, determines different cognitive and optical changes so not all remedies and approaches can be generalised and considered positive for everyone. In some cases, it has indeed been noted that decisive measures and modifications of light have benefited some subjects while generating problems for others (Boyce 2011). In a truly human centric lighting approach we must therefore consider each individual, their residential context and their daily activities.

The first category of problems is related to the reduction in the performance of the visual system, which can have consequences on the ability to relate to the environment and its objects, but also to access information written on paper or other media. The use of lighting that can adapt to different visual needs makes it possible to improve the psychophysiological performance of the elderly, with respect to the needs of everyday life, with differences that can be significant (Izsó et al. 2009).

Assuming that the current standards for interior lighting are defined for normally-sighted humans, a first approach is to consider the visual stimulus for the elderly or the visually impaired, using the relative visual performance (RVP) defined by Rea and Ouellette (1991). For instance, by increasing the size and contrast of the visual task, the visual performance of the elderly can be improved. The size refers to the font size of the written text, which can be larger or enlarged with other systems. The increase in contrasts offsets the light dispersion that occurs in the lens of the elderly and is obtained by acting on the luminance ratios, which in turn depend on the reflection factors of the materials and on the illuminance. In addition to the luminance contrasts, work can be done on surfaces, using colours to increase chromatic contrasts. This leads to a better vision of the objects in the environment, but also to a better perception of the space so as to facilitate the movements inside it.

Lighting for the elderly must be designed differently in terms of SPD and spatial distribution of light. Considering the RVP, attention is placed on the visual activity in relation to the age of the individual and the quantity of light. An increase in illuminance levels improves accuracy and speed in reading, but also the ability to move more safely in the interior and discriminate colours. For this last purpose, it is preferable to use high colour rendering sources. However, there are also divergent needs: such as, on the one hand, to limit the wavelengths below 450 nm, because blue light can annoy the elderly (Laufer et al. 2009), but on the other hand to have a contribution of short wavelengths necessary for a correct stimulation of the circadian system in interior spaces. Contribution necessary also when the elderly has been subjected to lens replacement surgery (Giménez et al. 2016).

### 6.3.1 Lighting and 3D Perception

The relationship between the individual and the surrounding environment is mediated by the senses. The well-known American psychologist Gibson (1966) has described the senses as five perceptual systems. These are the orientation system, the auditory system, the tactile system, the taste-smell system and the visual system. Thanks to the information received from the environment, the brain can handle aspects such as spatial orientation, posture and movement. The most important system for the relationship with the environment, however, is the visual system, which allows us to determine the forms, dimensions and distances of the structures that surround us. Along with the direct vision of the environment there is also the spatial representation that each individual creates in their brain, also outside the field of vision, using non-visual information (Intraub 2012). This mental construction of the environment is based on memory and on a learning process able to integrate spatially and temporally the visual inputs coming from every direction. The mental image of space also influences the movements of the individual in the three-dimensional space (Shioiri et al. 2018).

The overall perception of depth, horizontality and verticality of the surfaces is at the basis of the awareness of the three-dimensional space that surrounds us. Previc (1998) has proposed a model consisting of four behavioural realms to describe three-dimensional perception:

- The peripersonal, related to visuomotor operations in near-body space.
- The focal extrapersonal, related to visual search and object recognition.

- The extrapersonal action, related to orienting in a topographically defined space.
- The extrapersonal ambient, related to orienting in earth-fixed space.

Each of these is associated with different areas of the cerebral cortex that combine to generate the three-dimensional perception of the environment around us. However, the ambient extrapersonal realm is the most important for the perception of our position and for the orientation with respect to the main surfaces of space, acting both on the posture and on the movement of the human body in relation to the surrounding environment. The reference planes of an environment and the way in which they are connected by forming edges and vertices, provide the individual with the main visual information on the environmental structure. The perception of verticality and horizontality is the result of the integration of this visual information with that coming from the vestibular system. The light reflected by surfaces is the medium that carries this information from the surrounding environment to our eyes and, since reflection depends on the material characteristics of the surface, but also on the light that illuminates it, lighting becomes a determining element that can highlight the relationships between surfaces and the three-dimensional perception through the luminance and chromaticity contrasts.

### *6.3.2 Lighting and the Stability of Posture*

With reference to the risk of falls of the elderly, related to the stability of posture, several researches have highlighted the importance of the visual signal (Brooke-Wavell et al. 2002) and the amount of lighting (Rugelj et al. 2014) to promote a less swaying and more stable posture. A low level of lighting leads to a more swaying and less stable posture, which further worsens in the darkness. The presence of a more recognisable visual signal is in fact an element that contributes in a decisive way, along with the vestibular system, in sending to the brain the information necessary for the control of the motor system governing the posture. This effect applies to both young and older people, but several researches have shown that older people are more sensitive to a visual signal deficiency and this causes greater instability in the posture (Prieto et al. 1996; Poulain and Giraudet 2008). In particular Lord and Menz (2000), Lord and Dayhew (2001) have proposed that visual deficits in the elderly, with the consequent lack of the visual signal, would be a contributing factor to the increase in the probability of falling.

From the application standpoint Figueiro et al. (2008a, 2012b) have highlighted that the visual cues derived from lighting affect the control of the posture in the movement and in the action of standing up from a sitting position. In particular, the amount of ambient light is important but, after sunset, it can compromise the quality of sleep, negatively affecting the circadian system. For this reason, light signals have been experimented on door jambs, with LED strips, to facilitate the posture and nocturnal orientation of the elderly, showing that these systems are favourably accepted and even preferred compared to conventional night lights, which are usually installed at the level of the electrical sockets. Another study (Figueiro et al. 2011) has analysed the relationship between different lighting conditions and the gait and the stride of the elderly. The three tested lighting conditions included: a classic general ceiling lighting, night lights available on the market to be installed in electrical sockets and night lights integrated with floor lights, made with laser, outlining the route. The test, carried out on 24 elderly subjects, has shown that the safest walk occurs in the presence of the first daytime lighting condition. While the preferred night lighting condition has been the one with the light guides outlining the route. In this nocturnal condition, the elderly have shown a faster gait and a more constant stride length compared to the condition with only the night lights. Therefore, in the presence of low light, the addition of visual light signals facilitates the stability of the elderly and can potentially reduce the risk of falling.

### *6.3.3 Lighting and Psychophysiological Issues*

Another aspect of fundamental importance is the relationship between the lighting that can be present in homes, ALFs and NHs, in relation to the physiological and behavioural changes that occur with ageing, such as SAD and the alterations of the circadian system, with consequent sleep disorders, which affect the immune system and therefore the general well-being of the individual (Yadav et al. 2017). In this sense, several studies, illustrated in Chaps. 2 and 3, have amply demonstrated the importance of lighting to adjust the circadian system and improve the psychophysical well-being of individuals.

An important issue is that, as illustrated in Chap. 4, in many types of interior spaces natural light may be not sufficient for a correct circadian stimulation. Unfortunately, the artificial lighting that is currently designed and used in interior spaces is mainly able to respond to the visual requirements prescribed by the standards and generally pursues energy saving purposes. The circadian purposes of artificial lighting and the fact that it can integrate the lack of natural light in interior spaces remain outside today's lighting design criteria. In addition, the artificial lighting that is designed today is conceived in a static mode, while it should have dynamic features that make it possible, during the day, to vary the amount, the SPD and maybe even the direction, in a similar way to natural light.

Based on these premises, Figueiro et al. (2008b) have proposed a 24-h lighting program for the elderly in interior spaces, with the aim of better supporting the effects of ageing of the visual system, the circadian system and the neurological system. The system has been tested in elderly subjects living in eight rooms of an ALF, assessing their quality of sleep before and after the installation of the new lighting system. The principles behind the lighting scheme of this system are based on the circadian stimulus factor (CS) proposed by Rea et al. (2012), and can be summarised as providing:

- Circadian light during the day. Created by guaranteeing to the cornea at least 400 lx of natural light or artificial light with a CCT of 6,500 K.
- Non-circadian artificial light after sunset. Obtained with values lower than 100 lx at the cornea and a maximum of 2,700 K for the CCT.
- Night light with clear visual signals to promote posture and movements resulting from nocturnal awakenings.

The tests carried out have shown that after the installation of this lighting scheme the elderly saw an improvement in the quality of sleep, with fewer daytime naps and more sleep during the night. In addition, the participants have declared to prefer the new lighting scheme both day and night, for the awakenings and the movements between the bedroom and other spaces. Basically, the lighting scheme tries to bring into the interiors a circadian lighting stimulation similar to that one would have outside.

The best lighting for the elderly is that of natural light that would be achievable spending many hours a day outdoors. While on the one hand proper exposure to the natural light by the elderly should certainly be encouraged, this poses problems both at home and in ALFs or NHs. This is due to lifestyles to which the elderly have become accustomed over time and also to problems of psychophysical health. Moreover, the latitude and climatic conditions typical of many industrialised countries make it difficult to live outdoors, even for a few hours a day, in cold seasons. Therefore, the importance of good life habits of the elderly and of the characteristics of artificial light becomes evident (Czunyi and Craib 2017).

From the application standpoint, White et al. (2013) have conducted a review of research on architectural lighting in the living environments of the elderly. In the study, they have focused on the issues of circadian rhythm, sleep quality and well-being, in relation to the lack of natural light in interior spaces and artificial lighting design strategies. The analysed elements have been the amount of light, the duration of exposure, the SPD and the illumination timing. In this study too they have highlighted the importance of dynamic lighting strategies, based on alternating schemes of circadian lighting, aimed at reproducing the natural light trend over a period of 24 h. Two other aspects highlighted by the study are the importance of nocturnal darkness and the careful assessment of the circadian needs of staff working in shifts in NHs.

### *6.3.4 Therapeutic Effects of Lighting*

In designing lighting for the elderly, it is also necessary to consider the typical pathologies afflicting this age group. Going beyond the aspects concerning a correct circadian lighting in interior spaces, some researches have begun to evaluate the therapeutic aspects of lighting for diseases typical of the elderly. A study conducted by Sloane et al. (2008) emphasised the fact that, although much research has been done in the last decades on the therapeutic effects of light, there is no unanimous consensus on the use of lighting as an integration or replacement of the usual pharmacological therapies used for sleep disorders and depression in the elderly. This is despite the drugs being not always effective and sometimes also having unwanted side effects. One aspect that should not be underestimated is that light used as therapy is something widely available and is not patentable, therefore of little interest as a research or marketing topic for companies. The study also indicates the doses of artificial white light that would be necessary to obtain positive effects, proposing a minimum of exposure on the cornea of 2 h at 2,500 lx or half an hour at 10,000 lx. Natural light outdoors on a cloudy day is equivalent to ~2,500 lx, while on a sunny day it can reach 100,000 lx. The study also emphasises the need to pay the utmost attention to avoiding glare phenomena and proposes commercial products to implement this type of light stimulation (BIO-LIGHT 2018; Lumie 2018).

As illustrated in Chap. 2, research has long shown the effectiveness of light therapy, which is widely used in the Nordic countries, in the treatment of SAD. Instead, less attention has been given to other depressive disorders, particularly in the elderly. In an evaluation review of previous research considered of high level, Tuunainen et al. (2004) have shown that moderate positive effects of light therapy have been found also in the treatment of non-seasonal depressions, as by now claimed by many other researchers as well (Golden et al. 2005; Benedetti et al. 2018). In the absence of a unanimous consensus on the use of light therapy for the treatment of depression, however, there is evidence that the effect of light therapy for the treatment of depression is anyway greater when used in addition to drug therapy (Benedetti et al. 2003; Martiny et al. 2005). Some studies focused on the general elderly population have found moderate positive effects of light therapy in the treatment of non-seasonal depressions (Yamada et al. 1995; Sumaya et al. 2001; Tsai et al. 2004). While in other studies, carried out on elderly patients with dementia, no light effects on depression have been found, perhaps also because this pathology is difficult to be evaluated objectively in people suffering from senile dementia.

The various forms of senile dementia are another topic that requires a reflection on the most appropriate light for the affected elderly. Indeed, a field-based analysis of 24-h light exposure patterns, on a group of healthy elderly people and another one affected by DAT and related forms of dementia, has discovered that the subjects of the latter group are on average exposed to lower light levels throughout the winter, with possible negative effects on sleep quality (Figueiro et al. 2012a). For the elderly with dementia it is difficult to propose light therapy using portable devices, therefore Sloane et al. (2007) have experimented with the use of a white ambient light, with an intensity of 2,500 lx and low luminance to avoid glare, on 66 elderly people suffering from dementia living in NHs. This lighting intervention has been carried out, in the communal areas of the residences, with three different timings, in the morning or in the evening for 2.5–3 h, or throughout the day for 8.5 h, comparing it with simple standard lighting. The activities of patients have been monitored through the use of wrist actigraphs, noting that the amount of night-time sleep has increased in the elderly exposed to lighting in the morning or throughout the day, compared to those exposed to evening light or standard lighting.

From an application standpoint, the elements that must be taken into consideration in the lighting design for the elderly with dementia can therefore be summarised in three aspects that lighting should facilitate (Torrington and Tregenza 2007; Hanford and Figueiro 2013):

- Ensuring a normal 24-h light and dark cycle. Day lighting should facilitate the normal circadian cycle and night lighting should be limited in order not to affect the circadian system.
- Supporting social activities in ALFs and NHs. In communal areas and transition spaces, light must facilitate wayfinding by differentiating the areas used for different activities.
- The perception of the place. The sense of disorientation of the patient with dementia, due to the loss of short-term memory, must be countered by furnishings and lighting that are consistent over time and facilitate the recognition of the interior spaces frequented by the patient, whether public or private.

### *6.3.5 Lighting in the Elderly's Homes*

Despite the analyses presented in the previous paragraphs, the residential lighting present in the vast majority of private homes as well as in ALF and NH structures is often scarcely studied and almost never adapted to the needs of an ageing population, according to a correct human centric lighting approach.

According to some application studies (Rizzo 2007; Noell-Waggoner 2006; Gochenour and Andersen 2009; Rossi et al. 2017) the residences in which the elderly live often suffer from numerous lighting problems. Indeed, they present inadequate solutions in terms of photometric distribution on corridors, stairs, ramps and paths, but also low levels of illuminance, which do not respect the rules, in correspondence of visual tasks that require attention such as, for example, reading, cutting/sewing and selection of objects. Lighting is often different between different rooms and transition areas, causing problems for the visual system of the elderly, which has lost some of its ability to adapt to different light levels with ageing. Moreover, glare problems often arise due to the direct vision of lighting sources and the few possibilities of control by users, which are limited to the simple switching on and off of artificial light or the control of natural light with curtains and blinds. It is known that the elderly are not very inclined to make changes in their homes, either because they are creatures of habit or because they often have limited financial resources (Gilderbloom and Markham 1996). In addition to this, among the possible changes that can be made to the house, lighting is not considered useful because its beneficial effects on housing conditions is not perceived, even in situations of visual deficits (Percival 2007).

Some authors argue that to be considered acceptable in homes, circadian lighting solutions should be affordable, convenient, accessible and easily feasible, i.e. simple to install in normal homes. But also simple to understand, use and control. They must clearly communicate their added value for end users, also in terms of customisation according to the actual needs of the user (Slay 2002; Percival 2007). Finally, the fact that the user can participate consciously and actively, i.e. be involved in the decision-making process concerning the project, facilitates the approval of new lighting solutions.

## 6.4 Results of the Research

The analysis illustrated above shows that designing a correct circadian lighting in residential areas for the elderly, but also for a broader population, cannot be based only on the currently available standards, which essentially meet three design objectives:

- Facilitating the vision of normally-sighted individuals.
- Providing a comfortable vision in relation to the expected visual task.
- Pursuing energy saving goals.

To achieve true human centric lighting, the design must be based on an integrated approach that, in addition to the elements listed above, takes into consideration the human factor in relation to the light available and the environment considered, in this case the residential one. The additional characteristics that must be considered for a circadian lighting design are the following:

- Dynamism: the ability to change the luminous performance over time, according to the model of natural light and/or based on the different activities carried out in the spaces. This dynamic feature refers to changes in the amount of light, SPD and directionality of the light flux.
- Efficiency and economy of the system: the ability to customise the product within a family of products, making it possible to change lighting performance as needed. This is why the lighting product must be designed with interchangeable parts that can guarantee minimum performance, intermediate performance or high performance.
- Customisation: the ability of the system to be controlled automatically, by means of pre-set and timed scenes, or manually by the end user, be this an elderly person or a healthcare worker of an ALF.

Because of the lack of lighting products for this type of applications, the research has taken into consideration both the aspects of lighting design and those of lighting product design, based on the availability of new LED lighting technologies. In addition, some essential characteristics for improving the quality of life of the elderly in their home or within an ALF have been deduced. Indeed, some of the fundamental elements, within the psychological principle of self-determination (Inghilleri 1996), are the concept of de-hospitalisation and the feeling of maintaining one's own freedom and independence, both in one's living environment and in the less familiar environment of the ALF. For this reason, the lighting project must be conceived, in

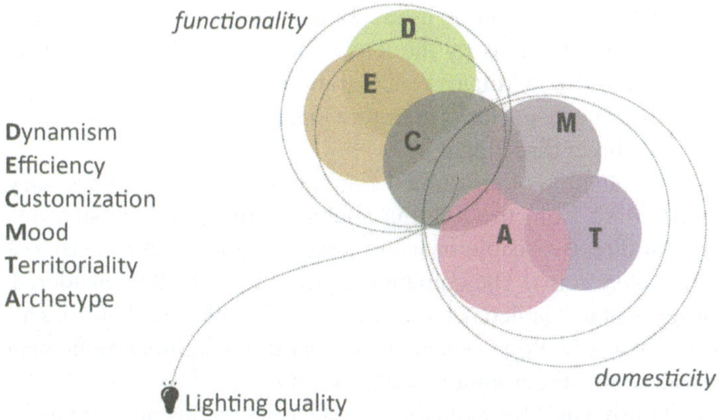


**Fig. 6.1** Fundamental characteristics of the project for guaranteeing lighting quality

some of its formal and functional characteristics, as a useful factor for the recovery of domesticity, even in ALFs, through the possibility of:

- Changing the mood through the atmosphere of the environment, i.e. making a space more welcoming with warm light (low CCT) or more detached with cool light (high CCT).
- Evaluating the formal and iconic aspect of the luminaire that must recall a domestic archetype.
- Providing the possibility of interaction and transformation of light, which must be functional to the recovery of territoriality and the personalisation of space.

These concepts are represented in Fig. 6.1. After establishing the design method, the research has defined the typical spaces in which the elderly may reside, in homes and ALFs, identifying three main areas of intervention: the bedroom or hospital room, which can be single or double, the dining room or refectory and the living room or communal area dedicated to social activities.

### *6.4.1 The Lighting Requirements*

The development of a lighting project for the elderly must primarily focus on the applications of artificial light for their specific needs. In particular, choices must be made on how to direct the light, which type of light to use, where to place it, which area to illuminate, which effects to create, which shadows to determine and which objects to illuminate more than others.

As a consequence of the visual deficits of the elderly, as explained above, it is necessary to increase the lighting levels on the visual task, to facilitate reading and other manual activities and improve the perception of colour. To this end, light sources with a higher CCT can be used, compensating for the decreased ability of transmitting short wavelengths by the lens (Cheng et al. 2011). The use of higher levels of illuminance also facilitates concentration and increases alertness. Increasing lighting for predictable visual tasks therefore means increasing the contrast of details and colour, using direct directional light.

Through the evaluation of the unified glare rating (UGR), it is necessary to design avoiding direct glare from the lighting sources. This means positioning them so that they are not directly visible or using indirect lighting and microprismatic low luminance optical systems. The containment of glare must be considered also with respect to the eyes of a disabled person sitting on a wheelchair, therefore at a distance of about 100 cm from the floor, especially as regards the lighting of the under-cabinet in kitchens and all the direct lighting components.

The issue of the balance between direct and indirect light components is important to ensure that users have a correct perception of space. If on the one hand a totally indirect lighting is very comfortable because it avoids glare effects, on the other it generates a feeling of uniformity, with low contrasts and flattened depths. A space very uniformly illuminated on all surfaces tends to generate a Ganzfeld-like effect, which reduces the effects of light stimulation on brain activation and the cognitive performance of users. On the other hand, the presence of direct light components or, in any case, of more intense areas of indirect light on the walls, generates a less uniform and more stimulating environment. In this last situation the brightest areas of the visual field, direct or indirect, are perceived as sources of light and can recall the perceptible differences of natural light filtering through the trees with inhomogeneous effects. These concepts refer to the current of biophilic design (Kellert et al. 2008), which proposes a reconnection to the natural environment in the design of artificial human spaces. According to the theory of Wilson (1984), human beings would indeed have an innate tendency towards natural spaces. In interiors spaces the inhomogeneities of the light in nature can be evoked by indirect inhomogeneous lighting and also with projections of coloured lights on the ceiling and on the walls (USAI Lighting 2016). In line with the attention restoration theory (ART) formulated by Kaplan and Kaplan (1989), the topic of bio-inspired design is also aimed at the ability to recover from mental fatigue and to improve concentration, which are considered facilitated by the vision of natural contexts (Berman et al. 2008; Stevenson et al. 2018).

With regard to the distribution of lighting levels, an appropriate distribution of light in interior spaces must be specifically designed, defining two types of areas: those with a highly uniform illuminance and those in which the attention is focused with different and greater quantities of light, using a layered approach. Light uniformity does not mean that lighting must be made equal in all areas, because light must also be an element that marks the difference between rooms with different functions, highlights their use and guides facilitating the wayfinding. In order to facilitate visual adaptation, the project should also conceive transition zones with intermediate illuminance levels, leading from very bright areas to those with lower illuminance levels. In order to reduce the risk of falling, transition spaces must provide light guides that facilitate posture and movement at night (Figueiro et al. 2012b).

**Table 6.1** The lighting requirements increased to facilitate the vision of the elderly

| Application and activity | Average illuminance levels | | | Uniformity Avg/min | UGR max | $R_a$ |
|---|---|---|---|---|---|---|
| | $E_{floor}$ | $E_{work\text{-}plane}$ | $E_{eye}$ | | | |
| *– Bedroom* | | | | | | |
| General | 100 lx | | 60 lx | 3:1 | 19 | >80 |
| Desk | | 800 lx | 75 lx | 3:1 | | |
| Reading in the bed | | 400 lx | 100 lx | 3:1 | 19 | >80 |
| Chairs | | 400 lx | 50 lx | 3:1 | | |
| TV watching | 60 lx | | 30 lx | 3:1 | | |
| Medical examination | 300 lx | 500 lx | 150 lx | 3:1 | 16 | 90 |
| *– Dining room* | | | | | | |
| General | 100 lx | | 40 lx | 5:1 | | |
| Formal | | 100 lx | 40 lx | 4:1 | 22 | >80 |
| Informal | | 200 lx | 80 lx | 4:1 | 22 | >80 |
| Breakfast | | 400 lx | 100 lx | 5:1 | 22 | >80 |
| Lunch | | 200 lx | 40 lx | 3:1 | 22 | >80 |
| Dinner | | 200 lx | 40 lx | 3:1 | 22 | >80 |
| *– Living room* | | | | | | |
| General | 60 lx | | 60 lx | | | >80 |
| Table games | | 400 lx | 150 lx | | | >80 |
| Reading, studying | | 1,000 lx | 400 lx | | | >80 |

Depending on the changes to the visual and cognitive systems occurring with age, illustrated in the previous paragraphs, it is essential to increase the levels of general lighting and focus it primarily on visual tasks and activities that require concentration (Rea and Ouellette 1991). In particular, the illuminance levels should be able to vary according to age and be greater than those provided in the standards and guidelines for interior lighting (EN 12464-1:2011; CIBSE 2013; ANSI/IES 2017). Based on these assumptions, the lighting requirements necessary for the elderly and the visually impaired have been defined in the three envisaged areas of intervention. In Table 6.1 are summarized the lighting requirements, increased to facilitate the vision of the elderly in residential spaces. The last column on the right shows the chromatic rendering index ($R_a$).

## 6.4.2 Luminaires Requirements

Compliance with the lighting design requirements described above involves the definition of the luminous intensity distribution (LID), i.e. the photometric solid, of the luminaires that must be used in the three areas of intervention, namely the bedroom, the dining room and the living room. The light emission characteristics defined by the LID depend on the requirements defined so far for the design research, namely:

- Creating a comfortable lighting without glare: making sure that the LED light sources are not visible, checking the angles of the light beams and evaluating the UGR in the various useful positions in the environment.
- Having an important contribution of indirect lighting: by using ceiling mounted luminaires, controlling the light emitted by LEDs with mini-reflectors and integrating LED sources with coloured emission. In addition, making it possible to control the light for different sectors in order to also obtain dynamic lighting variations.
- Having different direct lighting for the various residential areas changing only some elements of the luminaire: diffused, through a diffusing screen, or with a controlled light beam, through micro-prismatic systems with a wide and narrow beam.
- The LED light engine must be able to produce, from the product in which it is inserted, an indirect flux of 5,000 lm and a direct flux of at least 1,500 lm. Moreover, it must make it possible to obtain different LIDs with the replacement of interchangeable elements for different applications.
- To achieve energy saving goals, the luminaire must have an efficiency of at least 80–85%.

### *6.4.3 The Designed Luminous Intensity Distributions*

For the evaluation of the indirect lighting component, a realistic reflection factor $\rho = 0.7$ has been assumed for the ceiling of the environments considered. Based on these assumptions, the LID has been defined for a suspended luminaire for the bedroom, single or double, capable of producing an average general lighting of 100 lx, with direct emission through a medium beam and a greater indirect emission on the ceiling of the room (Fig. 6.2). For reading in bed and activities on a possible desk, additional task lights are required.

A LID for a suspended luminaire has been designed for the dining room or refectory. This is able to produce a general lighting in the 100–150 lx range on the floor and an illuminance in the 400–500 lx range on the work surface, thanks to a diffused direct emission of the flux and an indirect emission component on the ceiling (Fig. 6.3). It is also expected that the luminous flux can be adjusted independently both in the direct and the indirect components.

A suspended luminaire, located above the table, has also been designed for the living room or communal room, to make the most of the direct emission components. The LID of this product configuration generates a luminous flux, able to obtain a general illuminance on the floor in the 100–150 lx range and a focused illumination directed on the work surface, able to produce an illuminance in the 600–700 lx range

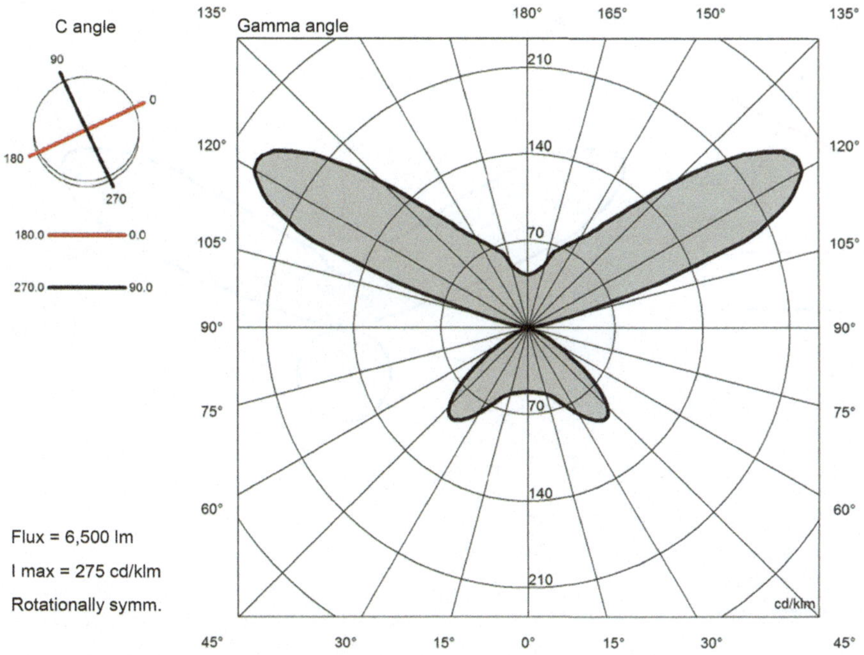


**Fig. 6.2** The LID designed for the bedroom or inpatient room

(Fig. 6.4). Also in this version of the product the luminous flux can be adjusted independently both in the direct and the indirect components.

### *6.4.4 Evaluation of the Luminous Intensity Distributions*

The photometric performance of the three configurations envisaged for the luminaire in the three selected environments has been then checked using the Dialux lighting CAD software (DIAL 2018). To this end, three-dimensional models have been created, typical of the bedroom (Fig. 6.5) and of the other environments.

In the *bedroom*, size 4 × 8.5 m, the compliance with the lighting requirements has been verified, as indicated by the calculations shown in Figs. 6.7 of the lighting in the room seen from two different positions and depending on the luminaire in the configuration designed for this environment, with indirect and direct emission. The calculation are represented with pseudocolour scales. The LID is represented in wire-frame. Figure 6.6 illustrates the calculation of the illuminances represented with isolines to evaluate the general lighting in the bedroom on the part of the floor adjacent to the window. A medium illuminance has been determined on the floor $E_m$

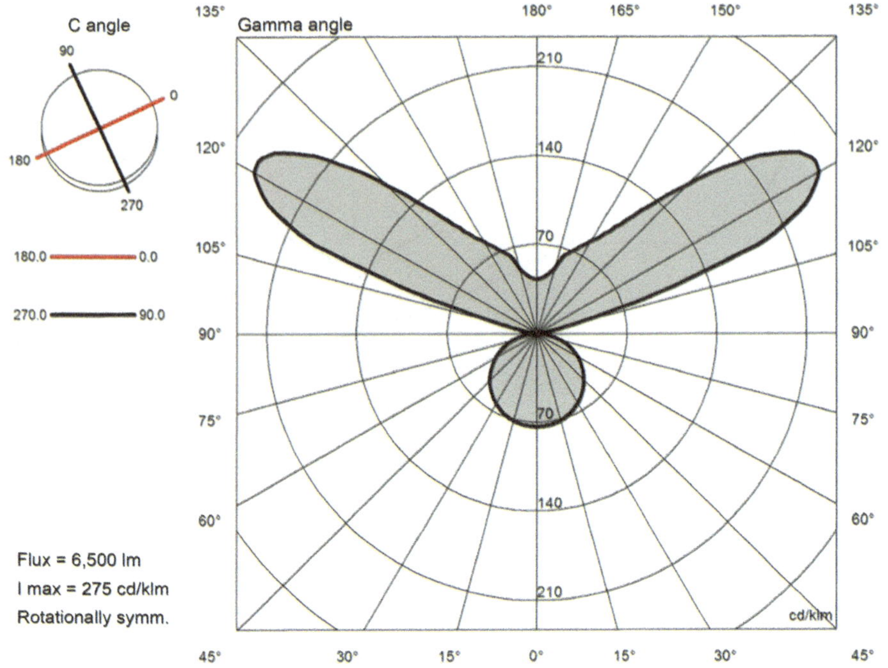


**Fig. 6.3** The LID designed for the dining room or refectory

= 108 lx, a minimum illuminance $E_{min}$ = 52 lx, a maximum illuminance $E_{max}$ = 138 lx and uniformity ratios $E_{min}/E_m = 0.48$ and $E_{min}/E_{max} = 0.38$.

In order to evaluate the visual comfort in the bedroom according to the designed luminaire, six typical places have been chosen that users can take in the room when they are awake (Table 6.2). Considering the typical user face orientation, the values of the UGR factor, calculated in these six places and orientations (Fig. 6.8), are well below the design limits established in Table 6.1. The lighting system therefore provides a high visual comfort thanks to the fact of having a higher flux upwards, to obtain indirect illumination by means of the ceiling, and a lower portion of lighting focused downwards, to guarantee the presence of shadows and contrasts functional to the vision of the elderly.

With the product configured for the *dining room*, the system has the same indirect emission as the other configurations but has a diffused direct emission. This version of the lighting system has been evaluated for a refectory of a NH or ALF, size 15 × 15.5 m, able to accommodate twenty square tables with four seats each. A suspended luminaire has been provided over each table. Using a lighting CAD software, an analysis of the general lighting levels has been carried out throughout the environment and is shown in Fig. 6.9 according to the luminaire configured with indirect and direct diffused emission. Lighting with only the indirect component has also been evaluated, in the hypothesis of switching off the direct light component to facilitate relaxation

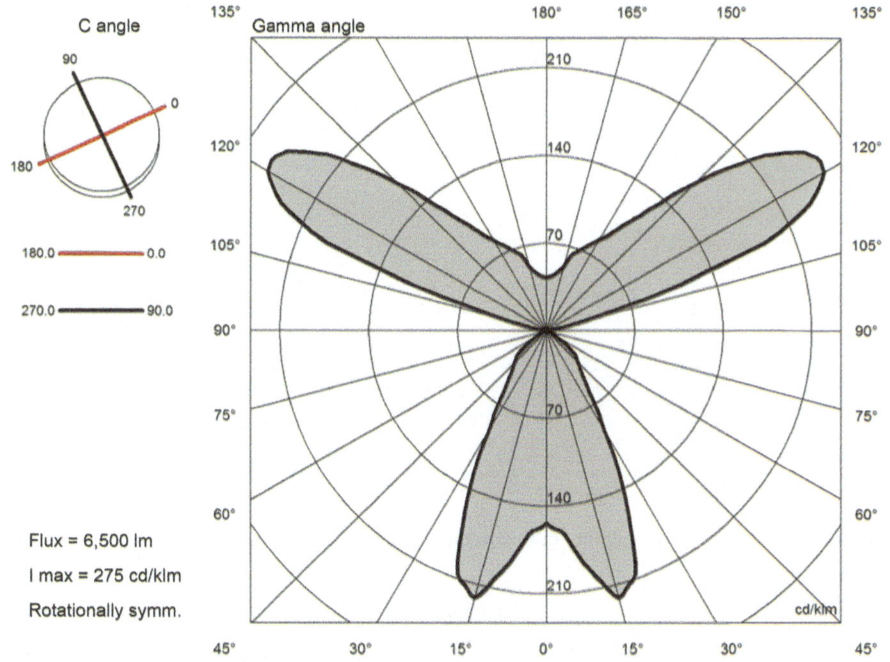


**Fig. 6.4** The LID designed for the living room or communal area

**Fig. 6.5** The model used for the evaluation of artificial lighting in the bedroom

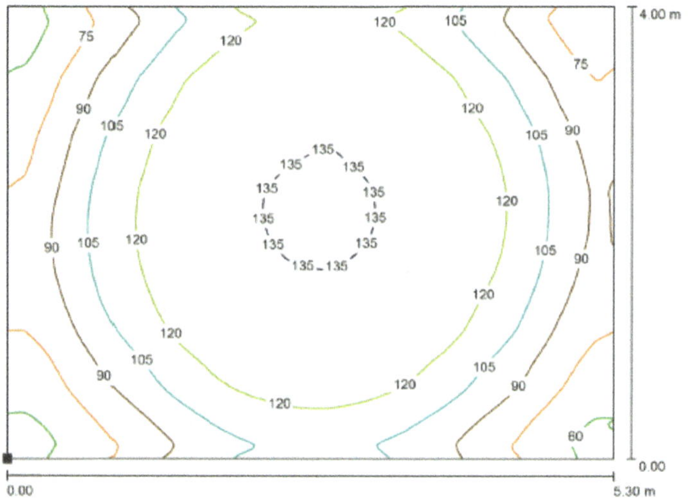


Fig. 6.6 Calculation of the lighting in Lux in the bedroom

**Table 6.2** Values of the UGR factor calculated in the bedroom in some typical user positions

| Ref. | Position | Angle | UGR |
|---|---|---|---|
| 1 | Standing, between the two beds, facing the desk | 90° | 15 |
| 2 | Lying on the bed, reading | 90° | 13 |
| 3 | Sitting on the bed, oriented towards the window | 0° | 17 |
| 4 | Sitting at the desk, facing the wall | 55° | <10 |
| 5 | Sitting at the desk, facing the beds | −110° | 20 |
| 6 | Standing, entering the room | 0° | <10 |

with lower illuminance levels. Fig. 6.10 shows the calculation of the lighting in the refectory according to the luminaire configured with only an indirect emission. Even in this condition the system still provides illuminance in compliance with the standards. Figure 6.11 shows the calculation of the lighting on the corner table with the lighting system in the dining room configuration with direct and indirect emissions. The small box shows the position of the table with respect to the refectory. A medium illuminance has been determined on the table $E_m = 447$ lx, a minimum illuminance $E_{min} = 384$ lx, a maximum illuminance $E_{max} = 482$ lx and uniformity ratios $E_{min}/E_m = 0.86$ and $E_{min}/E_{max} = 0.80$.

To evaluate the visual comfort the UGR factor has been calculated for three seats of a table placed in a corner with respect to the refectory and a seat of a table in the central area (Fig. 6.12). The position of the table in the corner has been chosen deliberately as it has been considered the worst both from the illumination standpoint, because it receives a lower light contribution from other luminaires, and

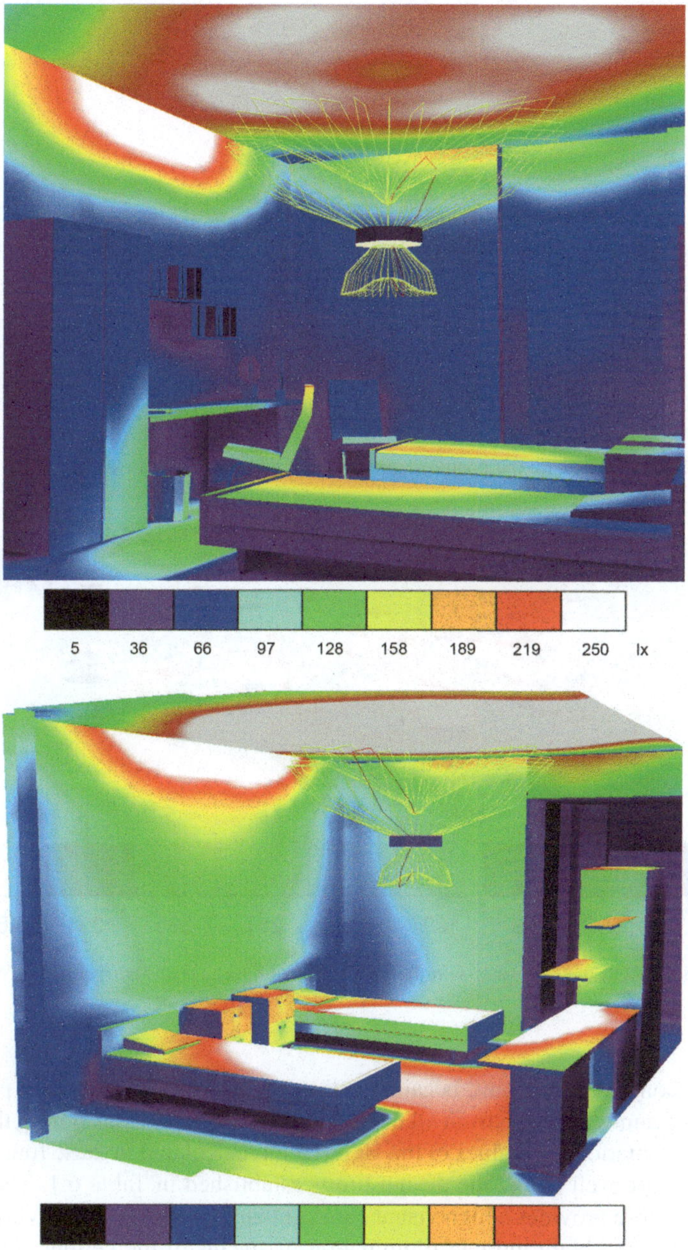


**Fig. 6.7** Calculation of the lighting in the bedroom seen from two different positions

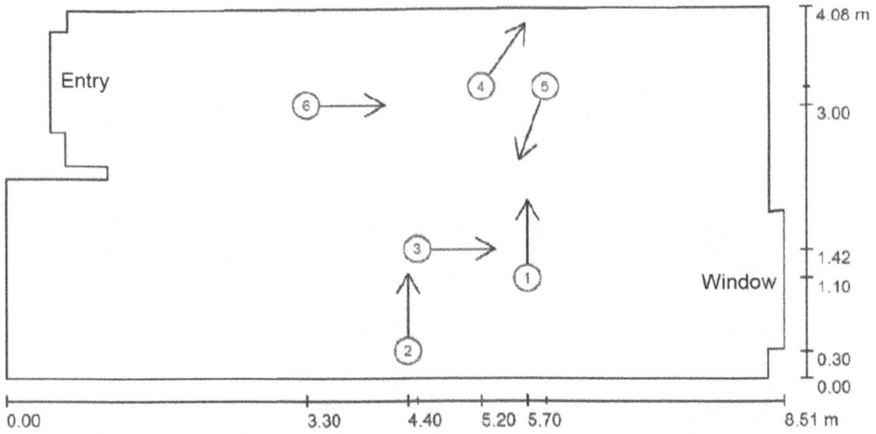


**Fig. 6.8** Bedroom plan with the positions and orientations chosen for the calculation of the UGR

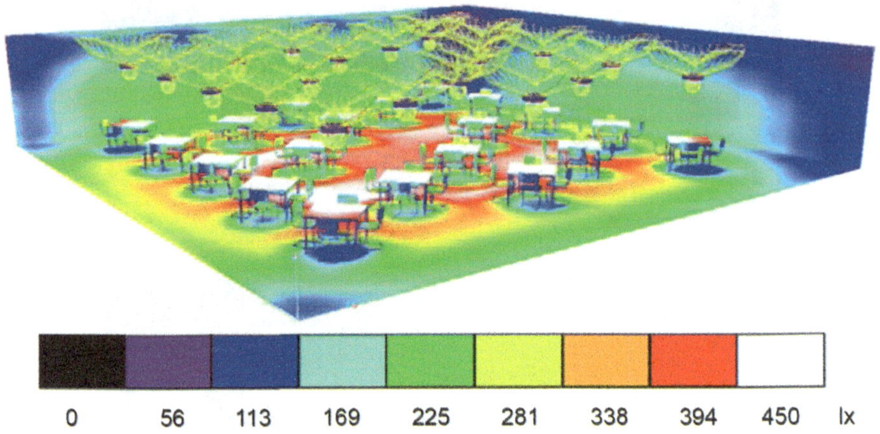


**Fig. 6.9** Lighting in the refectory with the luminaire configured with indirect and direct diffused emission

the visual comfort standpoint, because in this position there is a greater possibility of receiving annoying glare from all the luminaires installed. Considering the typical user face orientation, the values of the UGR factor, calculated in these four positions (Table 6.3), are well below the design limits established in Table 6.1. The lighting system therefore provides a high visual comfort thanks to the fact of having a higher flux upwards, to obtain indirect illumination by means of the ceiling, and a lower flux downwards.

Finally, the configuration of the luminaire designed for the *living room* has been evaluated, with a table for reading, playing or carrying out other activities with demanding visual tasks such as, for instance, sewing. This configuration has the same indirect emission as the previous ones but adopts an emission with a direct

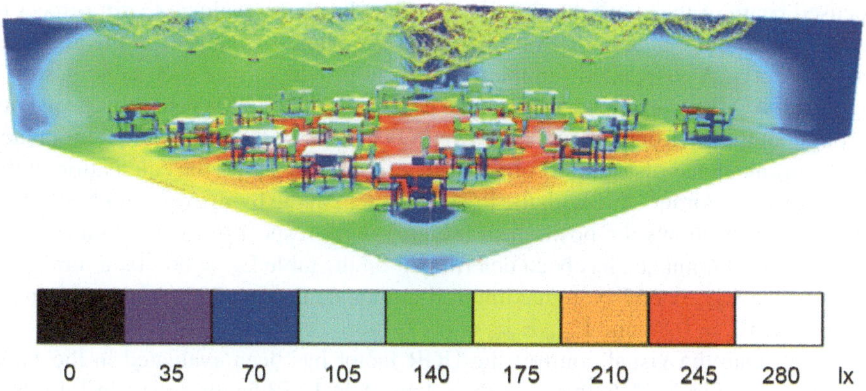


**Fig. 6.10** Lighting in the refectory with only a indirect emission

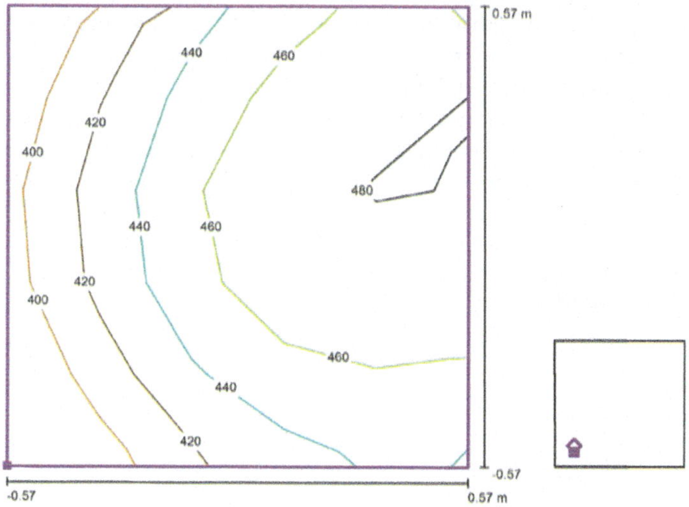


**Fig. 6.11** The Lux on the corner table with the lighting system in the dining room configuration

**Table 6.3** Values of the UGR factor with the lighting system in the dining room configuration

| Ref. | Position | Angle | UGR |
|---|---|---|---|
| 1 | Sitting at the table, facing the refectory | 90° | 18 |
| 2 | Sitting at the table, facing the wall | −134° | <10 |
| 3 | Sitting at the table, facing the refectory | 35.5° | 15 |
| 4 | Sitting at the table, facing the refectory | 35.5° | 13 |

focused beam of the spotlight type. The product has been evaluated with respect to a possible communal area of an NH or ALF, equipped with tables for the activities of the elderly. A suspended luminaire has been provided over each table. Using a lighting CAD software, an analysis of the general lighting levels has been carried out throughout the environment and is shown in Fig. 6.13. Furthermore, the illuminance has been calculated on a single corner table. Figure 6.14 shows the calculation of the lighting on the corner table with the lighting system in the living room configuration. The small box shows the position of the table with respect to the communal area. An average illuminance has been determined on the table $E_m = 665$ lx, a minimum illuminance $E_{min} = 557$ lx, a maximum illuminance $E_{max} = 725$ lx and uniformity ratios $E_{min}/E_m = 0.84$ and $E_{min}/E_{max} = 0.77$.

To evaluate the visual comfort the UGR factor has been evaluated in the same positions as in Fig. 6.12. In this case, the values of the UGR factor, shown in Table 6.4, are even lower than those of the product configured for the dining room and are well below the design requirements set out in Table 6.1. In this version the lighting system therefore provides a high visual comfort. This is possible thanks to the fact of having an upward flux, to obtain an indirect illumination by means of the ceiling and a significant portion of direct lighting concentrated downwards. The latter provides high levels of illuminance on the table (Fig. 6.14) and, thanks to the strong directionality, increases the visual contrasts to facilitate demanding tasks on the table and the concentration of the users.

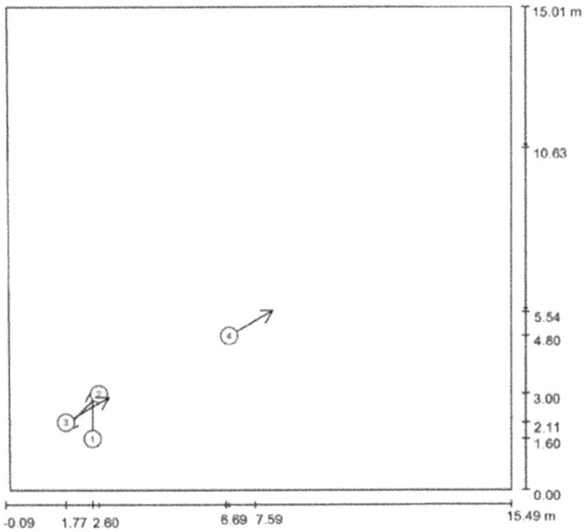


**Fig. 6.12** The four positions chosen within the refectory for the evaluation of the UGR factor

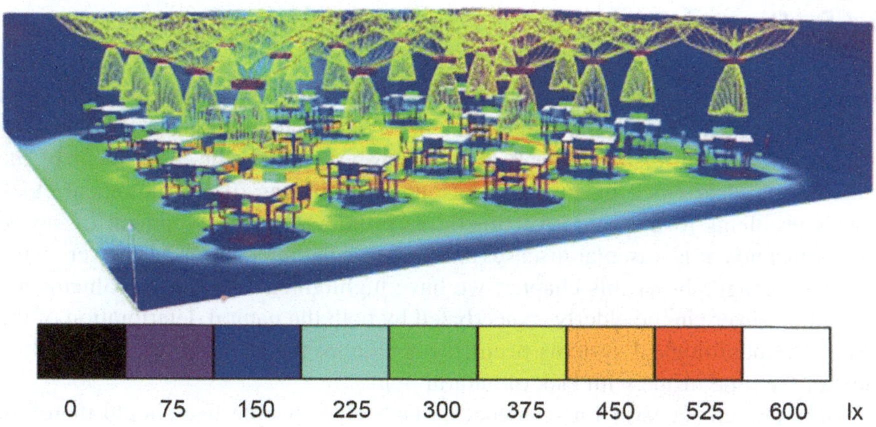


**Fig. 6.13** Lighting with the luminaire configured with indirect and direct focused emission

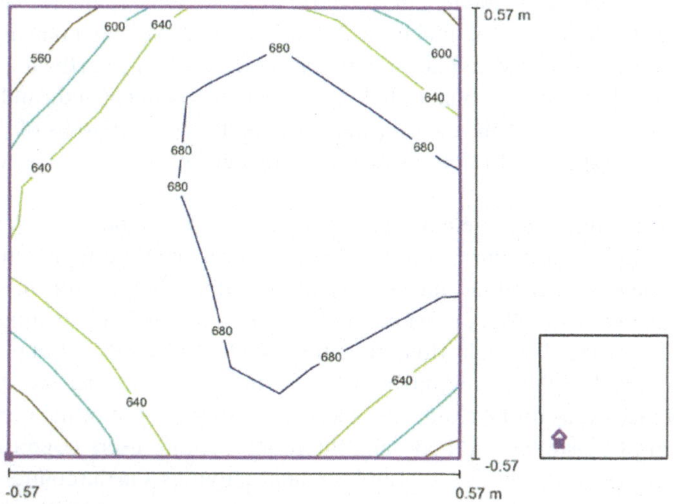


**Fig. 6.14** The Lux on the corner table with the lighting system in the living room configuration

**Table 6.4** Values of the UGR factor with the lighting system configured for the living room

| Ref. | Position | Angle | UGR |
|---|---|---|---|
| 1 | Sitting at the table, facing the refectory | 90° | <10 |
| 2 | Sitting at the table, facing the wall | −134° | <10 |
| 3 | Sitting at the table, facing the refectory | 35.5° | <10 |
| 4 | Sitting at the table, facing the refectory | 35.5° | <10 |

### 6.4.5 Dynamic Lighting

The studies illustrated in Chaps. 2 and 3 show a concrete relationship between light and people's health, particularly regarding the effects that lighting has on the quality of rest, on the control of alertness levels, but above all on the adjustment of the human biological clock in different seasons. The shifting of the biological clock causes problems to the cognitive functions, promotes the development of stress, metabolic and cardiovascular diseases and limits social interaction. Moreover, in the previous paragraphs of this chapter, we have highlighted how these problems are even more severe in the elderly, exacerbated by both the natural deterioration of the visual and neurological systems occurring with age, and by the habit of spending most of the time inside with lack of natural light.

An improved scenario for residences where older people live should therefore promote health by synchronising the light with the activities and circadian rhythms of the users in a 24-h natural cycle. It must also take into account both seasonal changes, which determine the length of the day, and weather changes, which affect the presence and duration of natural light. A residence that uses circadian lighting, i.e. a light that synchronises the human circadian system, contributes to the improvement of the quality of life of its inhabitants, helping them to rest better at night and to remain attentive and active during the day. It can also alleviate the symptoms of depression and promote interpersonal relationships, thus favouring the general health of the individual.

The circadian lighting system designed for these purposes must envision variable lighting over time and in relation to residential spaces, facilitating a perception of the environment as a familiar and natural place, rather than a uniform one, as it happens today in hospitals. A circadian lighting system must also be related to the natural cycles of the day-night rhythm, acting on the CCT and lighting levels. As already explained, lighting should be a stimulating factor during the day, in the most frequented areas of the residence, such as the living room or the dining room, with the aim of activating the body and the mind, making users cognitively more performing in order to promote social and personal activities. On the contrary, lighting should be a non-stimulating factor during the evening and at night, especially in areas used for watching TV or reading, be it the living room or bedroom. The system provides for the lighting scenarios to be changed automatically, according to the time, following a pre-set program that resets at 24:00. At the same time, the system has been designed to be flexible, in order to be managed manually by the user, through understandable interfaces, easy to handle and to use with the possibility to select three pre-set programs suitable for different types of activities.

Figure 6.15 shows the variation of the CTT according to the time for the four designed lighting programs. On the right, the diagram of the automatic lighting program and, on the left, the diagrams of the programs selectable by the user according to specific activities. To meet the requirements outlined above, the system provides a dynamic lighting performance, which changes over time and according to the different activities carried out, in the luminous flux, the CCT and the ratio between direct

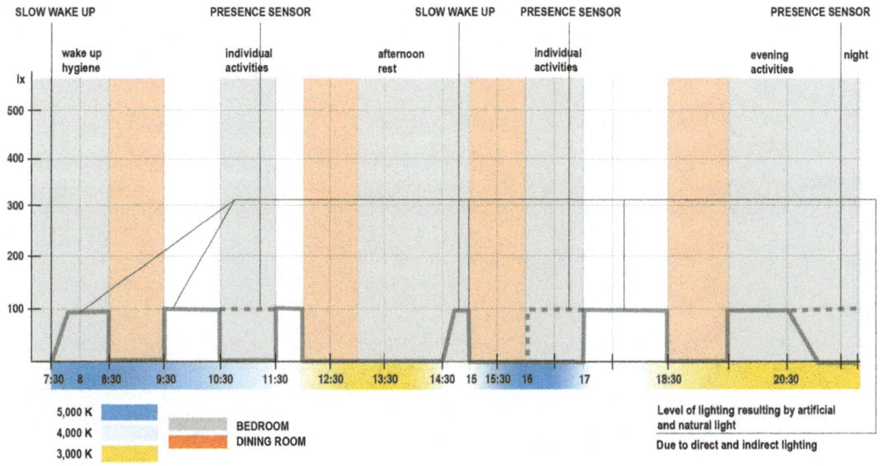


**Fig. 6.16** Lighting scheme applied to the bedroom

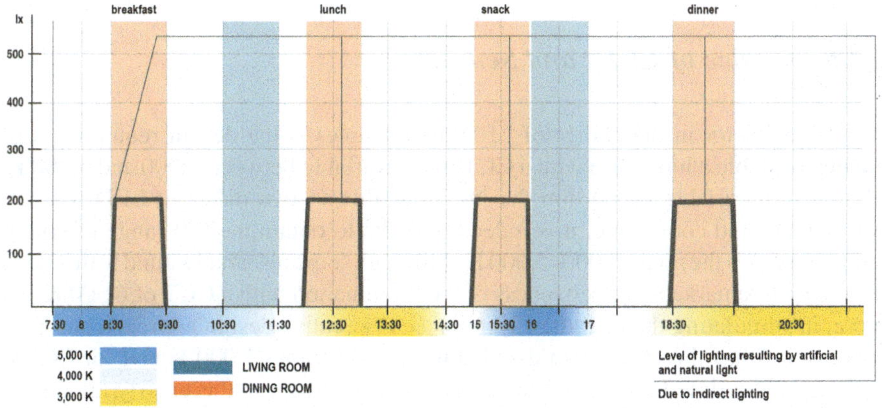


**Fig. 6.17** Lighting scheme applied to the dining room

The manually activated lighting programs last two hours, returning to the automatic program and anyway resetting at 24:00. The arousing program provides a high CCT of 5,000 K, and high illuminance, over 500 lx, with both indirect and direct lighting components. Over time, both the CCT and the illuminance are lowered, respectively to 4,000 K and 300 lx. The relaxing program starts from the end conditions of the arousing program to further lower the CCT to 3,000 K and the illuminance to 200 lx. Finally, the ART program uses intermediate light levels in the 200–300 lx range with white light, both indirect and direct, and an addition of dynamic coloured indirect light, in order to slowly animate the ceiling with coloured shades thus facilitating the recovery from mental fatigue.

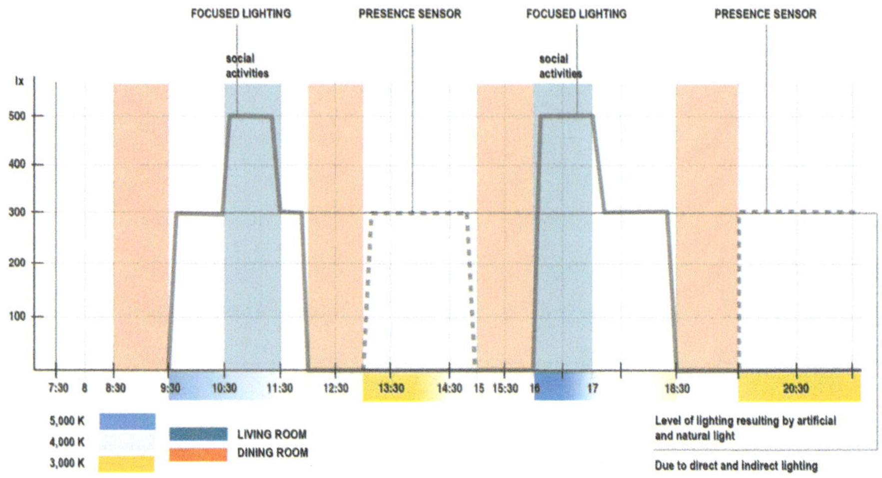


**Fig. 6.18** Lighting scheme applied to the living room

### *6.4.6 Choice of LEDs and Sensors*

A decisive factor in the choice of LED light sources is the design requirement of having available a luminaire with a CCT that is variable between 3,000 and 5,000 K. Mixing in a variable proportion the luminous flux of two different LED sources, warm white and cool white, it is indeed possible to obtain the full range of colour temperatures in the range 3,000–5,000 K. Although some researches cited in this book have shown remarkable effectiveness of the illumination with a CCT of 6,500 K and more, in stimulating the circadian system, a conservative design approach has been evaluated. It has therefore been decided not to go beyond 5,000 K and for limited periods of the day. Indeed, in the research on lighting for the elderly, alongside positive results in the use of very cool light, there are also discordant opinions on the pleasantness of short wavelengths for the elderly (Laufer et al. 2009). Moreover, in some limited cases, excessively stimulating effects have been highlighted on elderly people with senile dementia, for whom a very cool light can lead to increased agitation (Kuller and Wetterberg 1993; Terman and Terman 1999; Lyketsos et al. 1999). In addition to these motivations, in many countries the culture of interior lighting design is not inclined to the use of high CCT in residential spaces, also because a very cool light can bring to mind hospital environments.

Another parameter that has been taken into consideration is having a colour rendering index $R_a \geq 80$ with the possibility of also considering hybrid solutions with $R_a > 90$. This is to try to facilitate the perception of colours by the elderly. Finally, to pursue a general goal of energy efficiency, it has been decided not to consider LED light sources with an efficacy of less than 100 lm/W.

For these reasons, it has been preferred to screen LED products with medium-low power, identifying in Osram Duris® E5 E5 GW JDSMS1.EC one of the possible solutions, as it has a typical efficiency up to 130 lm/W, an $R_a = 85$ and an emission angle of 120°. Both the 3,000 K version (GW JDSMS1.EC-FSFU-5R8T-L1N2) and the 5,000 K version (GW JDSMS1.EC-FSFU-5H7I-L1N2) emit a medium luminous flux, in a selection of products, of 50.25 lm for a junction temperature Tj = 25°. Under operating conditions, a Tj = 75° is assumed, at which there is a 0.93 reduction factor in the flux and therefore an emitted flux of ~45 lm, with a power consumption of 0.38 W and a real efficacy just over 118 lm/W. Under these operating conditions, the LED operates at 67% of its maximum power to ensure high efficacy and long life. It would in fact be possible to increase the emitted flow, increasing the supply power, but this would decrease both efficacy and duration.

Based on this data and the project requirements decided for the luminaire, with a minimum efficiency of 80% and a 5,000 lm indirect and 1,500 lm direct flux, 139 LEDs are required for the indirect flux and 42 LEDs for the direct flux. This is the case for both channels provided for the CCT, i.e. a total of 181 LEDs with a CCT of 3,000 K and 181 LEDs with a CCT of 5,000 K. Obviously, these LEDs will never be turned on all together at full power. For instance, when the luminaire emits only the maximum 5,000 K CCT, only the 5,000 K LEDs will be switched on at 67% of their maximum power. The same rule applies to 3,000 K LEDs when the product emits this CCT at the maximum expected flux. Under these conditions, the maximum consumption of the light sources is of ~69 W but with an emitted total flux (indirect and direct) of 6,500 lm. The electronic driver is able to adjust the flux emitted by the two types of LEDs in order to mix them and obtain the intermediate CCTs, in the 3,000–4,000 K range.

Although the performance of the designed system offers already a high standard compared to traditional technologies, the colour vision needs of the elderly have also been taken into consideration. These require an increase in the colour rendering of the system which, however, does not go beyond an $R_a = 85$ because radiation components in the red area are missing in the spectrum of the emitted light. This is because the white LEDs are made with blue LEDs, covered with yellow phosphors that convert the light, but there is always the lack of a red light component. Therefore, within the research, the possibility of using a hybrid technology to increase colour rendering has been also assessed. In a solution of this type, the dynamic white light, with a variable CCT, is obtained by combining the light of a red or amber LED with that of a white LED, which has a spectral emission shifted to the green area of the chromaticity diagram. This white-green source is obtained by a blue LED coated with green phosphors, which has the advantage of having a very low conversion loss factor. This white-green LED is therefore able to produce, in combination with the blue emission of the LED, a very efficient light source. However, due to the strong green colour cast, it cannot be used alone as a source of white light.

In a two-channel hybrid lighting system capable of generating dynamic white through a series of white-green and red-amber LEDs, the CCT adjustment can be carried out within a range of 3,000–4,000 K because, by mixing at the ends the two light sources, a coloured light is obtained that can no longer be considered white. In addition, a chromaticity sensor is required to enable the LEDs driver to have an electronic feedback. This in order to suitably power supply the two LEDs channels for obtaining the desired CCT. As part of the research, two possible sensors useful for the experimentation have been chosen among several solutions.

The first sensor is the Avago Technologies ADJD-S311-CR999, with a very low cost, made with CMOS technology with coloured filters and featuring 4 channels: red, green, blue and white. This device is typically used in mobile devices, such as smartphones, to determine the quantity and chromaticity of the light that reaches them. It also has a sensitivity that can be electronically controlled to adapt to different applications.

The second sensor identified is a more complex, but also more sophisticated product, the AMS TCS3200. This sensor is based on a matrix of 64 photodiodes, of which 16 are for red, 16 for green, 16 for blue and 16 for white. The quantity that acts on the sensor is indeed irradiance. The three coloured channels, distributed with respect to the visible frequencies, allow the programmable converter to obtain precise information on the chromaticity of the light and therefore make it possible to evaluate the CCT of the light reaching the sensor. Furthermore, it is possible to program the full sensitivity scale to adapt to different orders of magnitude of the lighting.

Using therefore white-green and amber LEDs, a prototype has been created in which the light emitted by the two types of sources has been combined, thanks to a mixing chamber made of white diffusing material (Furukawa 2018). This is to avoid the phenomenon of coloured shadows that occurs when white light is created by adding sources of different colours. From the tests carried out with laboratory measurements, it has been possible to obtain a white light with a CCT in the 3,000–4,000 K range. In Fig. 6.19 are shown the relative SDPs measured in the hybrid lighting solution for the two different CCTs of 3,000 and 4,000 K. For the light produced by the prototype, with a CCT of 3,000 K, a colour rendering index $R_a = 93$ has been measured. In particular, the value of the red colour sample, usually critical for all LED sources, is $R_a = 69$. Furthermore, a value of LER = 350 lm/W has been calculated. In the case of white light at 4,000 K, a rendering of the red colour similar to the

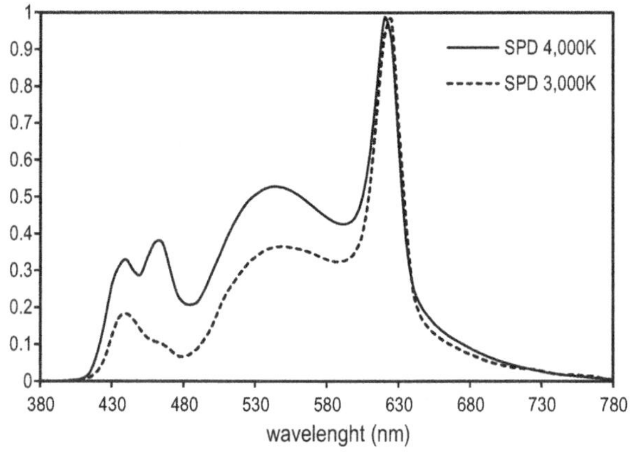


**Fig. 6.19** The relative SDPs measured in the hybrid lighting solution

previous one has been obtained, while the colour rendering index has increased to $R_a = 94$. This increase has led to a slight decrease in the efficacy of the system that, for this CCT, has a LER = 321 lm/W. And this is a general rule: although the LED light sources have raised the efficiency bar in the production of luminaires, an increase in colour rendering always entails a lower energy efficiency of the system.

### *6.4.7 Luminaire Concepts*

As illustrated, the three configurations envisaged for the lighting system are aimed at adapting to the main areas of a residence for elderly people, located in the various rooms, based on the analysis of potential activities carried out by users during the day. The choice of each luminaire is defined according to its luminous performance, its geometric and dimensional characteristics and its positioning within the environment of an ALF. The de-hospitalisation issue, in this context, is very important because it is aimed at maintaining or recovering the sense of independence and self-esteem of the elderly. The possibility of modifying the luminous atmosphere, making it warmer and more personal, but also the formal recall to the archetype of the domestic luminaires and the possibility of establishing a relationship of ritual interaction with the lighting atmosphere, are all factors aimed at recovering the sense of domesticity and identity also in a context different from the home. Therefore, the luminaires have been specifically designed for aesthetic customisation, making it possible to offer different alternatives and be able to adapt to the preferences of the elderly and the specific residential environment. At the same time, the development of a luminaire with domestic characteristics makes it also suitable for the redesign of lighting in the homes of the elderly in the ageing in place context.

Two different light engines have been developed to be inserted into different luminaires, adaptable to different environments and contexts. They have been conceived to perform different lighting functions, direct and indirect, in the same luminaire, which in turn responds to the different lighting functions envisaged by the lighting schemes presented in the previous paragraphs.

In order to pursue the above mentioned objectives, the developed light engines have been implemented with a series of LEDs with warm white (WW) and cool white (CW). In this way, dynamic effects are obtained by varying the CCT and the amount of light. In a more advanced version, the light engine can also be equipped with a red, green and blue LED system to create coloured light effects with indirect light emission on the ceiling. Through these light engines it has been therefore possible to implement the dynamic, adaptive and customisable lighting programs, designed for the bedroom, dining room and living room, in a residence for the elderly.

Both light engines have been designed to fit mainly into suspended luminaires, circular in shape on an aluminium support used to dissipate the heat produced by the LEDs. The metal core printed circuit board (MCPCB), on which the surface mounted device (SMD) LEDs are electrically connected, is joined to this heat sink. The use of a metal core PCB is aimed at maximising the heat exchange between the LEDs and the heat sink, to guarantee the durability and efficiency of the light engine. In both cases, possible variations have been envisaged for the use of hybrid technology, with the aim of obtaining a high colour rendering along with a good efficiency.

In the *first light engine*, the PCB is a single ring and the LEDs are therefore all positioned on the same side of the heat sink and face downwards. Figure 6.20 shows a sectional view of the first developed light engine. The main elements are: 1 heat sink, 2 outer ring of LEDs for indirect lighting, 3 inner ring of LEDs for direct lighting, 4 transparent screen, 5 thermoformed reflector, 6 MCPET diffusing material, 7 light mixing chamber, 8 diffuser or microprismatic refractor. While Fig. 6.21 illustrates a plan view of the first light engine: the highlighted elements are: 1 outer ring of LEDs for indirect lighting, 2 inner ring of LEDs for direct lighting, 3 LEDs for white dynamic technology CW and WW, 4 possible hybrid technology with amber and white-green LEDs, 5 possible external ring of RGB LEDs for indirect coloured light, 6 possible chromaticity sensor.

A thermoformed circular reflector is designed to reflect upwards the light produced by the LEDs located on the outer ring of the PCB to obtain indirect light. Among the LEDs, in this outer ring of the PCB, LEDs with variable colour (RGB) can also be provided, with power circuits separated from the others, in order to create colour shades in the indirect light. The central part of the PCB, on the other hand, contains LEDs that introduce light into a mixing chamber with diffusing surfaces to achieve a direct illumination. A chromaticity sensor can also be inserted in this mixing chamber, if one wants to use the hybrid technology for the LEDs (Fig. 6.21). The output of the mixing chamber can be a diffuser, to obtain a diffused direct illumination, or a microprismatic refractor, to obtain a direct focused light, with a narrower beam. The light engine can then be customised with different lampshades so as to obtain products that can be adapted to different contexts, with reference to the topics of domesticity and de-hospitalisation. Figure 6.22 illustrates possible customisations

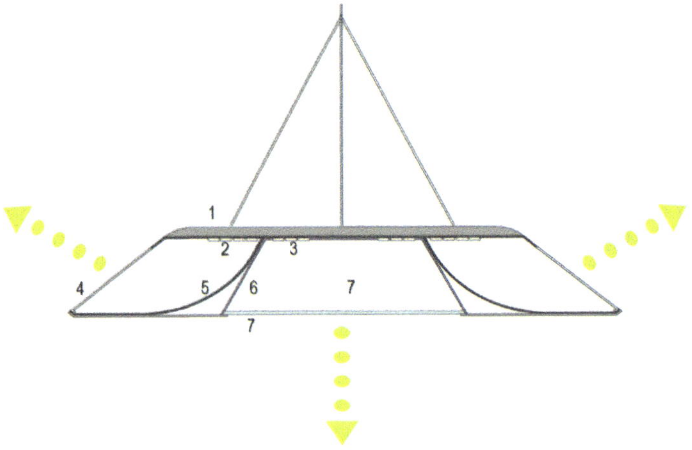


**Fig. 6.20** Sectional view of the first developed light engine

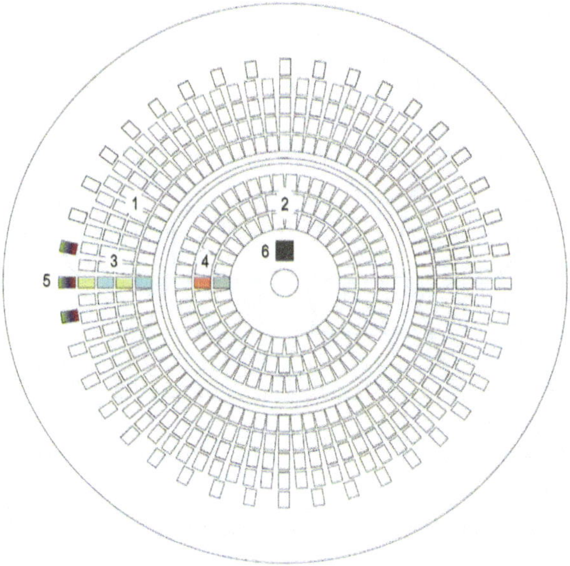


**Fig. 6.21** Plan view of the first light engine

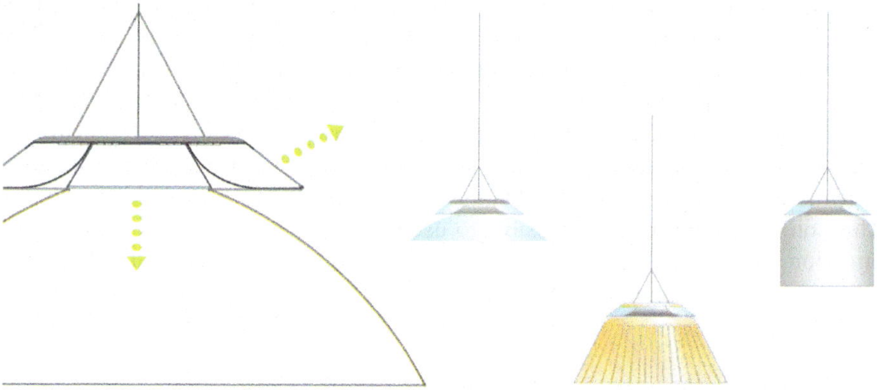

**Fig. 6.22** Possible customisations of the first light engine

of the first light engine useful for creating luminaires that can be integrated into different environments and contexts.

In the *second light engine* there are two separate ring PCBs on the two sides of the heat sink, upwards and downwards. Figure 6.23 shows the sectional view of the second developed light engine. The main elements are: 1 transparent screen, 2 thermoformed reflector, 3 upper ring of LEDs for indirect lighting, 4 lower ring of LEDs for direct lighting, 5 toroidal light mixing chamber, 6 heat sink, 7 MCPET diffusing material, 8 diffuser or microprismatic refractor, 9 air flow. The heat sink

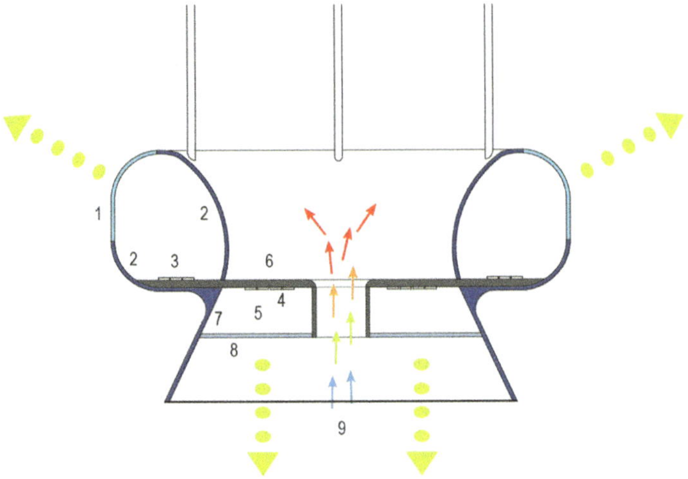


**Fig. 6.23** Sectional view of the second developed light engine

has a central hole joined with an empty cylinder, to facilitate cooling, thanks to the movement of hot air upwards due to the chimney effect. The larger ring of LEDs is located on the upper side, inside a toroidal mixing chamber, for the purpose of obtaining indirect illumination. The mixing chamber, facing upwards and on the side, is closed by a transparent screen to prevent the entry of dust and insects. The LEDs on the inner ring of the PCB are instead facing downwards, towards a second mixing chamber that, on its output, can have a diffuser, to obtain diffused direct lighting, or a microprismatic refractor, to obtain a focused direct light with a narrower beam. This solution could make it possible to use the hybrid technology for direct light emission. Figure 6.24 illustrates possible customisations of the second light engine useful for creating luminaires that can be integrated into different environments and contexts.

## 6.5 Final Considerations

When it comes to artificial lighting, the four elements around which design research revolves are human beings, the light sources, the surfaces of the environments to be illuminated and the ecosystem. But it has not been always like that. Only in recent years, indeed, the human factor has begun to be considered not only for the aspects of visibility and comfort with respect to lighting, but also for the aspects concerning psychophysiological well-being.

The research presented in this chapter is the part implemented at the Politecnico di Milano in the context of a larger industrial research project, co-funded by the Lombardy Region and the Italian Ministry of Education and Research (MIUR),

**Fig. 6.24** Possible customisations of the second light engine

together with industrial partners. The main partner of the industrial research project is the company Light Contract srl of the FLOS SpA Group; other participating companies are Design Group Italia srl and D'Alesio & Santoro srl. The research has been presented in this chapter in a design-oriented perspective, with dissemination purposes, omitting an excessive treatment of the technological aspects and other details that are covered by non-disclosure agreements, however illustrating all the main elements taken into consideration during the research activities.

A positive aspect of this research has been that of having tackled with a multidisciplinary approach the theme of lighting in reference to people, in particular the elderly and those suffering from low vision, analysing and testing the new technologies for solid-state lighting, but also paying attention to the issue of energy consumption of artificial light.

Another positive aspect of the research has been the technological transfer carried out towards the industrial partners with respect to the main topics that have been treated, namely:

- Raising awareness regarding the topic of circadian lighting as a key element in achieving true human centric lighting.
- The identification of new application areas for which to create new product categories for new possible market areas.
- The definition of the needs of specific users of lighting services, the elderly and the visually impaired, who are not currently considered by international standards.
- The development of some possible guidelines for artificial lighting, for the categories of users mentioned above, both in terms of lighting and product design.
- Laboratory analysis and experimentation of the technologies available at the time of the research.

Thanks to the development of the blue LED from which the white light has been derived, LEDs for artificial lighting have started to appear on the market at the beginning of the third millennium with products that could be considered little more than gadgets, which have not really been taken into consideration by the main companies in the sector. Ten years later, the situation had radically changed, with innovations regarding light sources that have followed each other at a rapid pace, every six months, and the difficulty of luminaires' manufacturers to keep up with this technological innovation process.

Today, the long and impressive wave of this technological innovation process, which has radically changed the lighting industry in a few years, has begun to slow down. We are moving towards a stabilisation of new technologies and manufacturers in the sector face the need to identify new application areas for research and development, in which to make the most of the longer life and flexibility of Solid State Lighting (SSL) technologies. The topic of human centric lighting, as also highlighted by the European association of manufacturers (Lighting Europe 2018), is considered a key element in future innovation processes.

## References

Addis, V. M., DeVore, H. K., & Summerfield, M. E. (2013). Acute visual changes in the elderly. *Clinics in Geriatric Medicine, 29*(1), 165–180. https://doi.org/10.1016/j.cger.2012.10.009.

Anni Azzurri. (2008). Case di riposo Residenza San Faustino. Available at: https://anniazzurri.it/residenza-anni-azzurri-san-faustino. Accessed June 10, 2018.

ANSI/IES. (2017). *ANSI/IES RP-29-16 Lighting for hospitals and healthcare facilities*.

Bauer, M., et al. (2015). An evaluation of Snoezelen(®) compared to "common best practice" for allaying the symptoms of wandering and restlessness among residents with dementia in aged care facilities. *Geriatric Nursing, 36*(6), 462–466. https://doi.org/10.1016/j.gerinurse.2015.07.005.

Bell, S. (2006). Visual deficits and their relationship to falls/fractures. *The Art of Geronursing, 12*(2), 1–4.

Benedetti, F., et al. (2003). Morning light treatment hastens the antidepressant effect of citalopram: A placebo-controlled trial. *The Journal of Clinical Psychiatry, 64*(6), 648–653.

Benedetti, F., et al. (2018). Evidence for the efficacy of bright light therapy for bipolar depression. *The American Journal of Psychiatry, 175*(9), 905–906. https://doi.org/10.1176/appi.ajp.2018.18020231.

Bengtson, V. L., & Settersten, R. (Eds.). (2016). *Handbook of theories of aging* (3rd ed.). New York: Springer.

Berman, M. G., Jonides, J., & Kaplan, S. (2008). The cognitive benefits of interacting with nature. *Psychological Science, 19*(12), 1207–1212. https://doi.org/10.1111/j.1467-9280.2008.02225.x.

BIO-LIGHT. (2018). *Bright light therapy. Dawn simulators. BIO-LIGHT light boxes and accessories*. Available at: http://www.bio-light.com/lightsproducts.html. Accessed June 6, 2018.

Boyce, P. (2011). *Research discussion paper n. 5 matching lighting to sight loss*. Thomas Pocklington Trust. Available at: https://www.pocklington-trust.org.uk/project/matching-lighting-to-eye-conditions/. Accessed June 7, 2018.

Boyce, P. (2014). *Human factors in lighting* (3rd ed.). Boca Raton: CRC Press.

Brooke-Wavell, K., et al. (2002). Influence of the visual environment on the postural stability in healthy older women. *Gerontology, 48*(5), 293–297. https://doi.org/10.1159/000065252.

Brundle, C., et al. (2015). The causes of falls: Views of older people with visual impairment. *Health Expectations, 18*(6), 2021–2031. https://doi.org/10.1111/hex.12355.

Campbell, V. A., et al. (1999). Surveillance for sensory impairment, activity limitation, and health-related quality of life among older adults–United States, 1993–1997. *MMWR CDC Surveillance Summaries, 48*(8), 131–156.

Chan, S. W., et al. (2010). The effects of multisensory therapy on behaviour of adult clients with developmental disabilities—A systematic review. *International Journal of Nursing Studies, 47*(1), 108–122. https://doi.org/10.1016/j.ijnurstu.2009.08.004.

Cheng, W., et al. (2011). The effect of LED lighting on color discrimination and preference of elderly people. *Human Factors and Ergonomics in Manufacturing & Service Industries, 26*(4), 483–490. https://doi.org/10.1002/hfm.20333.

Chéron, E. J. (2011). Elderly consumers in Japan: The most mature "silver market" worldwide. In P. Haghirian (Ed.), *Japanese consumer dynamics* (pp. 65–90). London: Palgrave Macmillan UK. https://doi.org/10.1057/9780230302228_4.

CIBSE. (2013). *LG09 lighting guide 09: Lighting for communal residential buildings*.

Collier, L., & Jakob, A. (2017). The multisensory environment (MSE) in dementia care: Examining its role and quality from a user perspective. *HERD: Health Environments Research & Design Journal, 10*(5), 39–51. https://doi.org/10.1177/1937586716683508.

Cronin-Golomb, A., et al. (1991). Visual dysfunction in Alzheimer's disease: Relation to normal aging. *Annals of Neurology, 29*(1), 41–52. https://doi.org/10.1002/ana.410290110.

Czunyi, L., & Craib, D. (2017). Lighting the way to independent living: Preventative methods for senior health inspired by daylight. In W. Chung & C. S. Shin (Eds.), *Proceedings of Advances in Affective and Pleasurable Design*. Springer.

DesMarais, C. (2014, June 19). 6 companies profiting from an aging population. *Inc.com*. Available at: https://www.inc.com/christina-desmarais/6-companies-profiting-from-an-aging-population.html. Accessed March 22, 2018.

DIAL. (2018). *DIALux*. Available at: https://www.dial.de/en/dialux/. Accessed October 8, 2018.

Duchi, F., et al. (2018). Design of a multisensory room for elderly people with neurodegenerative diseases. In *Proceedings of World Congress on Medical Physics and Biomedical Engineering* (pp. 207–210). https://doi.org/10.1007/978-981-10-9023-3_37.

Elliott, D. B. (1987). Contrast sensitivity decline with ageing: A neural or optical phenomenon? *Ophthalmic and Physiological Optics, 7*(4), 415–419. https://doi.org/10.1111/j.1475-1313.1987.tb00771.x.

EN 12464-1:2011. *Light and lighting—Lighting of work places—Part 1: Indoor work places*.

Eurostat. (2012). *Active ageing and solidarity between generations*. Portrait of the European Union. (Eurostat Statistical Books). https://doi.org/10.2785/17758.

Eurostat. (2015). *Demography report—3/2015 edition*. Analytical web note. https://doi.org/10.2767/769227.

Evans, J. R., et al. (2002). Prevalence of visual impairment in people aged 75 years and older in Britain: Results from the MRC trial of assessment and management of older people in the community. *British Journal of Ophthalmology, 86*(7), 795–800. https://doi.org/10.1136/bjo.86.7.795.

Figueiro, M. G., & Rea, M. S. (2005). New research in the light and health field is expanding the possibilities for LED lighting in healthcare environments. In *CIE Midterm Meeting Conference Proceedings* (pp. 1–11). Leon (ES): CIE.

Figueiro, M. G., et al. (2008a). A novel night lighting system for postural control and stability in seniors. *Lighting Research & Technology, 40*(2), 111–126. https://doi.org/10.1177/1477153507084198.

Figueiro, M. G., et al. (2008b). Developing architectural lighting designs to improve sleep in older adults. *The Open Sleep Journal, 1*(1).

Figueiro, M. G., et al. (2011). Lighting and perceptual cues: Effects on gait measures of older adults at high and low risk for falls. *BMC Geriatrics, 11*, 49. https://doi.org/10.1186/1471-2318-11-49.

Figueiro, M. G., et al. (2012a). Field measurements of light exposures and circadian disruption in two populations of older adults. *Journal of Alzheimer's Disease, 31*(4), 711–715. https://doi.org/10.3233/jad-2012-120484.

Figueiro, M. G., et al. (2012b). Lighting for improving balance in older adults with and without risk for falls. *Age and Ageing, 41*(3), 392–395. https://doi.org/10.1093/ageing/afr166.

Furukawa. (2018). *Microcellular foamed reflector MCPET*. Furukawa Electric Co., Ltd. Available at: https://www.furukawa.co.jp/mcpet/english/feature.htm. Accessed June 12, 2018.

Gaymu, J., et al. (2006). Determinants of the living arrangements of older people in Europe Déterminants des modes de vie des personnes âgées en Europe. *European Journal of Population/Revue européenne de Démographie, 22*(3), 241–262. https://doi.org/10.1007/s10680-006-9004-7.

Gibson, J. J. (1966). *The senses considered as perceptual systems* (1st ed.). Boston: Houghton Mifflin Co.

Gilderbloom, J. I., & Markham, J. P. (1996). Housing modification needs of the disabled elderly: What really matters? *Environment and Behavior, 28*(4), 512–535. https://doi.org/10.1177/0013916596284005.

Giménez, M., et al. (2016). Melatonin and sleep-wake rhythms before and after ocular lens replacement in elderly humans. *Biology, 5*(1), 12. https://doi.org/10.3390/biology5010012.

Gochenour, S. J., & Andersen, M. (2009). Circadian effects of daylighting in a residential environment. In *Proceedings of Lux Europa 2009*.

Golden, R. N., et al. (2005). The efficacy of light therapy in the treatment of mood disorders: A review and meta-analysis of the evidence. *The American Journal of Psychiatry, 162*(4), 656–662. https://doi.org/10.1176/appi.ajp.162.4.656.

Gupta, D., & Chen, P. P. (2016). Glaucoma. *American Family Physician, 93*(8), 668–674.

Hanford, N., & Figueiro, M. G. (2013). Light therapy and Alzheimer's disease and related dementia: Past, present, and future. *Journal of Alzheimer's Disease, 33*(4), 913–922. https://doi.org/10.3233/jad-2012-121645.

Haslam, R. A., et al. (2001). What do older people know about safety on stairs? *Ageing & Society, 21*(6), 759–776. https://doi.org/10.1017/s0144686x01008364.

Hedrick, S. C., et al. (2009). Assisted living pilot program health outcomes. *Journal of Aging and Health, 21*(1), 190–207. https://doi.org/10.1177/0898264308328929.

Hendrick, A. M., Gibson, M. V., & Kulshreshtha, A. (2015). Diabetic retinopathy. *Primary Care: Clinics in Office Practice. (Primary Care Ophthalmology), 42*(3), 451–464. https://doi.org/10.1016/j.pop.2015.05.005.

Huff, P. (2017, February 13). 15 home-based business ideas that serve senior citizens. *Home Business Magazine*. Available at: https://homebusinessmag.com/businesses/business-opportunities/silver-and-gold-senior-services/. Accessed March 22, 2018.

Inghilleri, P. (1996). Luci, emozioni, comportamento. In Lighting fields (Ed.), *Carlotta de Bevilaqua*. Artemide Editoriale Lotus (Vol. 2, pp. 6–17).

Intraub, H. (2012). Rethinking visual scene perception. *Wiley Interdisciplinary Reviews. Cognitive Science, 3*(1), 117–127. https://doi.org/10.1002/wcs.149.

Izsó, L., et al. (2009). Psychophysiological, performance and subjective correlates of different lighting conditions. *Lighting Research & Technology, 41*(4), 349–360. https://doi.org/10.1177/1477153509336798.

Jackson, G. R., & Owsley, C. (2003). Visual dysfunction, neurodegenerative diseases, and aging. *Neurologic Clinics, 21*(3), 709–728. https://doi.org/10.1016/s0733-8619(02)00107-x.

Jenkins, A. J., et al. (2015). Biomarkers in diabetic retinopathy. *The Review of Diabetic Studies: RDS, 12*(1–2), 159–195. https://doi.org/10.1900/rds.2015.12.159.

Kaplan, R., & Kaplan, S. (1989). *The experience of nature: A psychological perspective* (1st ed.). Cambridge: Cambridge University Press.

Kellert, S. R., Heerwagen, J., & Mador, M. (2008). *Biophilic design: The theory, science and practice of bringing buildings to life*. Hoboken: Wiley.

Kendig, H., et al. (2017). Preferences and predictors of aging in place: Longitudinal evidence from Melbourne, Australia. *Journal of Housing for the Elderly, 31*(3), 259–271. https://doi.org/10.1080/02763893.2017.1280582.

Knopman, D. S., & Petersen, R. C. (2014). Mild cognitive impairment and mild dementia: A clinical perspective. *Mayo Clinic Proceedings, 89*(10), 1452–1459. https://doi.org/10.1016/j.mayocp.2014.06.019.

Kuller, R., & Wetterberg, L. (1993). Melatonin, cortisol, EEG, ECG and subjective comfort in healthy humans: Impact of two fluorescent lamp types at two light intensities. *Lighting Research & Technology, 25,* 71–80. https://doi.org/10.1177/096032719302500203.

Laham, M. T. S. (2015, September 14). This is really when old age begins. *Huffington Post*.

Laufer, L., et al. (2009). Psychophysiological effects of coloured lighting on older adults. *Lighting Research & Technology, 41*(4), 371–378. https://doi.org/10.1177/1477153509336803.

Lighting Europe. (2018). *Human centric lighting*. Available at: https://www.lightingeurope.org/human-centric-lighting. Accessed October 12, 2018.

Livingston, G., et al. (2014). A systematic review of the clinical effectiveness and cost-effectiveness of sensory, psychological and behavioural interventions for managing agitation in older adults with dementia. *Health Technology Assessment, 18*(39), 1–226, v–vi. https://doi.org/10.3310/hta18390.

Löfqvist, C., et al. (2013). Voices on relocation and aging in place in very old age—A complex and ambivalent matter. *The Gerontologist, 53*(6), 919–927. https://doi.org/10.1093/geront/gnt034.

Lord, S. R. (2006). Visual risk factors for falls in older people. *Age and Ageing, 35*(suppl_2), ii42–ii45. https://doi.org/10.1093/ageing/afl085.

Lord, S. R., & Menz, H. B. (2000). Visual contributions to postural stability in older adults. *Gerontology, 46*(6), 306–310. https://doi.org/10.1159/000022182.

Lord, S. R., & Dayhew, J. (2001). Visual risk factors for falls in older people. *Journal of the American Geriatrics Society, 49*(5), 508–515. https://doi.org/10.1046/j.1532-5415.2001.49107.x.

Lumie. (2018). *Wake-up, SAD and skincare light therapy*. Lumie. Available at: https://www.lumie.com. Accessed June 6, 2018.

Lyketsos, C. G., et al. (1999). A randomized, controlled trial of bright light therapy for agitated behaviors in dementia patients residing in long-term care. *International Journal of Geriatric Psychiatry, 14*(7), 520–525. https://doi.org/10.1002/(sici)1099-1166(199907)14:7%3c520:aid-gps983%3e3.0.co;2-m.

Mantravadi, A. V., & Vadhar, N. (2015). Glaucoma. *Primary Care: Clinics in Office Practice. (Primary Care Ophthalmology), 42*(3), 437–449. https://doi.org/10.1016/j.pop.2015.05.008.

Martiny, K., et al. (2005). Adjunctive bright light in non-seasonal major depression: Results from patient-reported symptom and well-being scales. *Acta Psychiatrica Scandinavica, 111*(6), 453–459. https://doi.org/10.1111/j.1600-0447.2005.00532.x.

Maseda, A., et al. (2018). Multisensory stimulation and individualized music sessions on older adults with severe dementia: Effects on mood, behavior, and biomedical parameters. *Journal of Alzheimer's disease: JAD, 63*(4), 1415–1425. https://doi.org/10.3233/jad-180109.

Mendez, M. F., Cherrier, M. M., & Meadows, R. S. (1996). Depth perception in Alzheimer's disease. *Perceptual and Motor Skills, 83*(3), 987–995. https://doi.org/10.2466/pms.1996.83.3.987.

NCIPC. (2016). *WISQARS (Web-based Injury Statistics Query and Reporting System)*. Available at: https://www.cdc.gov/injury/wisqars/index.html. Accessed March 22, 2018.

NEI. (2015). *Facts about age-related macular degeneration*. National Eye Institute. Available at: https://nei.nih.gov/health/maculardegen/armd_facts. Accessed March 24, 2018.

Noelker, L. S., & Browdie, R. (2014). Sidney Katz, MD: A new paradigm for chronic illness and long-term care. *The Gerontologist, 54*(1), 13–20. https://doi.org/10.1093/geront/gnt086.

Noell-Waggoner, E. (2006). Lighting in nursing homes—The unmet need. In *Proceedings of the 2nd CIE Expert Symposium Lighting and Health*. Ottawa: CIE.

Nugent, P. M. (2013). *What is senile miosis?* Psychology Dictionary. Available at: https://psychologydictionary.org/senile-miosis/. Accessed March 24, 2018.

OECD. (2012). *The future of families to 2030*. Organisation for Economic Co-operation and Development.

Ong, S. R., et al. (2018). Physical activity, visual impairment, and eye disease. *Eye, 32*(8), 1296–1303. https://doi.org/10.1038/s41433-018-0081-8.

Peek, S. T. M., et al. (2014). Factors influencing acceptance of technology for aging in place: A systematic review. *International Journal of Medical Informatics, 83*(4), 235–248. https://doi.org/10.1016/j.ijmedinf.2014.01.004.

Percival, J. (2007). *Lighting the homes of people with sight loss: An overview of recent research*. VisionUK. Available at: https://www.visionuk.org.uk/lighting-the-homes-of-people-with-sight-loss-an-overview-of-recent-research-pdf/. Accessed June 7, 2018.

Petersen, R. C., et al. (1999). Mild cognitive impairment: Clinical characterization and outcome. *Archives of Neurology, 56*(3), 303–308. https://doi.org/10.1001/archneur.56.3.303.

Poulain, I., & Giraudet, G. (2008). Age-related changes of visual contribution in posture control. *Gait & Posture, 27*(1), 1–7. https://doi.org/10.1016/j.gaitpost.2007.02.007.

Previc, F. H. (1998). The neuropsychology of 3-D space. *Psychological Bulletin, 124*(2), 123–164. https://doi.org/10.1037/0033-2909.124.2.123.

Prieto, T. E., et al. (1996). Measures of postural steadiness: Differences between healthy young and elderly adults. *IEEE Transactions on Biomedical Engineering, 43*(9), 956–966. https://doi.org/10.1109/10.532130.

Pruchno, R. A., & Rose, M. S. (2000). The effect of long-term care environments on health outcomes. *The Gerontologist, 40*(4), 422–428. https://doi.org/10.1093/geront/40.4.422.

Rea, M. S., & Ouellette, M. J. (1991). Relative visual performance: A basis for application. *Lighting Research & Technology, 23*(3), 135–144. https://doi.org/10.1177/096032719102300301.

Rea, M. S., et al. (2012). Modelling the spectral sensitivity of the human circadian system. *Lighting Research & Technology, 44*(4), 386–396. https://doi.org/10.1177/1477153511430474.

Renesas. (2013). *Evolution of information technology focused on aging society creates China's big market*. Renesas Electronics. Available at: https://www.renesas.com/us/en/about/edge-magazine/global/10-chinas-market.html. Accessed March 22, 2018.

Rizzo, P. (2007). Lighting for universal design. *Ultimate Home Design, 07*, 50–59.

Robison, J. T., & Moen, P. (2000). A life-course perspective on housing expectations and shifts in late midlife. *Research on Aging, 22*(5), 499–532. https://doi.org/10.1177/0164027500225003.

Rossi, M., Casciani, D., & Musante, F. (2017). Chromatic and spectral quality of lighting for self-dependent elderly people well-being: A case study. In *13th AIC Congress 2017 Being Color With Health*. Jeju Korea: KSCS.

Rowles, G. D., & Bernard, M. (Eds.). (2013). *Environmental gerontology: Making meaningful places in old age* (1st ed.). New York: Springer.

Rugelj, D., Gomiščck, G., & Sevšek, F. (2014). The influence of very low illumination on the postural sway of young and elderly adults. *PLoS ONE, 9*(8), e103903. https://doi.org/10.1371/journal.pone.0103903.

Sabia, J. J. (2008). There's no place like home: A hazard model analysis of aging in place among older homeowners in the PSID. *Research on Aging, 30*(1), 3–35. https://doi.org/10.1177/0164027507307919.

Schaie, K. W., & Willis, S. (Eds.). (2015). *Handbook of the psychology of aging* (8th ed.). Boston: Academic Press.

Scuffham, P. A., et al. (2002). The incidence and cost of injurious falls associated with visual impairment in the UK. *Visual Impairment Research, 4*(1), 1–14. https://doi.org/10.1076/vimr.4.1.1.15635.

Shioiri, S., et al. (2018). Spatial representations of the viewer's surroundings. *Scientific Reports, 8*(1), 7171. https://doi.org/10.1038/s41598-018-25433-5.

Slay, D. H. (2002). Practice note-home-based environmental lighting assessments for people who are visually impaired: Developing techniques and tools. *Journal of Visual Impairment & Blindness, 96*(2), 109–115.

Sloane, M. E., Owsley, C., & Alvarez, S. L. (1988). Aging, senile miosis and spatial contrast sensitivity at low luminance. *Vision Research, 28*(11), 1235–1246. https://doi.org/10.1016/0042-6989(88)90039-9.

Sloane, P. D., et al. (2007). High-intensity environmental light in dementia: Effect on sleep and activity. *Journal of the American Geriatrics Society, 55*(10), 1524–1533. https://doi.org/10.1111/j.1532-5415.2007.01358.x.

Sloane, P. D., Figueiro, M. G., & Cohen, L. (2008). Light as therapy for sleep disorders and depression in older adults. *Clinical Geriatrics, 16*(3), 25–31.

Staal, J. A., et al. (2007). The effects of Snoezelen (multi-sensory behavior therapy) and psychiatric care on agitation, apathy, and activities of daily living in dementia patients on a short term geriatric psychiatric inpatient unit. *International Journal of Psychiatry in Medicine, 37*(4), 357–370. https://doi.org/10.2190/pm.37.4.a.

Steinfeld, E., & Maisel, J. (2012). *Universal design: Creating inclusive environments* (1st ed.). Hoboken: Wiley.

Stevens, G. A., et al. (2013). Global prevalence of vision impairment and blindness: Magnitude and temporal trends, 1990–2010. *Ophthalmology, 120*(12), 2377–2384. https://doi.org/10.1016/j.ophtha.2013.05.025.

Stevenson, M. P., Schilhab, T., & Bentsen, P. (2018). Attention restoration theory II: A systematic review to clarify attention processes affected by exposure to natural environments. *Journal of Toxicology and Environmental Health, Part B, 21*(4), 1–42. https://doi.org/10.1080/10937404.2018.1505571.

Sumaya, I. C., et al. (2001). Bright light treatment decreases depression in institutionalized older adults: A placebo-controlled crossover study. *The Journals of Gerontology: Series A, 56*(6), M356–M360. https://doi.org/10.1093/gerona/56.6.m356.

Tanner, B., Tilse, C., & de Jonge, D. (2008). Restoring and sustaining home: The impact of home modifications on the meaning of home for older people. *Journal of Housing for the Elderly, 22*(3), 195–215. https://doi.org/10.1080/02763890802232048.

Taylor, P., et al. (2009). *Growing old in America: Expectations vs. reality*. Pew Research Center. Available at: http://www.pewsocialtrends.org/2009/06/29/growing-old-in-america-expectations-vs-reality/. Accessed April 15, 2018.

Terman, M., & Terman, J. S. (1999). Bright light therapy: Side effects and benefits across the symptom spectrum. *The Journal of Clinical Psychiatry, 60*(11), 799–808.

Tinetti, M. E. (2003). Preventing falls in elderly persons. *New England Journal of Medicine, 348*(1), 42–49. https://doi.org/10.1056/nejmcp020719.

Todder, D., Levartovsky, M., & Dwolatzky, T. (2016). Measuring the effect of multi-sensory stimulation in the Snoezelen room on sleep quality of Alzheimer patients using actigraph. *Harefuah, 155*(12), 727–730.

Torrington, J. M., & Tregenza, P. R. (2007). Lighting for people with dementia. *Lighting Research & Technology, 39*(1), 81–97. https://doi.org/10.1177/1365782806074484.

Tsai, Y.-F., et al. (2004). The effects of light therapy on depressed elders. *International Journal of Geriatric Psychiatry, 19*(6), 545–548. https://doi.org/10.1002/gps.1125.

Turner, P. L., & Mainster, M. A. (2008). Circadian photoreception: Ageing and the eye's important role in systemic health. *The British Journal of Ophthalmology, 92*(11), 1439–1444. https://doi.org/10.1136/bjo.2008.141747.

Tuunainen, A., Kripke, D. F., & Endo, T. (2004). Light therapy for non-seasonal depression. *The Cochrane Database of Systematic Reviews, 2*(2), CD004050. https://doi.org/10.1002/14651858.cd004050.pub2.

USAI Lighting. (2016). *Biophilic lighting's role in biomimicry*. Available at: http://www.usailighting.com/biophilic-lightings-role-in-biomimicry. Accessed June 7, 2018.

Varma, R., et al. (2016). Visual impairment and blindness in adults in the United States: Demographic and geographic variations from 2015 to 2050. *JAMA Ophthalmology, 134*(7), 802–809. https://doi.org/10.1001/jamaophthalmol.2016.1284.

Vilela, V. C., et al. (2017). What do Cochrane systematic reviews say about non-pharmacological interventions for treating cognitive decline and dementia? *Sao Paulo Medical Journal, 135*(3), 309–320. https://doi.org/10.1590/1516-3180.2017.0092060617.

Voleti, V. B., & Hubschman, J.-P. (2013). Age-related eye disease. *Maturitas, 75*(1), 29–33. https://doi.org/10.1016/j.maturitas.2013.01.018.

Weert, J. C. M. V., et al. (2005). Behavioral and mood effects of Snoezelen integrated into 24-hour dementia care. *Journal of the American Geriatrics Society, 53*(1), 24–33. https://doi.org/10.1111/j.1532-5415.2005.53006.x.

West, S. K., et al. (1997). Function and visual impairment in a population-based study of older adults. The SEE project. Salisbury eye evaluation. *Investigative Ophthalmology & Visual Science, 38*(1), 72–82.

White, M. D., Ancoli-Israel, S., & Wilson, R. R. (2013). Senior living environments: Evidence-based lighting design strategies. *HERD: Health Environments Research & Design Journal, 7*(1), 60–78. https://doi.org/10.1177/193758671300700106.

WHO. (2002). *Active ageing: A policy framework*. World Health Organization. Available at: http://www.who.int/ageing/publications/active_ageing/en/. Accessed September 18, 2017.

Wijk, H., et al. (2002). Colour perception among the very elderly related to visual and cognitive function. *Scandinavian Journal of Caring Sciences, 16*(1), 91–102. https://doi.org/10.1046/j.1471-6712.2002.00063.x.

Wilson, E. O. (1984). *Biophilia* (Reprint ed.). Cambridge, MA: Harvard University Press.

Wysocki, A., et al. (2015). Long-term services and supports for older adults: A review of home and community-based services versus institutional care. *Journal of Aging & Social Policy, 27*(3), 255–279. https://doi.org/10.1080/08959420.2015.1024545.

Yadav, A., Verma, P., & Singh, S. (2017). Going beyond the limits: Effect of clock disruption on human health. *Biological Rhythm Research, 48*(5), 693–700. https://doi.org/10.1080/09291016.2017.1345428.

Yamada, N., et al. (1995). Clinical and chronobiological effects of light therapy on nonseasonal affective disorders. *Biological Psychiatry, 37*(12), 866–873. https://doi.org/10.1016/0006-3223(94)00221-n.

# Chapter 7
# Conclusions and Possible Guidelines for Circadian Lighting Design

## 7.1 Circadian Light Deficiencies in Interior Spaces

In Chap. 4 of this book, we have dealt with the topic of lighting actually present in interior spaces, focusing on the circadian effects of light. Our analysis has considered three typical internal environments of everyday life, in which people spend many hours: a workplace, a residence and a school. In the three situations, the natural light present, even when supplemented by artificial light, was not able to guarantee correct circadian stimulation, except in limited cases. In the following paragraphs we analyse the reasons for the lack of circadian light in the three interior spaces considered, comparing the results with other similar researches.

### *7.1.1 Poor Circadian Lighting in Workplaces*

At the end of the research in the library of Marcallo con Casone it has been possible to determine that the poor circadian stimulation in the workplace is a consequence of the ambient light inside the building, which therefore depends on three main factors:

- The first factor is the SPD of artificial light sources (Fig. 4.12), which has a low content of blue light and a real fixed colour temperature of 3,824 K, significantly lower than that of natural daylight.
- The second factor is the lack of openings to the outside, aggravated by the fact that occupants tend to limit the contribution of natural light through the use of Venetian blinds. This is done to avoid the annoying glare generated by the contrast between the average luminances of the surfaces present in the environment, on average less than 30 cd/m$^2$ and the luminance of the openings to the outside that is at least two orders of magnitude greater. Figure 7.1 shows the luminance in the visual field of workers in the office area at 19:00.
- The third factor is the contribution of the wall structures, furnishings and all the objects in the library that, with their spectral reflectance, contribute to modify



M. Rossi, *Circadian Lighting Design in the LED Era*, Research for Development,
https://doi.org/10.1007/978-3-030-11087-1_7

**Fig. 7.1** Marcallo con Casone library. Luminance in cd/m$^2$ in the visual field of workers

the SPD and CCT of the ambient light, decreasing the value of the latter with respect to that emitted from artificial light sources. This can also be seen from the image-based measurement of CCTs, present in the visual field of workers at 16:00 in the office area dedicated to the VDU, made with the Konica Minolta CA2000 videophotometer (Fig. 7.2). The values measured in the visual field are on average lower than the 3,824 K measured on artificial light sources.

The topic of the quantity and quality of natural light present in the offices has been the subject of research for many years. Boubekri et al. (1991) have assessed the way in which the size of windows and the degree of penetration of natural light can affect the mood and the job satisfaction level. In their study, they have found that the size of windows would not have a significant impact on the emotional state or on the degree of user satisfaction, but the level of penetration of natural light can affect workers who have windows to one side of their visual field.

A more general research, designed to assess stress, the level of satisfaction and well-being at work, in relation to natural light and windows, has been carried out by Leather et al. (1998). This research, which has analysed a sample of 100 workers, found instead a direct relationship between the degree of penetration of natural light in the workplace and the level of satisfaction and general well-being.

In 2001, the National Research Council Canada (NRC) published a scientific report on the state of research to relate windows to the well-being of people in the workplace (Farley and Veitch 2001). In this review, a number of investigation fields have been taken into consideration, but in particular environmental psychology. In

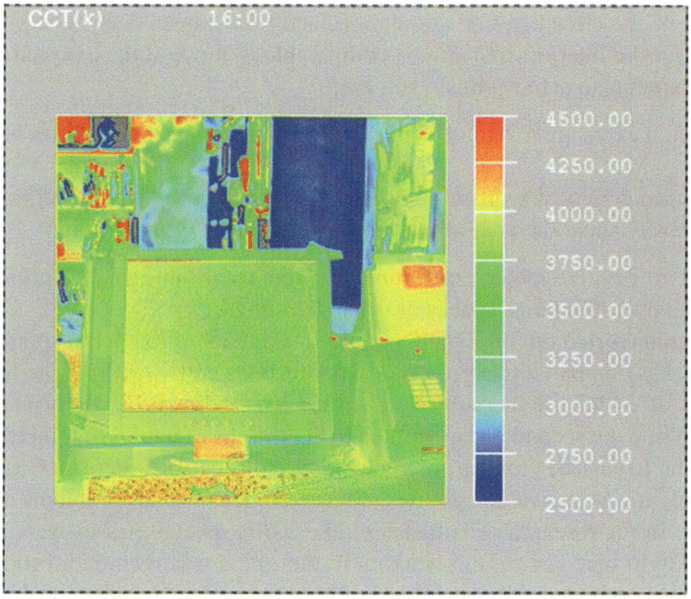


**Fig. 7.2** Marcallo con Casone library. The CCT in the visual field of workers

general, workers prefer the view of natural environments outside the window, rather than artificial or urban ones. This makes it possible to improve the state of well-being and the level of interest and job satisfaction; the view on natural landscapes would also have a positive effect on the recovery times of patients after surgery. However, all this would not involve also an increase in the productivity of workers and students.

Since the topic of circadian lighting has begun to interest researchers and designers, much research has been done to assess the quality of artificial lighting in interior spaces, often highlighting negative and rarely positive aspects on the circadian qualities of artificial light. However, there are few researches in which only the aspect of natural light present in the offices has been assessed. Among these, two research stand out, Andersen et al. (2012) and Mardaljevic et al. (2014), which have primarily tried to define a working methodology for the evaluation of the NIF effects of daylight in the interiors.

In an interesting case study by Bellia et al. (2014b), the natural light present in three offices has been analysed in two different seasons: winter and summer. These offices are located on the seventh and last floor of a building in the city of Naples. They are small, suitable for a single worker and have large windows compared to the size of the office. They are therefore interior rooms that, even if inserted in an urban context with a very high building density, enjoy maximum exposure to natural light, both in terms of position and shape, with lighting levels in the order of thousands of lux. These offices require artificial light only during limited hours in winter. Indeed, using the circadian photometry model of Rea et al. (2012) during the daytime hours, the values of the circadian CS stimulation factor have been determined. These are very high for the interiors and almost comparable to those of the external ones. The two key elements to achieve this result are:

- the optimal exposure to natural light, without other adjacent buildings of the same height;
- the reduced size of the office compared to the windows, which implies that the user is always near the area of maximum penetration of natural light.

These two positive factors, unfortunately, represent more an exception than the norm in the overall interior spaces of urban buildings.

A research carried out on 13,296 people working in a French transport company near Paris, has compared two groups of workers (Leger et al. 2011). The first group, composed of 34.9% of the subjects, worked in conditions of exclusive artificial lighting. While the second group worked in conditions of partial or total exposure to natural light. In the first group, a greater incidence of sleep disorders such as insomnia and fatigue on waking were unequivocally detected, as well as problems of daytime sleepiness. In the research of Hubalek et al. (2010) special glasses were used with which the light received by the workers in the office was recorded. From the data collected they deduced a direct relationship between the light received during the day and the quality of sleep on the following nights.

In the recent research conducted by Figueiro et al. (2015), the authors have recorded and collected the exposure to natural light of workers in a public office building, specifically designed to provide the occupants with the maximum contribution of natural light. Data have been collected in winter and summer to assess the effects on mood and sleep quality of workers. However, it has been surprisingly found that the CS factor during daylight hours has been generally lower than 0.2 in both winter and summer. With higher values in summer, but still less than 0.3, considered by them as the minimum effective level for circadian stimulation useful for a correct sleep timing. In a subsequent research Figueiro et al. (2018a) have introduced additional artificial lighting, in three US federal government buildings used as offices, with the aim of increasing the CS factor up to values greater than 0.3. From the results of the subjective tests to which the workers have responded, it has been shown that they have reported feeling more vital, energetic and alert in the days when they have been able to take advantage of the additional circadian artificial lighting.

This is a demonstration of the fact that artificial lighting, if properly designed, can compensate for the lack of natural light in the interior, both for the visual effects and for the NIF effects of light.

### *7.1.2 Poor Circadian Lighting in Residences*

The results of the research on natural lighting in the flat located in Milano, presented in Chap. 4, have shown a poor circadian stimulation. Only the kitchen/living room and the study are lit enough to activate the circadian system in the afternoon over the year, with the exception of winter and overcast conditions. In relation to this affirmation, it is necessary to further explore the topic of the observer's positioning within the rooms: indeed, in most of the cases in which the circadian activation occurs through natural lighting, the subject is oriented towards the window, in a very close position and with a look perpendicular to the same. This is most evident in the study, given that the desk is positioned in front of the window. However, in other observation points, further away from the windows and with different orientations, the subjects are not effectively activated, not only in overcast sky conditions, but also in clear and intermediate sky conditions. Therefore, the NIF effect of natural lighting occurs extensively and effectively only close to natural light sources and therefore only if the observer is directly exposed to light. On the other hand, if there is a diffused and indirect light condition in the more distant and remote areas of the rooms, the effectiveness of natural light decreases both for the elderly and for the young. As a consequence of this observation, the research has shown that in some specific sky conditions, in some periods of the year and far from observation points close to the windows, the lighting is not effective not only for the circadian cycle of an elderly observer but even for the young one. Therefore, a closer positioning with respect to natural light sources or a supporting artificial lighting system would be useful for visual and non-visual effects for any user.

As a further analysis of the comparisons of the simulation and metrological evaluation of circadian natural light, present in the considered residential environment, we can conclude that a correct stimulation of the circadian system in the interior occurs only in limited periods of time, in favourable weather conditions and with an appropriate exposure of the interior rooms. These conditions are in fact insufficient for a correct timing of the circadian cycle during the year. A correct stimulation of the human circadian system occurs only in situations where there is a clear sky and the sun is conveniently oriented with respect to the considered interior environment. Also the use of the rooms is a critical factor, as it would always be appropriate to have a sufficient dose of circadian light in the morning, but also in the afternoon, as it happens in the exteriors.

The problem of the unpleasant glare of natural light, due to the contrasts between the external natural light and the reduced luminance of the interior, is a consequence of the often-limited size of the windows compared to the rooms to be illuminated. This often leads to the use of simple methods for filtering natural light, such as curtains, which have the secondary effect of reducing the natural circadian light that can enter through the windows. This issue has been also present in the analysis of natural light available in the Marcallo con Casone library, where Venetian blinds have been used to reduce glare due to natural light entering through the windows. In addition, curtains are also often used to ensure the privacy of residents, regardless of the amount of natural light present in the exterior.

It should be considered that the case study assessed, despite being a flat located in the city, is still placed in a favourable position that allows the access of the sun on both sides, east and west, on which the windows are positioned. The flat has also an excellent exposure to natural light without surrounding urban obstacles, so it is an optimal case, compared to the average of the buildings present in large cities. Another topic is the size of the windows. In the case of existing buildings, this is a given data. In the case of those being planned, it can obviously be increased, with an indubitable benefit in terms of the amount of natural light that can enter the interior rooms. However, this factor also depends on elements related to project culture and climate. It should also be remembered that larger windows can also lead to energy surcharges for the air conditioning of buildings. However, this topic goes beyond the purposes of this book.

The results of this research fully confirm some of the hypotheses formulated in a preliminary study, only made through a simulation, in 19th century terraced houses in Boston (Gochenour and Andersen 2009). Also that study has considered the geometry of the living spaces, the windows, the orientation, the curtains, the characteristics of the surfaces and the possible positions of the observer. The authors have proposed that:

- The greatest stimulation occurs when a window is observed directly and progressively decreases when turning the gaze in the opposite direction.
- In the direct observation of the window the stimulation level decreases with the distance from the window.
- When the gaze is not directed towards the window, the level of stimulation depends on the reflection factor of the surfaces in the environment.
- Compared to the time factor, the optimal solution would be the one that makes it possible to have the largest amount of natural light in the rooms in accordance with the activity pattern of those living there.
- Work spaces should be placed in front of windows and higher floors are to be preferred, for longer-lasting activities, as they receive more natural light than lower ones.

A logical observation is that it would be desirable to promote outside activities and exposure to natural light on a regular basis for elderly and young subjects who have no motor limitations and are still active. And this often happens in the warmer seasons and with favourable weather conditions. However, elderly people, and often young too, are more reluctant to leave their homes and adopt a more sedentary lifestyle in unfavourable weather conditions and during the cold seasons, even if they are still active and without motor limitations. Unfortunately, it is precisely in these conditions that natural lighting would still be able to correctly stimulate the circadian system, but only outside, since it cannot sufficiently penetrate in the homes to facilitate the NIF effects of both elderly and young people.

### *7.1.3 Poor Circadian Lighting in Schools*

Although the study made on the two classrooms of the school building in Milan, Italy presented in Chap. 4 is limited to the case considered, it is important to remember that the measurements have been made in the daily period of almost maximum brightness with an overcast sky. However, this is a weather condition that often occurs in the city considered during the autumn and winter seasons (Table 4.6), when the duration of the daytime hours is also shorter and there is a continuous use of the school, except for a short winter interval for the Christmas holidays. As shown by the aerial view (Google Maps 2017), the school is certainly a classic case of a building inserted in a context of high urban density, with other neighbouring buildings in the surrounding areas and equipped with large windows in the classrooms that should contribute to a high input of natural light. However, it is clear from the observations made that, although the lighting inside the classrooms, thanks to the support of artificial light, complies with the European standard for interior lighting EN 12464-1:2011, full circadian stimulation, similar to that on the outside, is only available in limited positions in times of maximum insolation. This condition is however difficult to achieve, given that it has been obtained with the Venetian blinds completely raised, differently from what usually takes place inside the classroom.

A final aspect to highlight is that the contribution of artificial light, although meeting the standard for a correct vision in the interior spaces, is practically irrelevant to the circadian light in the classrooms, which would be particularly important from early morning (8:00) when lessons begin, but also during the entire school day.

This research carried out in a school of Milan is in line with the results of previous research conducted by Bellia et al. (2013) in a classroom of the Faculty of Architecture of the University of Naples Federico II, on the sixth floor of a building at a latitude of 40.85°. The classroom has three large windows facing west on the long side of 10.08 m and has a width of 4.76 m in the north-facing part where the teacher is located, 6.18 m in the central part and 5.56 m in the south-facing rear, with a capacity of 52 students. The measurements have been made from three positions, typical of the seated students, taken longitudinally with respect to the classroom at a distance of ~1.9 m from the west wall. Therefore, in a relatively favourable position for natural light with respect to the windows, since the farthest rows of desks is instead at ~4.75 m from the windows. A fourth measurement point has been placed at the teacher's desk at ~1.2 m from the west wall.

Measurements have been made from 9:30 to 16:30, at 1 h intervals, on two winter days (January 26 and February 2, 2012). On the first day, the sky has been clear while on the second it has been overcast. At that time of year, direct sunlight can enter from the windows in a limited time frame of three hours (13:00–16:00). The measured values, with reference to the typical position of the students' eyes, have been the SPD, the CCT and the vertical illuminance $E_{v,eye}$. Three possible vertical inclinations of the head have been also considered, from 0° with the gaze turned towards the teacher, up to 45° with the gaze turned to the work surface, for a writing action, and finally evaluating the average value of these three angles. The measurements have been conducted with natural light only and then only with the artificial lighting present in the classroom, after sunset and with the curtains drawn. In addition, a session of measurements with overcast sky has been also implemented with the addition of artificial lighting.

In the measurements carried out during the day with a clear sky, illuminance values to the eyes have been recorded in the interval of 18-662 lx, with an average of

228 lx, calculated in the 8 h of measurements, in the four positions and with respect to the three angles of the head. The CCT has been measured in the interval between 3,250 and 6,788 K, with an average of 4,587 K and with values significantly lower than the CCT of the sky at the same times. The CS circadian stimulus factor has been observed in the interval between 0 and 0.57, with an average of 0.31, slightly higher than the minimum threshold considered useful for the stimulation of the circadian system (Fig. 4.19).

In the day with an overcast sky, the natural light has been obviously different in quantity and SPD compared to that with a clear sky and this has been confirmed by the measurements. Furthermore, measurements at 16:30 have been not carried out due to the absence of natural light. In particular, the eye illuminance values have been in the range of 5–281 lx, with an average of 87 lx. The CCT has been measured in the interval between 3,629 and 5,246 K, with an average of 4,341 K and with values closer to the CCT of the sky at the same times. The CS circadian stimulus factor has been in the range 0–0.5, with an average of 0.11, well below the minimum threshold for stimulating the circadian system.

The artificial lighting system present in the classroom is based on linear fluorescent lamps with CCT 3,000 K. However, from the measurements made to the eyes, the CCT is always lower than 2,553 K, with an average value in the four positions and compared to the three inclinations of the head equal to 2,525 K, due to the inter-reflections of light on the internal surfaces that lower the CCT. The eye illuminances are significantly lower than those evaluated under natural light conditions and are in the range between 90 and 223 lx, with an average value of 154 lx. It should be noted that there are less than half the horizontal illuminances in the range between 300 and 400 lx in compliance with the European standard for interior lighting EN 12464-1:2011. The CS circadian stimulus factor has been recorded in the range 0–0.21, with an average of 0.08, well below the minimum threshold for stimulating the circadian system.

The last test has evaluated the effect of adding artificial light to natural light, adding up the SPD measurements made for the two light conditions taken individually. In this condition, a CS circadian stimulus factor has been recorded in the interval between 1 and 0.53, with an average of 0.31, slightly higher than the minimum threshold considered useful for the stimulation of the circadian system and similar to that obtained with natural light and a clear sky.

A limitation of this research, recognised by the authors themselves, is not having evaluated the innermost positions of the classroom, more distant from the windows. In conclusion, the authors have highlighted two main aspects. The first consists in the fact that it is always better to make the measure of the SPD on the eyes of the observers, rather than the simple illuminance value and the expected value of CCT given by the characteristics of the natural and artificial light sources. This is because the SPD on the eyes is significantly modified by inter-reflections of light on the internal surfaces. The second aspect highlighted is that the addition of artificial light is also necessary in day-time conditions to support both the visual and the light effects of NIF, albeit the latter to a limited extent.

As a further observation, it can be noted that the case study considered is at a latitude and in an urban condition, the sixth floor in a seven-storey building (Google Maps 2012), better than that of the school in Milan analysed in the previous paragraph. However, even in clear sky conditions in winter, the average of CS values is slightly higher than the minimum circadian stimulation level and, in any case, significantly lower than that found outside. The reasons for this lack of circadian light are attributable also in this case to an insufficient amount of natural light entering from the windows and to changes to the SPD of light due to reflections on the internal surfaces. Finally, even in the best condition, with direct access to sunlight in the afternoon, it is assumed that users in the classroom make use of curtains to avoid glare, thus further decreasing the contribution of natural light and consequently the CS circadian stimulus.

## 7.2 The Colour Issue

In Chap. 4, we have seen that the CCT of the light reaching the observer's eyes is not the same as that of the natural and artificial light sources present in an interior space. Indeed, the SPD of ambient light is always influenced by the chromatic reflection characteristics of surfaces and objects present in an interior space. The NIF effects due to ambient light therefore depend both on the light sources and on the materials present in an interior space, that is on the entire visual signal that arrives to the eyes of the human observer.

Light and colour are two aspects connected to each other in the process of human visual perception. Until a few years ago, lighting was designed considering only white light. The introduction of LEDs as lighting sources has definitely changed this perspective. Today, when we talk about colour in the field of lighting design we refer to the possibility of using coloured light sources. However, there is an area of design, the colour material finish (CMF) design, in which the concept of colour is treated by referring to the chromatic appearance and other surface properties of objects. These are the wall surfaces present in human spaces, internal or external, as well as industrial and craft products. Turning the attention to interior design, it is a fact that the CMF designer and the lighting designer rarely collaborate in the design process for a common visual-perceptive result. This is an open research topic in the field of interior design that goes beyond the scope of this book. However, there are two aspects that deserve to be treated: the first is the colour rendering of artificial light sources and the second is the pleasantness of the white light's shades.

### *7.2.1 The Open Issue of Colour Rendering*

The concept of colour rendering of a light source is meant as the ability of the light to allow the user to have a visual perception of the colours of objects in a correct way, comparing it with natural light or an ideal artificial light source. Being able to see the colours properly means to better distinguish hues, saturation levels and ultimately all colour contrasts. An illumination with a high chromatic rendering therefore facilitates visual perception and this is even more important for older people who, due to the ageing of their visual system, have a lower ability to distinguish colours and contrasts (Wijk et al. 2002).

The CIE has defined the colour rendering index (CRI) as the effect of a test light source on the chromatic measurement of objects with respect to the chromatic measurement under a reference illuminant. The current version of the method is described in the publication CIE 013.3-1995. The calculation of the CRI, indicated with the $R_a$ index, is based on the direct comparison of a set of 8 colour samples, each illuminated by two sources, the one under examination and the reference one, with the same CCT. The reference source is chosen from two possible sources that are ideal according to the CCT of the light source to be tested. If the latter is less than 5,000 K, the emission of the ideal black body of physics (Plackian emitter) of the same CCT is used as reference source. Otherwise, an illuminant of the D series (daylight) with the same CCT is used. The calculation then involves determining the chromatic differences, in the CIE 1960 UCS colour space, of the eight measured colour samples, when illuminated by the source under consideration with respect to the reference source. From these differences we obtain the eight rendering indices $R_i$ ($i = 1$–$8$) of each of the eight colours and the value $R_a$ is obtained from their arithmetic average. The value calculated in this way, between 0 and 100, describes the colour fidelity level of a light source. The two light sources that have the highest colour rendering ($R_a = 100$) are natural light and incandescent lamps, since their SPD is very similar to that of the reference illuminants.

The eight colour samples used are taken from the Munsell (1915) colour atlas, with medium saturation and so as to uniformly divide the circle of all the colours provided by the atlas. Six additional samples have been then added, of which the first four are red, yellow, green and blue with maximum saturation, the fifth is the pink colour of the Caucasian skin and the last is the colour of green leaves. With these colours, additional indices are sometimes calculated to better evaluate the colour rendering of fluorescent and LED light sources, since they are generally lacking in the red area of the visible spectrum. For instance, the $R_9$ index is used to evaluate the rendering on intense reds and $R_{13}$ to evaluate the rendering on human skin.

Even before the advent of LEDs in the lighting sector, the $R_a$ of the CIE has been criticized as it is not really able to provide complete information on the real colour rendering of many artificial light sources (Fumagalli et al. 2015). A technical committee of the same CIE, the TC 1-62 of Division 1—Vision and Color, has established that the colour rendering index $R_a$ is generally not applicable to establish a rating from the standpoint of the colour rendering of light sources, when in the set of the sources considered there is also a white light LED source (CIE 177:2007). The same committee has also recommended the development of a new and different index to describe the colour rendering.

As part of the research in this sector, alternative techniques have been proposed to determine the colour rendering of light sources. The methods suggested by Judd

(1967) and Thornton (1974) propose some subjective preference criteria for colours. Smet et al. (2010) have proposed a method based on psychophysical tests of visual similarity, based on the chromatic memory of known objects, when illuminated with the test source. Also the method of Bodrogi et al. (2011) is based on psychophysical tests regarding the perception of the difference of 17 colour samples, illuminated with both the test and the reference sources. The colour quality scale (CQS) index proposed by Davis and Ohno (2010) is based on 15 colour samples, chosen with high saturation, and uses a mathematical method different from the $R_a$ to determine an index between 0 and 100. Moreover, unlike the $R_a$, this system does not penalise light sources that cause an increase in colour saturation. The method proposed by Hashimoto et al. (2007) considers that a light source that provides an increase in the saturation of the objects, or an increase in the ability of colour discrimination, has a positive impact on the perception of colour quality. The same principle is also used in the method of Freyssinier and Rea (2010), in which a colour gamut index is proposed, alongside the $R_a$ index of CIE.

In a context where a lot of research had been proposing different methods, a new standard was presented in the USA in 2015 to describe colour rendering (IES TM-30-15). The principles behind this standard state that to determine if a light source facilitates colour vision:

- At least 2 indices are needed, colour fidelity $R_f$ and colour gamut $R_g$ instead of just one index ($R_a$).
- The colour space used for the calculations must be based on the appearance of the colours (CIECAM02) rather than on their measurement (CIE 1960 UCS).
- A graphic representation is needed to show how the light source changes the hue and saturation of colours.
- The colour samples used in the tests must be more than the 8 colours taken from the Munsell colour atlas and must be uniformly distributed in the spectrum.

The colour fidelity index $R_f$ is determined based on the hue variations, due to the test light, of 99 colours instead of the 8 of the CIE method and has a similar meaning, i.e. a value equal to 100 indicates maximum fidelity in the vision of the hue. The colour gamut index $R_g$ is determined based on 16 colour samples and indicates if a light source increases or decreases the saturation level of the perceived colours: a value lower than 100 indicates a loss of colour saturation, while a value greater than 100 indicates that the light source increases the perception of colour saturation. This standard also changes the method by which the reference light source is determined. The transition between the incandescent source (black body) and the daylight is no longer abrupt at 5,000 K but occurs with a gradual change between 4,500 and 5,500 K. This method is more complex to calculate than the $R_a$ of the CIE, but a software that makes it possible to determine its indices and plot its charts is freely available online (Sylvania 2018).

A series of tests conducted at Lab. Luce of the Politecnico di Milano measured a LED module, the Osram Prevaled PL-CORE-Z4-S43. By measuring the SPD emitted from this source, shown in Fig. 7.3, it is possible to observe that this LED uses a hybrid technique to obtain white light, with a CCT = 3,227 K and $R_a$ = 89. The

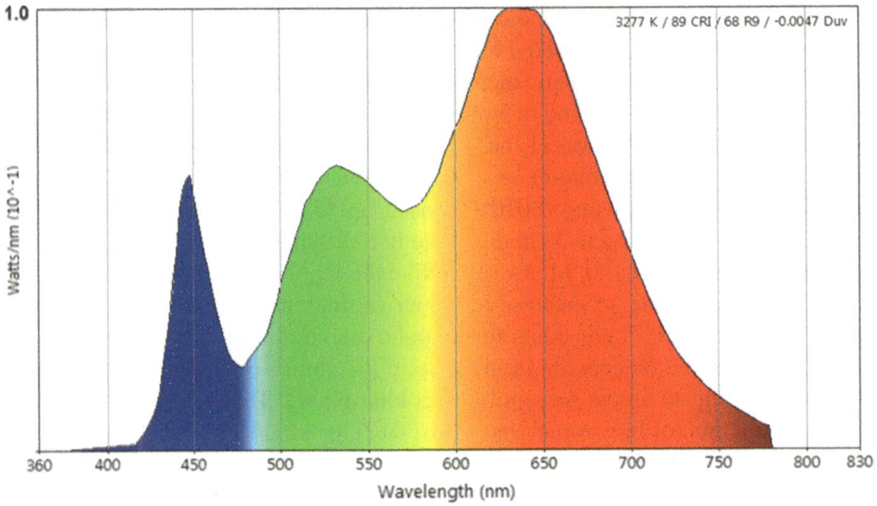


**Fig. 7.3** SPD measured for Osram Prevaled PL-CORE-Z4-S43

colour fidelity indices $R_f = 90$ and $R_g = 109$ have been then determined. This means that this light source has a good colour fidelity, as the hue change is very small, but also increases the perceived saturation of the colours ($R_g > 100$) highlighting them, with a beneficial effect, for all applications in which a greater colour contrast can be useful to facilitate vision. This information can be seen in Fig. 7.4, which shows the colour distortion icon of this source on the CIE LAB colour chart. The red broken line indicates the colours illuminated by the light source, while the blue one indicates the same colours illuminated by the reference source. When the red line is external to the blue one, it means that for that colour there is an increase in saturation. Conversely, when it is internal there is a decrease. When the vertices of the red line are rotated with respect to those of the blue line, it means that the illumination causes a hue variation, which increases with the rotation.

The availability of two indices and a graphic representation of colour rendering is essential for lighting design in an application-oriented perspective. Indeed, colour rendering of light sources is obtained at the cost of energy efficacy. Through the $R_f$ index, one can evaluate how much a light source changes the fundamental hue of colours, while using the $R_g$ index one can understand if lighting makes it possible to increase or decrease colour saturation. These two pieces of information can then be viewed in the colour distortion icon.

Subsequently, the CIE has published a new version of the colour rendering index (CIE 224:2017), the colour fidelity $R_f$ of the CIE. This index has been proposed as more accurate than the $R_a$ but the CIE has however recommended its use only for scientific purposes, without proposing its adoption also for the labelling of light sources on the market. This proposal by the CIE has been followed by the very recent ANSI/IES TM-30-18 standard that partially modifies the contents of TM-30-15 to

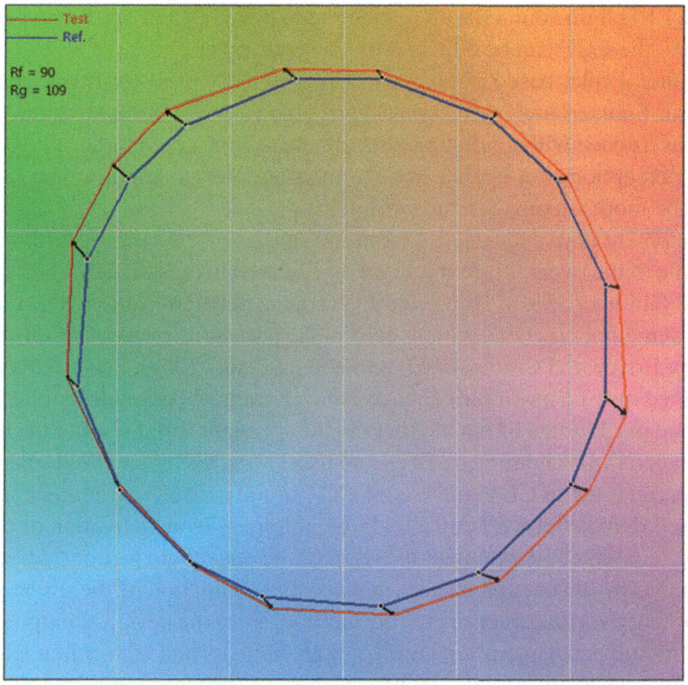


**Fig. 7.4** Colour distortion icon measured for Osram Prevaled PL-CORE-Z4-S43

calculate the colour fidelity index $R_f$ in the same way as the $R_f$ proposed by the CIE, with the aim of harmonising the method worldwide. In a note, the IES (2018) proposes, however, that the indices defined in the ANSI/IES TM-30-18 standard should be added to the $R_a$ of the CIE to replace it later in the light sources on the market. To date, almost all light sources on the market still show the colour rendering index $R_a$ of the CIE often referred to as CRI in their datasheet.

### *7.2.2 The Relationship Between Colour Temperature and Quantity of Light*

One of the topics most often addressed in the past for the lighting of workplaces is the relationship between the quantity of light and its colour temperature. This topic is equally important today, because the circadian stimulation of artificial light is greater with a CCT increase over 5,000 K. We should recall that until a few years ago white light sources on the market provided a limited choice regarding the available CCTs. It was customary to divide them into three categories: warm light up to 3,000 K, cool light over 5,500 K and intermediate light between these two values. Today, we can have white LED light sources with a wide range of CCTs and, in the case of tunable white LEDs, the CCT can be adjusted by the user.

An empirical rule, based on experience and widely used for many years is that with a warm light, of low intensity, it is possible to obtain an internal environment perceived as a space with a relaxing atmosphere, while a cool light of high intensity leads to a perception of a livelier and stimulating internal environment, a situation that could be more adequate in the workplace.

This statement derives primarily from the studies carried out by Kruithof (1941) in the middle of the twentieth century and represented in the curve that bears his name (Davis and Ginthner 1990). However, this topic is debated (Fotios 2001). Although the Kruithof method has been widely used in the past, controversial results have been obtained by Boyce and Cuttle (1990) and Viénot et al. (2009), because its curve has been obtained only on an empirical basis, since Kruithof has not described the evaluation method used for the liking, neither the independent variables, nor the number of subjects involved. By monitoring EEG and heart rate, Kuller and Wetterberg (1993), have compared a warm CCT and a cool CCT in the office environment. They have hypothesised that a cool CCT at 1,700 lx would provoke an activation of the central nervous system which, if extended throughout the day, could be considered visually uncomfortable by the occupants also in the social evaluation of the office space.

By adopting the consequences of this method, from the design standpoint, people should not be subjected to artificial light of high intensity and high colour temperature for too many consecutive hours. However, this contrasts with the characteristics of natural light in the exteriors, which we all consider pleasant.

Other more recent studies have shown different results. Boyce and Cuttle (1990) believe that the assessment of the workspace does not depend only on the CCT of the light but also on other natural and coloured elements that may be present. In a research conducted in Stockport (UK), in a call centre operating from 8:00 to 20:00, organised with working shifts, the effects of using lamps with a CCT of 17,000 K were compared with those of other lamps with a CCT of 2,900 K, with equal illuminance (Mills et al. 2007). This in order to compare results on the performance and well-being of 69 human subjects divided into two groups. This research came to the conclusion that high CCT lighting was well tolerated and seemed to lead to improvements in the feeling of well-being and work performance among the study participants. Similar results have been presented in the research of Viola et al. (2008), who have compared the difference between two colour temperatures, 4,000 and 17,000 K, on a sample of 104 office workers. The data collected from the research has confirmed that subjective improvements in mood, performance, alertness, reduction of daytime sleepiness and quality of night-time sleep have been observed with a high CCT.

However, the methods to achieve stimulating lighting can also be based on increasing the amount of light instead of the CCT, as demonstrated by Smolders et al. (2012); Smolders and de Kort (2014). In their studies, they used an illuminance with a CCT of 4,000 K and 1,000 lx at eye level. However, this leads to an increase in energy consumption and possible glare problems. This lighting was judged to be less pleasant in break and rest situations, although more suitable to work, even in situations of greater fatigue.

## 7.3 Possible Guidelines

In the field of lighting design, the concept of well-being has for many years been mainly addressed by making reference to the quality of vision: for instance, with research on direct and indirect glare control, on contrast rendering factor, and on contrast between the primary visual function area and the background. Users' satisfaction with lighting has also been put in relation to issues related to energy saving (Veitch and Newsham 1998). In the international standards, all these factors have been engineered and are currently used in lighting products and systems, through appropriate design requirements. However, current standards do not consider the NIF effects of light.

The producers of light sources have addressed the issue of vision quality by improving the characteristics of LED sources that are progressively replacing other light sources. Technological research has succeeded in making LEDs produce an ever-increasing value of light flux, with increasing energy efficacy, falling costs, performance and flexibility of use unthinkable for the previous light sources.

An argument often brought in contrast to the introduction of a circadian lighting design is that electric lighting in interior spaces has existed for over a hundred years, performs well its function for visual perception and does not generate problems to the health of users. This statement is however incorrect. It should rather be said that almost no one had posed, until a few years ago, the fundamental question of analysing, from a physiological standpoint, whether the light in interior spaces could create problems, in particular when there is a scarce contribution of natural light.

Chapters 2 and 3 of this book have presented many researches carried out in the last twenty years, which demonstrate the physiological effects and possible problems resulting from inadequate lighting. In recent studies, these negative effects were assessed even more effectively, by recording the exposure to light of people during the working day. A field research conducted by Figueiro and Rea (2014), in an office building recently renovated to maximise its energy efficiency and increase the use of natural light, has assessed the personal levels of daily exposure to light in summer and winter using the Daysimeter (Figueiro et al. 2013). The aim has been to evaluate the relationship between the total light received by people and the timing of their circadian system with reference to quality of sleep and mood. The results of the research show that in the summer there has been a predictable increase in exposure to light, but also a corresponding increase in the quantity and quality of sleep compared to winter, when the amount of light recorded daily had been lower. This study was then extended to five other buildings (Figueiro et al. 2017). The results have shown that people who receive adequate levels of circadian light in the morning have a more regular and good quality sleep than those who do not receive it. In addition, circadian light exposure throughout the working day improves the timing of circadian rhythm and decreases the propensity to depression.

All these elements provide design directions, for the construction and redevelopment of buildings, in order to increase the exposure of the occupants to circadian light, which can be natural light. However, this is often lacking and adequate circadian artificial lighting is therefore needed. We mention here, by way of example, what has been proposed by the Well building standard certification (IWBI 2014) introduced in Chap. 1.

From the review of the numerous researches cited in this book, we can now affirm with certainty that artificial lighting in interior spaces, in addition to facilitate vision, can also support the NIF effects of light. These can affect the level of alertness, mood, cognitive abilities, sleep-wake rhythm, immune system and ultimately the overall health of people.

Although not everything has yet been explained, in particular the effect of light exposure history (Higuchi et al. 2007; Figueiro et al. 2018b), the lighting requirements that come into play for the NIF effects are related to the time and duration of the exposure, the spectral distribution, and the quantity and the direction of the light. An element of complication is that these factors also depend on the environmental context (Bellia et al. 2014a, 2017), on the type of activity carried out, on the characteristics of the individual and on his/her level of fatigue.

There is also the ethical issue of evaluating the effect of artificial lighting, able to decrease melatonin and increase the state of vigilance (Boyce et al. 1997), on the health of people doing night work in interior spaces, with a consequent shift of the circadian rhythm (Åkerstedt and Wright 2009). Examples are doctors, nurses (Booker et al. 2018) and all those who do shift work with night shifts (Wang et al. 2011).

The combination of all these elements makes circadian lighting design an evolving research topic, more complex than the classic interior lighting design. However, some key elements that go beyond current standards can be considered.

In a situation where there is little availability of natural light, an artificial circadian lighting should be used, even if this requires more electricity than the illumination only aimed at vision. In the first place, we can define some key characteristics that lighting must have in order to be considered circadian:

- Artificial lighting in interior spaces should always be dynamic, with light quantity and CCT both variable during the day in a similar way to natural light.
- Lighting must ensure a contribution of short wavelengths (cool CCT) in at least two phases of the day: in the first half of the morning and in the first half of the afternoon.
- In periods of lighting with cool CCT, higher lighting levels should also be used.
- In moments of relaxation and before the evening, a warm CCT should be used with lower lighting levels.
- Currently, in the absence of specific standards, the quantification of natural and/or artificial circadian light $CL_A$ and of the consequent circadian stimulus CS, can be done with the model proposed by Rea and Figueiro (2016) and calculable by software tools: LRC (2018) and Sylvania (2018).
- The circadian stimulation of light must be greater in the morning and almost absent as the evening approaches.
- The evaluation of the real colour rendering of light sources should be carried out using a two-index system able to evaluate both colour fidelity and colour gamut.

A second element to be taken into consideration is that, in order to obtain an illumination with the above characteristics, there must be a smart dynamic lighting project, based on innovative products with special features:

- With the use of LEDs, LMS and smart systems, an increase in lighting levels can be managed in an energy-sustainable way thanks to the greater efficiency of light sources and to their smart control.
- A smart lighting system must include the use of sensors that are able to detect the characteristics of ambient lighting and human presence. Furthermore, in the future, the systems could also detect the position, physiological condition, quantity and quality of light received by the individual during the day.
- Devices for artificial circadian lighting must provide the possibility of adjusting the flow and the CCT emitted. Moreover, they must provide the possibility of controlling the direction of light, in order to vary the ratio between direct and indirect lighting, thus facilitating visual comfort.
- All products of the lighting system must be connected to an LMS and/or be part of the smart home via the Internet of Things.

Finally, bearing in mind that the NIF effects of light include circadian effects through the adjustment of melatonin, but also other direct effects on the nervous system, we must highlight the importance of the *human being* at the heart of the design process in relation to the *environment*:

- The lighting cycle to which people are exposed during the day can disrupt their circadian rhythm and negatively affect the duration and quality of sleep in the following days, with repercussions on health and productivity.
- The age of the individual must be considered in the design. Indeed, lighting for the elderly and the visually impaired involves requirements higher than that for normal-sighted people.
- The quantity and characteristics of the SPD of light can influence people's ability to maintain attention and good cognitive performance during the day, but also at night if they do night work.
- Exposure to light for night workers can reduce melatonin production and influence sleep quality, with long-term health effects.
- The quantification of the circadian effects of light must be done at the level of light that arrives on average to the user's eyes rather than on the work surface.
- The influences of the internal environment must be considered on the real CCT of the light that reaches the eyes, generally lower than that of the installed light sources.
- Lighting settings can influence people's assessment of lighting and interior spaces where they are located, while light settings preferred by individuals also show substantial individual-to-individual variations. It is therefore good to always provide an interface that makes it possible to customise the lighting.

As can be seen from these indications, the overall context must be considered. That is, the type of interior, the functions and the activities of the people who use it, remembering that, in any case, the appropriate lighting for each individual must always be guaranteed.

The fundamental principle is that human beings, with their needs for vision and circadian stimulation, must be placed at the centre of the design process. This should be the goal of human centric lighting, although the use of this term is controversial and debated (Houser 2018) along with the application of the criteria for a new way of understanding lighting (Rea 2016).

Natural light has variable directions, SPD and quantities during the day and humans are sensitive to these variations, which act like a timer, communicating to our body in which period of the day or season we are. Artificial light has made most of human daily activities independent of the availability of natural light. However, this change may result in an alteration of the circadian cycle for a large number of individuals, without their knowledge and with significant effects on their well-being and also on health, in the workplace and in residences.

Based on the research presented and analysed in this book, lighting design will increasingly have to deal with the aspects that can affect human well-being, in the design of both lighting systems and lighting products. Two world-renowned scholars such as Wout van Bommel, researcher of Philips Lighting and president of the CIE from 2003 to 2007, and Mark Rea, professor of cognitive science and director until 2017 of the Lighting Research Center of the Rensselaer Polytechnic Institute, for some years have been arguing that, thanks to research on the effects of light on human beings conducted in recent times, there are the conditions for lighting design to bring about a fundamental shift, taking care not only of the vision but also of the human psychophysical well-being in interior spaces.

## References

Åkerstedt, T., & Wright, K. P. (2009). Sleep loss and fatigue in shift work and shift work disorder. *Sleep Medicine Clinics, 4*(2), 257–271. https://doi.org/10.1016/j.jsmc.2009.03.001.

Andersen, M., Mardaljevic, J., & Lockley, S. (2012). A framework for predicting the non-visual effects of daylight—Part I: Photobiology-based model. *Lighting Research & Technology, 44*(1), 37–53. https://doi.org/10.1177/1477153511435961.

ANSI/IES TM-30-18. Method for evaluating light source color rendition.

Bellia, L., Pedace, A., & Barbato, G. (2013). Lighting in educational environments: An example of a complete analysis of the effects of daylight and electric light on occupants. *Building and Environment, 68,* 50–65. https://doi.org/10.1016/j.buildenv.2013.04.005.

Bellia, L., Pedace, A., & Barbato, G. (2014a). Indoor artificial lighting: Prediction of the circadian effects of different spectral power distributions. *Lighting Research & Technology, 46*(6), 650–660. https://doi.org/10.1177/1477153513495867.

Bellia, L., Pedace, A., & Barbato, G. (2014b). Winter and summer analysis of daylight characteristics in offices. *Building and Environment, 81,* 150–161. https://doi.org/10.1016/j.buildenv.2014.06.015.

Bellia, L., Pedace, A., & Fragliasso, F. (2017). Indoor lighting quality: Effects of different wall colours. *Lighting Research & Technology, 49*(1), 33–48. https://doi.org/10.1177/1477153515594654.

Bodrogi, P., Brückner, S., & Khanh, T. Q. (2011). Ordinal scale based description of colour rendering. *Color Research & Application, 36*(4), 272–285. https://doi.org/10.1002/col.20629.

Booker, L. A., et al. (2018). Individual vulnerability to insomnia, excessive sleepiness and shift work disorder amongst healthcare shift workers. A systematic review. *Sleep Medicine Reviews, 41,* 220–233. https://doi.org/10.1016/j.smrv.2018.03.005.

Boubekri, M., Hull, R. B., & Boyer, L. L. (1991). Impact of window size and sunlight penetration on office workers' mood and satisfaction: A novel way of assessing sunlight. *Environment and Behavior, 23*(4), 474–493. https://doi.org/10.1177/0013916591234004.

Boyce, P. R., & Cuttle, C. (1990). Effect of correlated colour temperature on the perception of interiors and colour discrimination performance. *Lighting Research & Technology, 22*(1), 19–36. https://doi.org/10.1177/096032719002200102.

Boyce, P. R., et al. (1997). Lighting the graveyard shift: The influence of a daylight-simulating skylight on the task performance and mood of night-shift workerst. *International Journal of Lighting Research and Technology, 29*(3), 105–134. https://doi.org/10.1177/14771535970290030501.

CIE 013.3-1995. Method of measuring and specifying colour rendering properties of light sources.

CIE 177:2007. Colour rendering of white LED light sources.

CIE 224:2017. Colour fidelity index for accurate scientific use.

Davis, R. G., & Ginthner, D. N. (1990). Correlated color temperature, illuminance level, and the Kruithof curve. *Journal of the Illuminating Engineering Society, 19*(1), 27–38. https://doi.org/10.1080/00994480.1990.10747937.

Davis, W., & Ohno, Y. (2010). Color quality scale. *Optical Engineering, 49*(3), 033602. https://doi.org/10.1117/1.3360335.

EN 12464-1. (2011). Light and lighting—Lighting of work places——Part 1: Indoor work places.

Farley, K. M. J., & Veitch, J. A. (2001). *A room with a view: A review of the effects of windows on work and well-being*. NRC Publications Archive—National Research Council Canada. Available at https://nparc.nrc-cnrc.gc.ca/eng/view/object/?id=ca18fccf-3ac9-4190-92d9-dc6cbbca7a98. Accessed June 12, 2018.

Figueiro, M. G., et al. (2013). Comparisons of three practical field devices used to measure personal light exposures and activity levels. *Lighting Research & Technology (London, England : 2001), 45*(4), 421–434. https://doi.org/10.1177/1477153512450453.

Figueiro, M. G., & Rea, M. S. (2014). Office lighting and personal light exposures in two seasons: Impact on sleep and mood. *Lighting Research & Technology, 48*(3), 352–364. https://doi.org/10.1177/1477153514564098.

Figueiro, M. G., et al. (2015). Daylight in office buildings: Impact of building design on personal light exposures, sleep and mood. In: *Proceedings of the 28th Session of the CIE*. Manchester: Commission Internationale de l'Eclairage.

Figueiro, M. G., et al. (2017). The impact of daytime light exposures on sleep and mood in office workers. *Sleep Health, 3*(3), 204–215. https://doi.org/10.1016/j.sleh.2017.03.005.

Figueiro, M. G., et al. (2018a). Circadian-effective light and its impact on alertness in office workers. *Lighting Research & Technology* (p. 147715351775000). https://doi.org/10.1177/1477153517750006.

Figueiro, M. G., Nagare, R., & Price, L. (2018b). Non-visual effects of light: How to use light to promote circadian entrainment and elicit alertness. *Lighting Research & Technology, 50*(1), 38–62. https://doi.org/10.1177/1477153517721598.

Fotios, S. A. (2001). Lamp colour properties and apparent brightness: A review. *Lighting Research & Technology, 33*(3), 163–179. https://doi.org/10.1177/136578280103300306.

Freyssinier, J. P., & Rea, M. S. (2010). A two-metric proposal to specify the color-rendering properties of light sources for retail lighting. In: *Tenth International Conference on Solid State Lighting. Tenth International Conference on Solid State Lighting, International Society for Optics and Photonics* (p. 77840V). https://doi.org/10.1117/12.863063.

Fumagalli, S., Bonanomi, C., & Rizzi, A. (2015). Experimental assessment of color-rendering indices and color appearance under varying setups. *Journal of Modern Optics, 62*(1), 56–66. https://doi.org/10.1080/09500340.2014.952694.

Gochenour, S. J., & Andersen, M. (2009). Circadian effects of daylighting in a residential environment. In: *Proceedings of Lux Europa 2009*.

Google Maps. (2012). Via Toledo, 402. Available at https://www.google.it/maps/place/Via+Toledo,+402,+80134+Napoli+NA/@40.8453674,14.2480479,184a,35y,39.49t/data=!3m1!1e3!4m5!3m4!1s0x133b085b133128f7:0xdf133fd4fae22f74!8m2!3d40.846769!4d14.2492768?hl=it.

Google Maps. (2017). Via Carlo Dolci, 5 Milano. Available at https://www.google.it/maps/place/Via+Carlo+Dolci,+5,+20148+Milano+MI/@45.4703496,9.1423935,190a,35y,39.44t/data=!3m1!1e3!4m5!3m4!1s0x4786c17766e55993:0xf7662030dc3e49b1!8m2!3d45.4719907!4d9.143046?hl=it. Accessed June 11, 2018.

Hashimoto, K., et al. (2007). New method for specifying color-rendering properties of light sources based on feeling of contrast. *Color Research & Application, 32*(5), 361–371. https://doi.org/10.1002/col.20338.

Higuchi, S., et al. (2007). Less exposure to daily ambient light in winter increases sensitivity of melatonin to light suppression. *Chronobiology International, 24*(1), 31–43. https://doi.org/10.1080/07420520601139805.

Houser, K. W. (2018). Human centric lighting and semantic drift. *LEUKOS, 14*(4), 213–214. https://doi.org/10.1080/15502724.2018.1501234.

Hubalek, S., Brink, M., & Schierz, C. (2010). Office workers' daily exposure to light and its influence on sleep quality and mood. *Lighting Research & Technology, 42*(1), 33–50. https://doi.org/10.1177/1477153509355632.

IES. (2018). PS-11-18: *IES position on TM-30-18, IES method for evaluating light source color rendition*. Available at https://www.ies.org/policy/position-statements/ps-11-18-ies-position-on-tm-30-18-ies-method-for-evaluating-light-source-color-rendition/. Accessed Nov 5, 2018.

IES TM-30-15. Method for evaluating light source color rendition.

IWBI. (2014). The WELL Building Standard, International WELL Building Institute. Available at https://www.wellcertified.com/en/explore-standard. Accessed May 14, 2018.

Judd, D. B. (1967). A flattery index for artificial illuminants. *Illuminating Engineering, 58*. Available at https://www.brikbase.org/sites/default/files/ies_023.pdf.

Kruithof, A. A. (1941). Tubular luminescence lamps for general illumination. *Philips Technical Review, VI*(3), 65–73.

Kuller, R., & Wetterberg, L. (1993). Melatonin, cortisol, EEG, ECG and subjective comfort in healthy humans: Impact of two fluorescent lamp types at two light intensities. *Lighting Research & Technology, 25,* 71–80. https://doi.org/10.1177/096032719302500203.

Leather, P., et al. (1998). Windows in the workplace: Sunlight, view, and occupational stress. *Environment and Behavior, 30*(6), 739–762. https://doi.org/10.1177/001391659803000601.

Leger, D. et al. (2011). Underexposure to light at work and its association to insomnia and sleepiness: A cross-sectional study of 13 296 workers of one transportation company. *Journal of Psychosomatic Research, 70,* 29–36. https://doi.org/10.1016/j.jpsychores.2010.09.006.

LRC. (2018). Web CS Calculator. Lighting Research Center. Available at https://www.lrc.rpi.edu/cscalculator/. Accessed July 20, 2018.

Mardaljevic, J., et al. (2014). A framework for predicting the non-visual effects of daylight—Part II: The simulation model. *Lighting Research & Technology, 46*(4), 388–406. https://doi.org/10.1177/1477153513491873.

Mills, P. R., Tomkins, S. C., & Schlangen, L. J. (2007). The effect of high correlated colour temperature office lighting on employee wellbeing and work performance. *Journal of Circadian Rhythms, 5,* 2. https://doi.org/10.1186/1740-3391-5-2.

Munsell. (1915). *Munsell color system. Color matching from Munsell color company*. Available at https://munsell.com/. Accessed Sept 5, 2018.

Sylvania, O. (2018). *LED color calculator*. Available at https://www.osram.us/cb/tools-and-resources/applications/led-colorcalculator/index.jsp. Accessed Nov 5, 2018.

Rea, M. S., et al. (2012). Modelling the spectral sensitivity of the human circadian system. *Lighting Research & Technology, 44*(4), 386–396. https://doi.org/10.1177/1477153511430474.

Rea, M. S. (2016). Opinion: On being PC. *Lighting Research & Technology, 48*(3), 266. https://doi.org/10.1177/1477153516644160.

Rea, M. S., & Figueiro, M. G. (2016). Light as a circadian stimulus for architectural lighting. *Lighting Research & Technology, 50*(4), 497–510. https://doi.org/10.1177/1477153516682368.

Smet, K. A. G., et al. (2010). Memory colours and colour quality evaluation of conventional and solid-state lamps. *Optics Express, 18*(25), 26229–26244. https://doi.org/10.1364/OE.18.026229.

Smolders, K. C. H. J., de Kort, Y. A. W., & Cluitmans, P. J. M. (2012). A higher illuminance induces alertness even during office hours: Findings on subjective measures, task performance and heart rate measures. *Physiology & Behavior, 107*(1), 7–16. https://doi.org/10.1016/j.physbeh.2012.04.028.

Smolders, K. C. H. J., & de Kort, Y. A. W. (2014). Bright light and mental fatigue: Effects on alertness, vitality, performance and physiological arousal. *Journal of Environmental Psychology (Light, lighting, and human behaviour), 39*, 77–91. https://doi.org/10.1016/j.jenvp.2013.12.010.

Thornton, W. A. (1974). A validation of the color-preference index. *Journal of the Illuminating Engineering Society, 4*(1), 48–52. https://doi.org/10.1080/00994480.1974.10732288.

Veitch, J. A., & Newsham, G. R. (1998). Lighting quality and energy-efficiency effects on task performance, mood, health, satisfaction, and comfort. *Journal of the Illuminating Engineering Society, 27*(1), 107–129. https://doi.org/10.1080/00994480.1998.10748216.

Viénot, F., Durand, M.-L., & Mahler, E. (2009). Kruithof's rule revisited using LED illumination. *Journal of Modern Optics, 56*(13), 1433–1446. https://doi.org/10.1080/09500340903151278.

Viola, A. U., et al. (2008). Blue-enriched white light in the workplace improves self-reported alertness, performance and sleep quality. *Scandinavian Journal of Work, Environment & Health, 34*(4), 297–306. https://doi.org/10.5271/sjweh.1268.

Wang, X.-S., et al. (2011). Shift work and chronic disease: the epidemiological evidence. *Occupational Medicine (Oxford, England), 61*(2), 78–89. https://doi.org/10.1093/occmed/kqr001.

Wijk, H., et al. (2002). Colour perception among the very elderly related to visual and cognitive function. *Scandinavian Journal of Caring Sciences, 16*(1), 91–102. https://doi.org/10.1046/j.1471-6712.2002.00063.x.

The manufacturer's authorised representative in the EU is Springer Nature Customer Service Centre GmbH, Europaplatz 3, 69115 Heidelberg, Germany. If you have any concerns regarding our products, please contact ProductSafety@springernature.com

Printed and bound by CPI Group (UK) Ltd, Croydon, CR0 4YY
30/09/2026
02225228-0009